Technician's Guide to Refrigeration Systems

John A. Corinchock

Library Resource Center
Renton Technical College
3000 N.E. 4th St.
Renton, WA 98056

McGraw-Hill

New York San Francisco Washington, D.C. Auckland Bogotá
Caracas Lisbon London Madrid Mexico City Milan
Montreal New Delhi San Juan Singapore
Sydney Tokyo Toronto

621
.56
CORINCH
1997

Library of Congress Cataloging-in-Publication Data

Corinchock, John A.
  Technician's guide to refrigeration systems / John A. Corinchock.
      p.   cm.
  Includes index.
  ISBN 0-07-013159-7 (hc)
  1. Refrigeration and refrigerating machinery.   I. Title.
TP492.C688 1997
621.5'6—DC20   96-42366
                                                                CIP

McGraw-Hill
A Division of The McGraw·Hill Companies

Copyright © 1997 by The McGraw-Hill Companies, Inc. All rights reserved. Printed in the United States of America. Except as permitted under the United States Copyright Act of 1976, no part of this publication may be reproduced in any form or by any means, or stored in a data base or retrieval system, without the prior written permission of the publisher.

1 2 3 4 5 6 7 8 9 0   DOC/DOC   9 0 1 0 9 8 7

ISBN 0-07-013159-7

The sponsoring editor for this book was April Nolan, and the production supervisor was Donald F. Schmidt. It was set in Century Schoolbook by North Market Street Graphics.

Printed and bound by R. R. Donnelley & Sons Company.

McGraw-Hill books are available at special quantity discounts to use as premiums and sales promotions, or for use in corporate training programs. For more information, please write to the Director of Special Sales, McGraw-Hill, 11 West 19th Street, New York, NY 10011. Or contact your local bookstore.

Information contained in this work has been obtained by The McGraw-Hill Companies, Inc. ("McGraw-Hill") from sources believed to be reliable. However, neither McGraw-Hill nor its authors guarantees the accuracy or completeness of any information published herein and neither McGraw-Hill nor its authors shall be responsible for any errors, omissions, or damages arising out of use of this information. This work is published with the understanding that McGraw-Hill and its authors are supplying information but are not attempting to render engineering or other professional services. If such services are required, the assistance of an appropriate professional should be sought.

This book is printed on acid-free paper.

*To my partner and love in life,
my dear wife Susan.*

Contents

Introduction   xi
Acknowledgments   xiii

Chapter 1. Careers in Refrigeration   1

  A Quick Look Back   1
  Mechanical Refrigeration   2
  Refrigeration Systems   3
  Working Conditions   4
  Basic Skills   5
  New Challenges   7
  Entering the Field   8

Chapter 2. Refrigeration Tools and Equipment   11

  Hand Tools   11
  Cutting and Bending Tools   12
  Flaring and Swaging Tools   14
  Brazing/Soldering Torch   16
  Specialty Tools   17
  Safety   33

Chapter 3. Basic Principles of Refrigeration   39

  Cold and Food Preservation   39
  What Is Refrigeration?   40
  Heat and Thermodynamics   42

Chapter 4. Refrigeration System Components   59

  Types of Control Devices   61
  Evaporators   75

viii   Contents

Suction Line	78
Compressors	80
Condensers	90

Chapter 5. Refrigeration Control Systems — 99

Control Switches	99
Pressure Regulators	106
Solenoid Valves	107
Controlling Refrigeration System Temperature	111
Control Applications	115
Control Systems for Capacity Reduction	128
Electronic Temperature Control Systems	134
Application Considerations	139

Chapter 6. Working with Tubing — 141

Types of Refrigeration Tubing	141
Cutting and Cleaning Tubing	143
Bending Tubing	143
Joining Tubing	144
Brazed and Soldered Connections	151
Troubleshooting	156
Repairing Tubing Leaks	157
Working with Tubing and Piping	157

Chapter 7. Modern Refrigerants — 169

Alternative Refrigerants	170
CFC/HCFC Phase-Out Regulations	174
Guidelines for Recycled and Recovered Refrigerants	176
Mixed Refrigerants	180
Applications for Alternative Refrigerants	181
Safety	182
Pressure-Enthalpy (Mollier) Diagrams	192

Chapter 8. Working with Refrigerants — 209

Recovering Refrigerant	209
Moisture and Contaminants	214
Purging Refrigerant Lines	215
Evacuation of Refrigeration Units	216
Charging the System	220
Retrofits with Alternative Refrigerants	227

Chapter 9. Understanding Refrigeration Electrical Systems   239

Electricity   239
Electrical System Components   249

Chapter 10. Electric Motors Used in Refrigeration   267

Induction Motor Design   267
Motor Strength   270
Shaded-Pole Motors   270
Split-Phase Motors   271
Permanent Split-Capacitor Motors   273
Capacitor-Start Motors   274
Capacitor-Start-Capacitor-Run Motors   275
Three-Phase Motors   276
Hermetic Compressor Motors   277
Electrical Troubleshooting   279
Mechanical Problems   284
Motor Replacement   286
Troubleshooting Hermetic Compressor Motors   288

Chapter 11. Domestic Refrigerators and Freezers   291

Domestic Refrigerator/Freezer Cabinet Designs   291
General Troubleshooting   309
Testing a Rebuilt System   322
Shutting Down Units   323

Chapter 12. Commercial Reach-In and Walk-In Units   325

Reach-In Refrigerators   325
Reach-In Freezers   329
Tips on Installing a Reach-In Unit   332
Maintenance   332
Service and Troubleshooting   333
Walk-In Refrigerators and Freezers   335
Tips on Controller Installation   344
Troubleshooting Refrigeration System Controls   350

Chapter 13. Supermarket Refrigeration Equipment   355

Supermarket Display Case Design   355
Multiplexed Refrigeration Systems   358
Parallel Compressor Operation   361
Defrost Methods   367
Subcooling Liquid Refrigerant   376

x Contents

Heat-Recovery Systems	377
Modern Control Systems	382
Computer-Controlled Systems	386
Installation Tips	389
Final Checks	395
Start-Up Procedures	396
Typical Serviceable Components	401

Chapter 14. Ice Makers, Water Coolers, and Food-Service Equipment   403

Ice-Making Equipment	403
Water Coolers	417
Food-Service Equipment	423

Chapter 15. System Troubleshooting   427

Compressor Failures	427
Evaporator Troubleshooting	435
Air-Cooled Condensers	436
Water-Cooled Condensers	436
TEV and Other Valves	437
Fans and Motors	437
Periodic Maintenance	437

Chapter 16. Calculating Refrigeration Loads   445

Refrigeration Loads	445
Equipment Selection	456

Glossary   459
Index   473

Introduction

Keeping goods and produce cold or frozen using mechanical refrigeration is a cornerstone of modern commercial and domestic life. The refrigeration industry that serves the food-service and other related industries is massive. Modern refrigeration systems are among the most reliable systems used in modern society. Few people give a second thought to the equipment and scientific principles that keep food and many other products fresh and preserved.

Technician's Guide to Refrigeration Systems contains practical information on how modern commercial and domestic refrigeration systems operate and the new refrigerants being used throughout the industry. It is a book that can serve as a primer for those considering a career in refrigeration or as a compact reference guide to apprentices and technicians already in the field.

Chapter 1 explains the careers available in the refrigeration service industry and details the skills and attributes that make up a successful technician. Chapter 2 covers the tools and equipment used in the trade.

In Chapter 3, the science of refrigeration is explained in straightforward terms that can be applied to practical applications. Chapters 4 and 5 concentrate on the physical components and control systems that are used in refrigeration systems.

Chapter 6 covers an essential skill for the refrigeration technician—working with tubing. Bending and brazing techniques, joint design, and the basics of laying out and installing tubing runs are all covered.

Chapter 7 is devoted to the most important topic in today's refrigeration industry—the new alternative refrigerants. By law, these alternative refrigerants must now replace older CFC/HCFC refrigerants that have been found to damage the earth's ozone layer. Important topics include refrigerant identification and numbering systems, CFC/HCFC phase-out regulations, and guidelines for recovering, recycling, reclaiming, and disposing of refrigerants. The latest information on retrofit procedures, the new refrigerant oils, and system compatibility concerns is presented. Safety is also a major topic. The final sections of Chapter 7 explain how to read and work with refrigerant pressure-enthalpy (Mollier) diagrams. These diagrams can be used to measure system performance and perform basic troubleshooting functions.

Chapter 8 concentrates on working with refrigerants including procedures for recovering refrigerant, purging refrigerant lines, evacuation of refrigeration units, and charging with new refrigerant. Additional information on retrofit procedures is presented.

Basic electrical skills are needed by all service technicians. Chapter 9 gives a solid overview of electrical principles as they apply to the refrigeration industry. Chapter 10 covers the types of electric motors used in compressors and fans.

Chapters 11 through 14 take a systems approach to the major types of equipment a refrigeration technician will work with on a daily basis. These chapters are devoted to domestic refrigeration units, commercial reach-in and walk-in units, supermarket refrigeration equipment, and ice makers, water coolers, and other food-service equipment. System design, defrost and control schemes, specialized components, maintenance procedures, and troubleshooting techniques are discussed for specific types of equipment. Much of the information deals with new electronic control systems for supermarket and other larger capacity equipment.

Chapter 15 covers system troubleshooting for a number of common components. Topics include handling compressor burnouts, evaporator and condenser troubleshooting, and servicing thermostatic expansion valves, solenoid valves, fans, and motors.

Chapter 16 explains another vital skill for refrigeration technicians—calculating refrigeration loads. Example calculations are given and data on equipment selection is included. The book closes with a glossary of common refrigeration terms.

John A. Corinchock

Acknowledgments

The author would like to thank the following companies for graciously supplying photographs, art, and information for this book.

Hussmann Corporation
Air-Conditioning and Refrigeration Institute
Fluoro Tech, Inc.
Watsco Components, Inc.
Robinair Division, SPX Corporation
MasterCool, Inc.
Fluke Corporation
DuPont Corporation
Copeland Corporation
Ranco Incorporated
Parker Hannifin Corporation
Sporlan Valve Company
Heatcraft Inc.
J.W. Harris Company
Tyler Refrigeration Corporation
Handy & Harman/Lucas-Milhaupt, Inc.
National Refrigeration Products
Tecumseh Products Company
Sears, Roebuck and Co.
Traulsen & Co., Inc.
Three Star Refrigeration Engineering, Inc.
Ram Freezers and Coolers Manufacturing Inc.

Paragon Electronics

Scotsman Ice Systems

Norlake Inc.

Elkay Manufacturing Company

True Refrigeration

American Society of Heating, Refrigerating, and Air-Conditioning Engineers

International Compressors Remanufacturing Association

A special thanks to Darla Fessler for her research help, Sheryl Doyle and Glenn Smith for their fine artwork, and Rick Paquette for his contributions to Chapter 11.

Chapter 1

Careers in Refrigeration

Today mechanical refrigeration forms one of the cornerstones of modern life. It ranks beside electricity, indoor plumbing, automobiles, and (unfortunately) television as a given convenience we rely on every day. The cold milk for your breakfast cereal, your deli sandwich at lunch, the steak you pick up for dinner, the kids' ice cream treat, and that late-night soft drink you grab at the corner convenience store or vending machine are all made possible by mechanical refrigeration (Fig. 1-1).

Mechanical refrigeration revolutionized the way we harvest, process, transport, store, and sell food products and other perishable items. Over 75 percent of the food we eat is refrigerated at some point in its preparation or distribution.

The refrigeration industry is a diverse field that offers great opportunities. The need for well-trained, dedicated service technicians eager to pursue careers in refrigeration service has never been greater.

A Quick Look Back

Until about one hundred and fifty years ago, nature was the only truly reliable refrigeration system. Thousands of years ago, enterprising Chinese merchants knew ice and snow made drinks more enjoyable in hot weather. So during the winter months they cut ice from frozen lakes and packed it in straw. Months later, thirsty shoppers paid a premium price for this valuable commodity.

The ancient Egyptians understood the cooling power of evaporation. They filled porous jars with water and set them on rooftops at sundown. As the water seeped through the walls of the porous jugs, the extremely dry dessert air quickly evaporated it. This created natural cooling, similar to raising your wet hand in the wind. By morning the water was chilled to a refreshing temperature. And on the coldest desert nights, a thin layer of ice might form.

Upper-class Greek and Roman citizens enlisted legions of slaves to haul snow down from the mountains and pack it into huge cone-shaped pits lined with straw and covered with thatched roofs. Alexander the Great lined trenches with snow to cool wine for his troops on the eve of battle. Nero loved frozen delicacies, and kept hundreds of slaves toiling to fill the snow trenches in his courtyard.

Centuries later, the cutting and sale of ice became big business in both Europe and North America. By the mid-nineteenth century, many companies in the northern United States had developed a lucrative ice trade. Yankee clipper ships, their cargo holds packed with ice insulated with sawdust, raced to ports in the south and abroad.

Mechanical Refrigeration

Artificial refrigeration grew out of the curiosity of dozens of nineteenth century scientists and engineers. In 1834, Jacob Perkins, an American

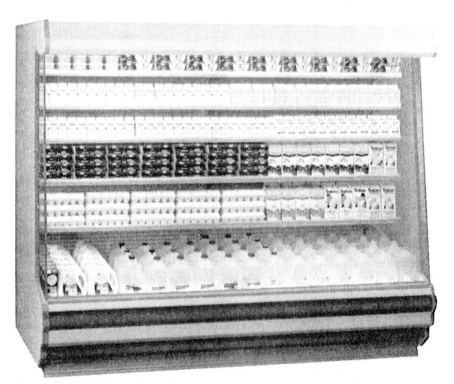

Figure 1-1 Mechanical refrigeration systems are the cornerstone for modern food processing and display. *(Hussmann Corporation)*

engineer living in England, invented and patented an ice maker based on a vapor-compression cycle. It proved successful in a number of installations.

Michael Faraday discovered the principles of absorption refrigeration in 1824. In 1858, Ferdinand Carre (of France) applied these principles to a practical refrigeration system. In 1885, a German engineer, Carl Linde, founded the first company to successfully produce and market ice-making machines based on absorption refrigeration.

By 1880, the United States patent office had issued over 3,000 patents on mechanical refrigeration systems. The technology was in place, but the public was not willing to embrace artificial ice. Some considered it unhealthy. What's more, cut ice was inexpensive and plentiful until 1890.

The winter of 1890 was exceptionally warm. Little or no ice formed on northern lakes and rivers. Out of necessity, people turned to artificial ice. Virtually overnight, a booming industry was formed. By the turn of the century, almost every town had an ice-making plant and every home had an ice box.

Commercial refrigeration had arrived, but one more important step had to occur before refrigeration entered domestic markets. As homes were wired for electric lights, they now had the power source needed to run their own refrigeration system. A small electric motor was all that was needed to run a vapor compression cycle. In 1918, Kelvinator introduced the first refrigerator aimed at the domestic market. The year's final sales figure of 67 gave little indication of what was to come.

By the end of the 1920s, the domestic refrigerator business was thriving. The Electrolux, an automatic absorption unit for home use, was introduced in 1927. The Monitor Top, the first sealed or "hermetic" home refrigerator was sold by General Electric the following year. The quick freezing of meats and vegetables to preserve them for long periods gained acceptance during the 1920s as well. The Great Depression of the 1930s and the Second World War slowed but could not stop the growth of the industry. In the propriety that followed the war years, home refrigeration became a reality for virtually all American households.

In the past twenty years, the incredible expansion of the fast-food and convenience store industries has placed literally millions of refrigeration systems into operation.

Refrigeration Systems

The refrigeration equipment you are called on to service can range from small self-contained units, such as a wall-mounted drinking fountain, to large multiplexed supermarket systems having multiple high-horsepower compressors, several branch circuits, and remote condensing equipment. Fine-tuning larger systems for optimum performance is an art in itself.

Domestic systems

Domestic refrigeration systems include household refrigerators and freezers. Many small businesses and companies also use residential-type refrigerators in lunchrooms, break rooms, and the like.

Commercial systems

Commercial refrigeration includes many, many applications, including: food processing, food service, and supermarkets and retail outlets.

Food processing. Refrigeration equipment is used extensively in the food-processing industries. Refrigerated storage rooms for fresh fruits and vegetables, refrigerated truck trailers, and supermarket meat-cutting rooms are just several examples of food-processing applications.

Food service. The food service industry includes restaurants, fast-food stores, public cafeterias, and cafeterias used in schools, hospitals, and other institutions. Equipment includes reach-in and specialty cabinets, roll-in cabinets, and walk-in coolers and freezers. Drink dispensers, slushie, ice cream, and frozen yogurt dispensers, and food service line equipment all contain refrigeration systems. Drinking fountains, coolers, and ice-making equipment form another branch of commercial equipment.

Supermarkets and retail outlets. About one-half of all food sold in supermarkets and retail stores requires refrigeration. Meats, fish, poultry, dairy products, fresh fruit, produce, deli items, ice cream, and frozen foods all require refrigeration. Display cases for meat and other items can be open design or closed service. Many units are often positioned in front of walk-in coolers so they can be loaded from the rear. Reach-in frozen food displays can be closed-door or island displays. Soft drinks and many other products that can be stored at room temperatures are often kept cold in reach-in display units for the convenience of customers.

Refrigeration cases are used by florists, medical laboratories, and many other industries that must keep products at controlled temperatures.

Working Conditions

Refrigeration technicians work in homes, supermarkets, convenience stores, bottle shops, office buildings, factories, food-processing plants, and anywhere food, beverages, and other produces must be kept cool or frozen. Varying work sites and types of systems encountered offer a new challenge every day.

Refrigeration technicians must be prepared to work outside in cold or hot weather. You should also be prepared to work inside a subfreezing walk-in locker on a 95°F summer's day. At times, the work area might be cramped or awkward to work in.

Technicians must practice safety with regard to electrical equipment, refrigerant handling, mechanical hazards, brazing and soldering torches, and basic hand tools.

In the refrigeration business the key word is *service*. A refrigeration system often preserves thousands of dollars worth of food or inventory. Because of this, most commercial customers expect their refrigeration repair service to be on call 24 hours a day, seven days a week. If you plan to work for or start a business aimed at commercial markets, then some late-night or weekend work is in your future.

Basic Skills

Refrigeration work combines the theory of scientific principles with numerous practical hands-on mechanical skills. A good refrigeration service technician is part scientist, pipe fitter, carpenter, electrician, and electronics technician. Many technicians are drawn to the field of refrigeration simply because it combines such diverse skills.

Today's technician must understand how and why a refrigeration system works. This involves knowing the basic vapor-compression cycle and how temperature and pressure affect modern refrigerants. Refrigeration technicians must understand and follow blueprints, design specifications, and manufacturer's installation instructions. You will install motors, compressors, condensing units, evaporators, and other components and connect this equipment to refrigerant lines, and electrical power sources. After making the connections, you must then charge the system with refrigerant and check it for proper operation.

Brazing and soldering skills are essential in making all types of refrigerant line connections and leak repairs.

You must also be familiar with ac power, electrical circuits, the operating characteristics of induction motors, and the basic principles of digital control systems.

Refrigeration service technicians use a variety of tools and equipment. These include simple hand tools, such as wrenches, pipe cutters, and benders and flaring tools. Acetylene torches and brazing equipment are needed to make pipe joints and metal-to-metal connections. Thermometers, manifold pressure gauges, manometers, leak detectors, and other specialized test instruments are used in troubleshooting refrigeration equipment. Electrical test equipment is also used to troubleshoot electric control circuits and electric motors and compressors. Refrigerant recovery and charging equipment is used to safely handle and store refrigerants (Fig. 1-2).

Installation work. Installating systems can be as simple as leveling and plugging in a self-contained reach-in cooler. On the opposite end of the scale, installation might involve installing roof-top condensers and cooling towers, fabricating refrigerant lines, installing control valves, connecting electrical components, and charging and testing the system.

The installation of walk-in coolers involves raising and joining insulated wall panels, installing doors, and setting all mechanical components in place. Some refrigeration companies specialize in manufacturing and installing prefabricated units.

Maintenance and repair work. When refrigeration equipment breaks down, service technicians diagnose the cause and make repairs. To find defects, they test parts, such as compressors, relays, thermostats, and timers. Newer units often include self-diagnostic systems that assist the technician in locating the problem.

Maintenance work includes inspection of various components and refrigerant lines to detect leaks and other faults. Technicians often

Figure 1-2 Recovery and safe handling of refrigerants are an important part of the service technician's job. *(Fluoro Tech, Inc.)*

adjust the system and its thermostatic controls to keep temperatures at specified levels.

Repair work involves fixing refrigerant leaks or changing faulty components, such as compressors, fan motors, solenoid valves, or pressure or temperature controls. Experience has shown that approximately 40 percent of all service calls involve electrical problems with motors and control circuits. Roughly 30 percent of service calls deal with locating and repairing a refrigerant leak, while the remaining 30 percent of work deal with mechanical problems, such as broken drive belts, cracked fan blades, and poorly sealed doors.

New Challenges

Today's refrigeration technician faces special challenges.

Refrigerants. For many years, the refrigeration industry used chlorofluorocarbon-based refrigerants (CFC) that were simply vented to the atmosphere when their service life was over. During the 1980s it became apparent that large amounts of CFCs in the atmosphere are contributing to the destruction of the earth's protective ozone layer. For this reason, the United States and many other nations have passed laws that will completely phase out the use of CFC refrigerants by the end of this century. Many refrigeration systems originally designed for CFC refrigerants are being retrofitted with newer, non-CFC refrigerants. In some cases this is a direct change. In others, compatibility of certain components becomes an issue and the service technician must perform additional work.

The venting of refrigerant to the atmosphere is no longer allowed. Refrigerants are now recovered from the system using special equipment that pumps the refrigerant into a storage cylinder. The used refrigerant is then cleaned and recycled. The job can be done using a self-contained system at the job site or the refrigerant can be sent to an outside reclaimation center. In either case, the technician is responsible for the safe handling of the refrigerant charge.

Electronic controls. Advances in electronics during the last 15 to 20 years have transformed control systems for many industries, and refrigeration is no exception. Computerized systems provide precise control of system temperature and other operating parameters on many newer commercial systems. These systems use electronic sensors and valves that communicate through digital direct-current voltage signals. Troubleshooting such systems requires a thorough understanding of how they operate and communicate. Many systems have unique methods of providing alarms and self-diagnostic information.

Entering the Field

There are two ways to enter the field of refrigeration: as an apprentice or as an on-the-job trainee. The method is usually determined by the extent of your knowledge in the field. High school graduates showing good mechanical aptitude and a willingness to learn are prime candidates for long-term apprenticeship programs.

Graduates from a secondary or post-secondary training program specializing in refrigeration are prime candidates for entry-level positions with refrigeration contractors and repair companies. They normally complete a short-term apprenticeship or on-the-job training period to prove their ability in job situations. This trial period is normally six months to a year.

Apprentice programs

Many refrigeration technicians start as helpers and acquire their skills by working for several years with experienced technicians. New workers usually begin by assisting experienced mechanics and doing simple jobs. They might carry materials or insulate refrigerant lines. In time, apprentices perform more difficult jobs, such as cutting and brazing tubing and checking electrical and electronic circuits.

Apprenticeship programs are run by unions and/or refrigeration and air-conditioning contractors. To be considered for the program, you must be a high school graduate and pass a mechanical aptitude test. Apprenticeships last four years and combine various work experience under qualified supervision with 144 hours of classroom study a year in related subjects, such as the use and care of tools, safety practices, blueprint reading and air-conditioning theory.

When hiring apprentices, employers prefer high school graduates with mechanical aptitude who have had courses in shop math, mechanical drawing, and blueprint reading. A basic understanding of microelectronics is becoming more important because of the increasing use of this technology in equipment controls. Good physical condition also is necessary because workers sometimes have to lift and move heavy equipment.

Job trainees. If the apprenticeship program is filled, the applicant might wish to enter the field as an on-the-job trainee. In this case the employer is contacted directly, or the state employment service might be used.

Vocational training

Many colleges, vocational schools, and technical training institutes offer one- or two-year programs in refrigeration technology and service.

The course of study normally includes classes in practical math, physics, and chemistry relating to refrigeration, mechanical drawing, and blueprint reading, which are the basics of electricity and electronic control.

A good post-high-school program will teach students refrigeration theory, plus the basics of equipment design, installation, maintenance, troubleshooting, and repair.

ARI/GAMA competency exams. The Air Conditioning and Refrigeration Institute in Arlington, Virginia, has also developed competency examinations to test the fundamental knowledge and basic skills of recent graduates of formal refrigeration training programs (Fig. 1-3). Only 60 percent of all students pass this exam. Those who do are listed in a directory distributed to interested employers in the field.

The completion of a post-high-school program or passing of competency exams is not absolutely necessary to success. Many successful technicians have no formal training other than that received during

Figure 1-3 ARI/GAMA compentency exams and manufacturer's certification programs help ensure that a technician's knowledge is accurate. *(The Air Conditioning and Refrigeration Institute)*

their apprenticeship. But employers increasingly prefer to hire trainees who are graduates of a formal refrigeration training program.

Refrigerant handling certification. The Environmental Protection Agency has set forth strict guidelines for the handling and disposal of refrigerants. Technicians must now receive special training in handling refrigerants and be certified to perform this work.

Continuing education. To keep up with changes in technology and to expand your skills, you should take advantage of training courses and educational materials offered by equipment manufacturers and by professional trade associations. Trade publications and magazines aimed at the refrigeration industry and the contracting business in general are good sources of up-to-date information.

Owning your own business

Most technicians entering the field work for another company to gain field experience. Good service technicians that keep up with the latest technology and service techniques often advance to positions as supervisors. Many top-level technicians eventually tackle the ultimate challenge and start their own businesses. There are thousands of successful refrigeration businesses in America.

To be successful in your own business, you must be a service- and sales-oriented person interested in dealing with and helping customers of all types. You must manage your time effectively and be disciplined in your work habits.

You must be able to work in an organized, accurate manner and realize the importance of small details that affect both profit and customer satisfaction. You must exercise good business sense and understand how policies and actions will affect your business and those who work for you.

The investment in time and effort will be great. But for some, the rewards and satisfaction of operating your own refrigeration service business can be tremendous.

Chapter

2

Refrigeration Tools and Equipment

Many of the tools and service techniques used to service and repair refrigeration systems are similar to those used to fix any mechanical or electro-mechanical system. As a refrigeration service technician, you must be well-versed in the safe use and care of common hand tools. You also will be called on to use digital multimeters, inductive ammeters, and other types of electrical test equipment on almost every service call.

Servicing modern refrigeration systems also requires special diagnostic and repair equipment unique to the refrigeration and air-conditioning industries. This chapter briefly outlines the tools and equipment used in the industry and covers basic safety concerns.

Hand Tools

To work safely and efficiently, you must have the right tools for the job. Professional service technicians use quality tools and keep them clean, organized, and in good working order. Hand tools made of heat-treated alloy steel should last the length of your career.

A tool chest and smaller tote tray are ideal for storing and carrying your basic hand tools. Store cutting tools, such as files, chisels, and hack saws, in separate compartments to avoid damaging the cutting edges. Keep tool sets, such as wrenches, Allen wrenches, and sockets together.

Screwdrivers and nutdrivers. The most common screwdrivers are the flat-blade and Phillips types. You should have several sizes of each. Nutdrivers combine the ease of use of a screwdriver with the superior gripping power of a socket. They are normally used on small nuts and bolts, and are handy for reaching odd locations.

Wrenches. To work on commercial and residential refrigeration systems, you should have complete sets of both USCS and metric wrenches. USCS or "standard" wrenches are most commonly sized in increments of $\frac{1}{16}$ of an inch. Metric wrenches are sized in increments of 1 mm.

Wrenches are either boxed or open-ended. A box wrench completely encircles a nut or bolt head and is less likely to slip and cause damage or injury. Combination wrenches have an open-end wrench on one end and a box wrench on the other, both sized the same. When working with nuts and bolts, never use a metric wrench or socket on a U.S. customary bolt or nut, or a customary wrench on a metric bolt. The wrench or socket will always be slightly oversized and will likely slip off or strip the nut.

Socket wrenches. Socket wrenches consist of two parts: the socket that fits over the nut or bolt and the handle or racket used to drive the socket. Your tool set should have a full set of both USCS and metric sockets combined with a ratchet handle and several extensions.

Flare-nut wrenches. To loosen or tighten refrigeration line fittings, special flare nut wrenches are needed. Using standard open-ended wrenches on soft metal-flared nut fittings will round the corners of the nut. Flare nut wrenches surround the nut and provide a better grip on the fitting. They also have a section cut out so that the wrench can be slipped around the refrigerant line and dropped over the flare nut.

Refrigeration ratchet wrenches. Special refrigeration reversible ratchet wrenches (Fig. 2-1a) are used to quickly open and close valve stems and fittings.

Allen wrenches. In refrigeration work, long-length Allen wrenches (Fig. 2-1b) are used to tighten the screws on fan blade bushings that hold the fan blade hub to the motor shaft.

Pliers/cutters. The most commonly used pliers are combination or slip joint pliers, long-nose pliers, and diagonal cutters. Slip joint pliers are a general gripping and pulling tool. Long-nose pliers are a must for tight spots. Diagonal cutters are needed to cut electrical wire. Vise-grip pliers are an indispensable tool for most refrigeration work, excellent for gripping small parts and also serving as an extra hand in unique service situations.

Mallets and hammers. In refrigeration work, a plastic, leather, or brass-faced mallet is used when tapping parts apart or aligning parts together. A carpenter's hammer is needed for mounting tubing and pipe supports and other odd jobs encountered in the field.

Cutting and Bending Tools

A number of tools are used to cut, clean, and bend metal tubing.

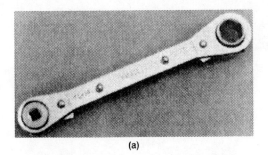
(a)

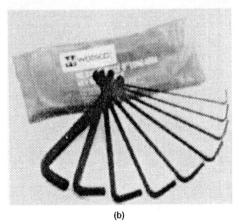
(b)

Figure 2-1 Refrigeration ratchet wrenches (a) and hex-head wrenches (b) are two types of specialty wrenches used in refrigeration work. *(Watsco Components Inc.)*

Hack saws. Hack saws are best for cutting hard copper tubing. A fine-cut blade is best (32 teeth per inch).

Tube cutters and pinchers. Wheel-type cutters clamp a cutting wheel against the tubing surface (Fig. 2-2a). As the tool is rotated around the tube, it gradually cuts through the tubing wall. A knob is used to tighten the tool's grip as the wheel cuts through the tubing. Tube pinchers (Fig. 2-2b) are used to close off tubing to keep dirt and moisture from entering the refrigeration system.

Reamers. Reamers are used to trim both the inside and outside of a cut tube.

Files. Files are used to deburr the cut ends of metal tubing or, in the case of small-diameter capillary tubes, to make the actual sizing cut. The most commonly used file for deburring or clean-up is the half-round or flat single-cut design.

Tube benders. Soft copper tubing can be bent by hand if the diameter is relatively large; however, it's best to use a tube-bending tool for tighter curves and to simply get more professional results. The sim-

Figure 2-2 Tube cutters (a) and a tube pinch-off tool (b). *(Robinair Division, SPX Corporation)*

plest version of this tool uses a tightly coiled steel spring that has an inside diameter equal to the outside diameter of the tubing. The tube is inserted into the spring and bent. This spring can also be used in reverse; that is, the spring can be inserted into a larger tube and the tube bent while the spring is inside it. Spring-type tube benders come in many sizes, and are often sold in sets.

A second type of bending tool contains a wheel that is used to bend the tubing (Fig. 2-3).

Flaring and Swaging Tools

Flared connections are used as an alternative to brazing on replaceable refrigeration components, such as filter-driers. Flared connections have specially designed tubing and fittings that make a leak-proof connection with the flare.

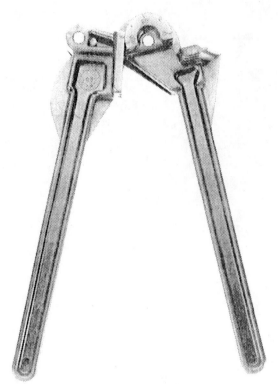

Figure 2-3 A lever-type tube-bending tool. *(Robinair Division, SPX Corporation)*

A flare can only be made with a specially designed tool (Fig. 2-4). A flared fitting is made up of the nut that fits over the tubing and a male threaded piece that will be held to the tubing by the nut. A correctly flared tube provides a vapor-proof seal.

A flaring tube has two main sections. The first is a flaring block to hold the tubing in place. The second is a cone-shaped flare that bends the tube to a 45° angle. Flaring tools can be used with many different tubing sizes. Proper use of a flaring tool is described in Chapter 6.

Swaging tools. These tools are used to prepare a tube end for an overlap-type joint. One type of swaging tool consists of an anvil block having different diameter holes and a set of corresponding punches. The tubing is mounted into the proper anvil opening and the punch is hammered down into the tubing to enlarge the tubing end to the correct size (Fig. 2-5). A second type of swaging tool is lever operated, and resembles a large riveting gun. The tube end is placed over the tools expander tip. Squeezing the lever closed expands the tube to the proper diameter.

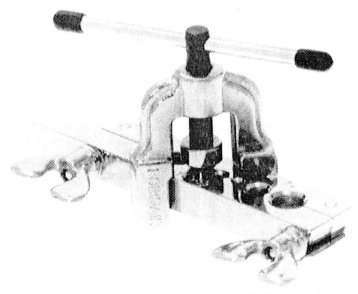

Figure 2-4 A tube-flaring tool. *(Robinair Division, SPX Corporation)*

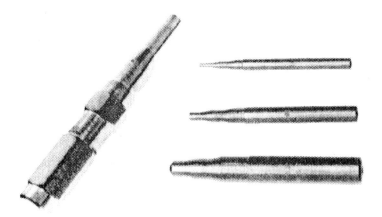

Figure 2-5 A swag joint-sizing tool set. *(Robinair Division, SPX Corporation)*

Brazing/Soldering Torch

Making soldered connections and repairs to tubing runs and evaporator and condenser coils requires the use of brazing torches. A portable oxy-acetylene torch or an air-acetylene torch provides the high temperatures necessary for silver brazing. Most can't reach the high temperatures needed for this process.

Mapp gas is a combination of propane and acetylene that can also be used as the fuel for brazing and soldering torches. Mapp gas is sold in small cylinders like a propane torch. The torch assembly has a built-in regulator for setting the high temperatures needed for brazing.

Specialty Tools

A number of unique specialty tools are designed for special jobs and situations.

Capillary tube cleaning kit. A capillary tube cleaning kit contains a portable hydraulic unit that can generate 500 psi of pressure, as well as a set of lead alloy wires. Each wire is several thousandths of an inch smaller than the diameter of a standard-sized capillary tube. A short piece of the appropriate wire is inserted $3/8$-inch into the capillary tube. The tubing is attached to the portable hydraulic unit, which forces the wire through the tube, pushing out any obstructions in the capillary tube. The wire simply falls into the evaporator after passing through the tube. It does not have to be removed and will cause no problems. If operating conditions indicate that the cleaning might not have been successful, the capillary tube should be replaced.

Fin combs. Fin combs are used to straighten and clean evaporator and condenser coil fins (Fig. 2-6a).

Inspection mirrors. The hand mirrors resemble dentist tools. They are ideal for inspecting tight spots and odd spaces (Fig. 2-6b).

Piercing valves. If a residential refrigerator or small commercial unit does not have a service valve for measuring system pressure or charging refrigerant, one can be added. Clamp-on piercing valves can be installed to the low-pressure side of the system and left on the unit permanently (Fig. 2-7).

Piercing valves are available in several different configurations, depending on the manufacturer. They are designed so that the two components of the valve are clamped to the tubing at an appropriate location. Once installed, the valve remains as a permanent addition to the system. A sharp pointed screw or needle, part of the valve assembly, pierces the tubing to allow access to the system. Some valves are designed with a combination piercing screw and valve, which is turned with an Allen wrench. Other designs have spring-loaded valves that automatically open when the charging hose is attached.

The valves are always supplied with caps that are placed on the threaded hose connection when servicing is completed. These caps prevent dirt from entering the valve. More importantly, they completely seal the valve and prevent refrigerant leakage in the event that the built-in sealing mechanism is faulty. Access valves should never be left uncapped.

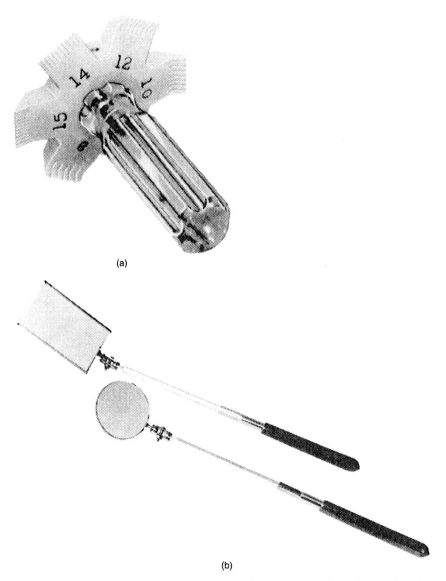

Figure 2-6 Fin combs (a) and inspection mirrors (b) are two specialty tools used in refrigeration work. *(Robinair Division, SPX Corporation)*

Thermometers

Thermometers are used to measure the operating temperature of various system components. Analysis of the temperature readings provides important information on the overall system performance, and can indicate whether a problem exists.

Several styles of thermometers are popular in refrigeration work (Fig. 2-8). The familiar mercury-filled glass tube model gives fast read-

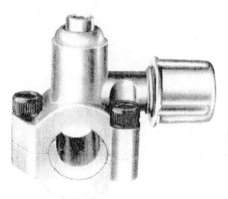

Figure 2-7 Clamp-on piercing valves provide access for pressure readings, recovery, evacuation, or charging. *(Watsco Components Inc.)*

ings, but it is hard to read and prone to breakage. Metal dial-stem thermometers read a temperature range of −40 to +120°F (−40 to +50°C) and are more durable than glass. Clamp-on thermometers easily attach to the tubing for an accurate reading.

Thermocouple and thermistor thermometers use electronics to record highly accurate temperature readings. Hand-held units with digital displays are ideal for many applications.

Manifold compound gauge set

The manifold gauge set is probably the refrigeration technician's most important piece of service equipment. It is used to measure the system's low- and high-side operating pressures when analyzing system operation or problems. The manifold gauge set is also used when purging and recharging the system, or when adding lubricating oil.

A manifold gauge set (Fig. 2-9) is basically a pair of pressure gauges mounted on a dual-valve assembly. One of the two gauges is a high-pressure gauge. This gauge typically reads up to 500 psi, and is used to measure the high-side (head) pressure of the refrigeration system.

The second gauge is a combination low-pressure and vacuum gauge. The low-pressure gauge measures up to 80 psi, and has a built-in retarder mechanism so that pressures up to 250 psi will not damage the gauge. The vacuum scale on the gauge measures up to a perfect vacuum of 29.9 inches of mercury and is used when the system is evacuated during a purging operation. (Some gauges are scaled to indicate vacuum in kPa.)

Many high- and low-vacuum gauges are equipped with additional scales calibrated to the evaporating temperatures of popular refriger-

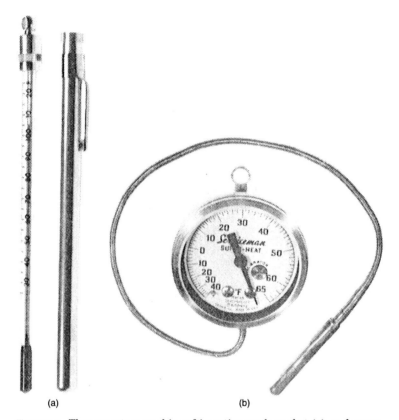

Figure 2-8 Thermometers used in refrigeration work: pocket (a), and superheat thermometer (b). *(Robinair Division, SPX Corporation)*

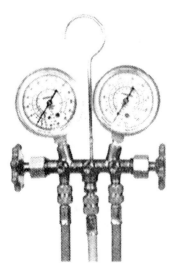

Figure 2-9 The manifold gauge set is an extremely important tool used to read system pressure. *(Robinair Division, SPX Corporation)*

ant types. These convenient scales eliminate the need for carrying pressure/temperature charts for these refrigerants.

Many newer gauges include a metric pressure scale (kg/sq cm) alongside of the standard psi scale. Sophisticated glycerine-filled gauges are designed to give a steady reading even under rapid pressure changes in a system.

The valve assembly has three separate hoses connected to it by ¼-inch SAE threaded fittings (Schrader fittings). They are the low-side, center, and high-side hoses. The manifold gauge set has two hand-operated valves. When opened, the valve(s) allow flow from the center hose to either or both outside hoses. The center hose can evacuate the system with a vacuum pump or charge the system from a refrigerant source. With the valves closed, each gauge with its hose is fully closed off from the center hose. The gauges will then record the existing pressure at their respective access valves.

Gauges and hoses are color coded for easy identification. The center hose might be yellow, the high-pressure hose might be red, and the low-pressure gauge and hose might be blue. Some manifold gauge sets use four hoses: two for evacuation and two for charging.

Manifold hoses are made of reinforced rubber. The ends of the hoses are ¼-inch SAE threaded fittings with center depressor pins to complement the charging ports of most systems. The threaded fittings are lined with rubber gaskets so that simple finger pressure can create a gas seal.

Special steel manifold gauge sets are designed for use with ammonia refrigeration systems. Unlike the standard brass sets, they will not corrode when exposed to ammonia-based refrigerants.

Manometer

A manometer is used to measure vapor pressures that are too low to be read accurately on a manifold gauge set. The pressure source pushes a liquid indicator up into a clear tube, against the force of gravity (Fig. 2-10). The reading can be made in inches of water or millimeters of mercury depending on the instrument being used.

In refrigeration systems, a manometer is typically used to measure air pressure drop across an evaporator or condenser coil. In ammonia-absorption refrigeration systems, the manometer is used to measure regulated natural or LPG gas pressure.

Vacuum pump

Vacuum pumps are used to remove air and moisture from a refrigeration system that has been opened/exposed to the atmosphere. A vac-

figure 2-10 Mercury manometers are used to accurately measure low pressure readings. (Robinair Division, SPX Corporation)

uum pump is a portable, self-contained unit. Both single-stage and double-stage models are used (Fig. 2-11). Pumps vary in terms of their drawing capacity and displacement. Smaller units might produce only 27.5 inches of vacuum (while 29.92 inches of mercury is a perfect vacuum) and displace less than 2 cubic feet per minute. Higher quality vacuum pumps measure vacuum in microns (one millionth of a meter) instead of inches, and can draw almost a perfect vacuum. Vacuum pumps that produce a vacuum of 10 microns or less ensure that all gas and moisture is removed from the system.

Larger displacement pumps (3 to 5 CFM) are often used when a larger refrigeration unit must be serviced or when many units need servicing at one time. These pumps are much faster than smaller units. Some vacuum pumps use a special oil that helps seal the unit while it draws a vacuum. This oil becomes contaminated over time reducing the unit's ability to draw a vacuum. It is important to replace the oil on a regular schedule.

Electronic vacuum gauge

An electronic vacuum gauge is needed to accurately measure the near-perfect vacuum created by larger, high-quality vacuum pumps. This gauge is a portable, battery-operated instrument that accurately mea-

Figure 2-11 A two-stage vacuum pump used for system evacuation to a deep vacuum. *(Robinair Division, SPX Corporation)*

sures vacuum levels as small as 50 microns or less (Fig. 2-12). Some gauges give readings in inches of mercury as well as microns.

Leak detectors and monitors

Because the internal system pressure in a refrigeration system exceeds atmospheric pressure, a refrigeration system will slowly lose its refrigerant charge due to leakage. However, most systems are manufactured so that the leakage is so minimal that it will not cause a significant problem over the lifetime of the equipment. The system should be charged before beginning a leak test. For larger leaks, the pressure need only be 5 or 10 psi.

Leak detectors are used for pinpointing specific leaks as well as for continually monitoring an entire area. Leak detection and monitoring is used to protect employees, conserve refrigerants, protect equipment, and reduce refrigerant emissions.

Leak detectors can be placed in two broad categories: leak pinpointers and area monitors. An instrument's sensitivity, detection limits, and selectivity should be considered before purchasing a monitor or pinpointer.

A *nonselective detector* detects any type of emission or vapor present, regardless of its chemical composition. These detectors are typically

Figure 2-12 Electronic vacuum gauges are used to accurately read extremely low vacuum levels. *(Robinair Division, SPX Corporation)*

quite simple to use, very durable, inexpensive, and usually portable. However, their inability to be calibrated, long-term drift, lack of sensitivity, and lack of selectivity limit their use for area monitoring. Some detectors are not sensitive enough for use with alternative refrigerant blends. Even detectors designed to detect a specific compound of the total blend, such as HCFC-22, might not be sensitive enough to detect enough HCFC-22 to confirm a leak.

A *halogen-selective* detector uses a specialized sensor that allows the monitor to detect compounds containing fluorine, chlorine, bromine, and iodine without interference from other species. The major advantage of such a detector is a reduction in the number of *nuisance alarms*—false alarms caused by the presence of some compound other than the target compound.

These detectors feature higher sensitivity than the nonselective detectors. (Detection limits are typically <5 ppm when used as an area monitor and <0.05 oz/yr when used as a leak pinpointer.) They are also usually easy to use, very durable, and easily calibrated.

The most complex, and also the most expensive, detectors are *compound-specific* detectors. These units are typically capable of detecting the presence of a single species without interference from other compounds.

Soap solutions. To detect larger leaks, leak-detection solutions are added to the system. These solutions produce bubbles at the source of the leak. If no bubbles appear, slowly increase the pressure until the bubbles start to form.

Halide torch leak detectors. A halide torch leak detector (Fig. 2-13) can be used to locate leaks as small as eight ounces a year. This detector consists of a propane torch with a specially designed burner head that has a flexible hose attached to it. The hose is moved about to locate the leak. The area should always be well ventilated when checking for leaks with a halide torch. There is the possibility of creating toxic phosgene gas.

As the hose draws the air needed to burn the propane, it also draws in any leaking refrigerant gas at the same time. A copper plate located inside the burner acts as a reactor. It is heated red-hot by the propane flame. A visual check of the propane flame indicates if there is refrigerant leaking from the system. If there is no leakage, the propane flame will be colorless. A small amount of leakage will color the flame green, while a large amount of refrigerant will turn the flame bright blue.

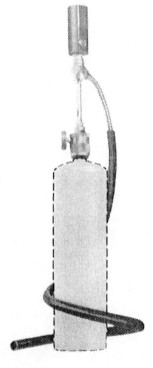

Figure 2-13 A halide torch refrigerant leak detector. *(Robinair Division, SPX Corporation)*

A bright blue flame is hazardous. It means that the detector is burning large amounts of refrigerant gas and is producing a toxic byproduct, phosgene gas. If the flame turns blue, immediately remove the torch from the leaking area. To further reduce the chances of producing phosgene gas, begin testing with a low charging pressure of 5 to 10 psi. Low pressures should reduce the chances of a high-volume leak. Also, if too much refrigerant enters the area around the leak, the flame will turn green before the torch locates the exact leak location. Ventilating the area using portable fans helps minimize these false readings so that the leak can be pinpointed.

Electronic leak detectors. To detect the smallest leaks (as small as ½ ounce a year), electronic leak detectors are used (Fig. 2-14). An electronic detector operates by sensing a change in the electrical conductivity of the air, or a change in its dielectric constant that occurs in the presence of commonly used refrigerant gases.

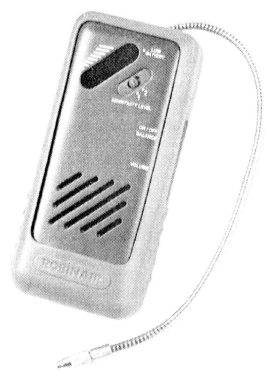

Figure 2-14 An electronic refrigerant leak detector. *(Robinair Division, SPX Corporation)*

Some detectors indicate the presence of refrigerant gas by emitting a low ticking noise. As the detector is moved closer to the source of the leak, the noise rises sharply.

Special circuitry in many electronic detectors allows the unit to compensate for ambient levels of refrigerant already in the surrounding air. This allows the instrument to detect a leak even when the surrounding air has been contaminated. Start with a small amount of pressure in the system and increase as needed. For very small leaks, the system can be gradually pressurized to a maximum of 150 psi using nitrogen or another inert gas.

UV light and fluorescent additives. Fluorescent additives have been used in refrigeration systems to pinpoint leaks for several years (Fig. 2-15). These additives are invisible under ordinary lighting but visible under ultraviolet (UV) light. Generally, these additives are placed in the refrigeration lubricant when the system is serviced, and a UV light is used to search for additive that has escaped from the system. Under UV light, the additive is usually a bright green or yellow.

As a leak pinpointer, fluorescent additives work very well because large areas can be rapidly checked by an individual. The recent introduction of battery-powered UV lights has made this task even simpler. Leak rates of less than 0.25 oz/yr can be found with the additives. The only drawback to the use of additives is that some leak sites might not be able to be seen because of cramped spaces.

The fluorescent additive should be tested for compatibility with the lubricant and refrigerant prior to use. The manufacturer of the fluo-

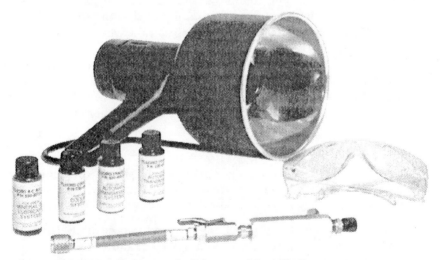

Figure 2-15 A UV light refrigerant leak detector. *(MasterCool)*

rescent additive can provide detailed information about which lubricants and refrigerants have been tested with which additives.

Inert gas supply

An inert gas, most commonly nitrogen, is used to purge the refrigeration system, or to pressurize the system to higher levels so leaks can be detected. Dry nitrogen, available in steel cylinders, is relatively inexpensive. Carbon dioxide can be used as a substitute gas.

A cylinder of nitrogen is pressurized to about 2000 psi, while carbon dioxide is 800 psi. The cylinders are equipped with regulators to reduce the pressure from these dangerously high levels to a safer, more controlled 150 psi or less.

Refrigerant recovery and recycling equipment

The no-longer-legal practice of venting used refrigerant into the atmosphere was neither the most cost effective, nor the most environmentally safe way to handle refrigerant disposal. CFC refrigerants contribute heavily to the depletion of the ozone layer. The expense of using new refrigerant whenever a system was contaminated also drove up the cost of ownership in commercial applications.

Refrigerant recovery and recycling systems provide an alternative to these problems. A recovery system safely removes refrigerant from the system without dissipating it into the atmosphere. The recovery system uses a compressor and condenser to remove the gaseous refrigerant from the system and convert it to a liquid that is pumped into a storage cylinder (Fig. 2-16). The used refrigerant can then be transported to a recycling center or proper disposal site.

Self-contained recycling units operate similarly to recovery units but also have the necessary filters and components to clean the refrigerant at its point of use. The purified liquid refrigerant is then stored in a container to be used again in the same unit or in another one.

Charging equipment

Refrigerant charge levels on many refrigeration systems are critical to the successful operation of the system. This is particularly true of systems using a capillary tube as a flow control device. (See Chapter 4.)

Charging cylinder. A charging cylinder is used to accurately charge a system with refrigerant. It consists of a tall, clear storage cylinder that is accurately marked with graduated scales (Fig. 2-17). These scales indicate in ounces the amount of refrigerant dispensed into the system.

Refrigeration Tools and Equipment 29

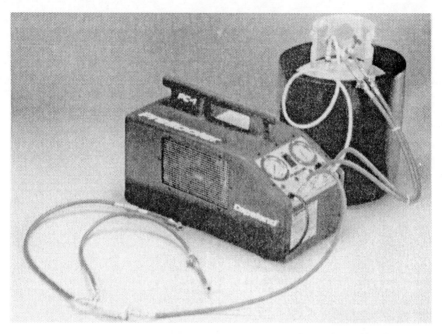

Figure 2-16 A refrigerant recovery unit used to remove and contain used refrigerants. *(Courtesy Copeland Corporation)*

The unit has a pressure gauge and charging and safety valves. Use of a charging cylinder is described in Chapter 8.

Electronic charging scales. Highly accurate electronic scales can be used to charge a system by dispensing a programmed amount of refrigerant by weight. In some states, refrigerant charging scales require certification by a NIST-approved laboratory.

Compressor oil charging pump. This is a special pump used to charge compressors with oil without the need of pumping the compressor down. A piston-type pump is often used to inject refrigerant oil into the compressor crankcase with the system fully charged with refrigerant (Fig. 2-18). It can typically be used at pressures up to 250 psi.

Low-pressure oil pumps are used to remove oil from the compressor. These pumps are equipped with a suction tube and a discharge hose connection.

Electrical test equipment

Almost half of all service calls made by refrigeration technicians involve an electrical problem. Understanding basic electrical concepts and working with electrical test equipment and tools are primary skills.

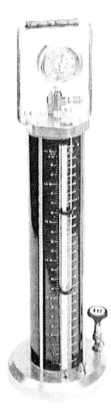

Figure 2-17 A refrigerant charging cylinder. *(Robinair Division, SPX Corporation)*

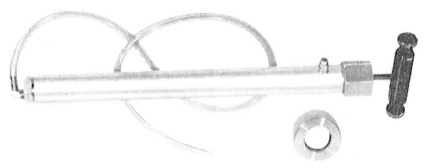

Figure 2-18 Refrigerant oil charging pumps are used to remove and charge compressor lubricating oil. *(Robinair Division, SPX Corporation)*

Circuit testers. Circuit testers or test lights are used to locate shorted or open electrical circuits. Low-voltage testers are used to troubleshoot 5- to 24-volt circuits commonly used in electronic control systems. A circuit tester resembles a stubby ice pick. Its handle is transparent and contains a light bulb. A probe extends from one end of the handle and a ground clip and wire extends from the other end. When the ground clip is attached to a good ground and the probe is touched to a live connector, the bulb in the handle lights up. If the bulb does not light, voltage is not available at the connector.

Continuity tester. A self-powered circuit tester is called a *continuity tester*. It is always used with the power off in the circuit being tested. A continuity tester looks like a regular test light, but it has its own small internal battery. When the ground clip is attached to the ground terminal of a component and the probe is touched to the feed wire, the bulb will light if there is continuity in the circuit. If an open circuit exists, the light will not illuminate.

Crimping/stripping tool. For electrical work, a combination tool that cuts and strips wire as well as crimps solderless connectors is recommended.

Ammeter. Ammeters are used to measure the rate of current flow in a circuit in amperes. An inductive ammeter, such as the one shown in Fig. 2-19 is used to measure alternating current. The meter's tongs are opened and slipped around the conductor. The current flow through the conductor generates a magnetic field that is read by the ammeter and converted into a digital readout in amperes. Always connect an ammeter in series with the circuit.

Wattmeter. A wattmeter measures the true wattage (power) being used by a motor or other electrical device. It automatically adjusts for

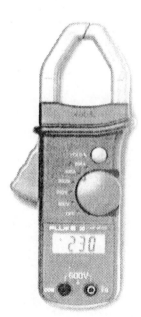

Figure 2-19 Inductive ammeters make measuring alternating current a simple task. *(Fluke)*

the power factor. A wattmeter is always connected in series with the circuit being tested.

Power factor meter. A power factor meter gives a direct reading of the power factor in an electrical circuit. The power factor is the ratio of the true wattage in the circuit as compared to the available wattage in the circuit. Knowing the power factor is useful when you are working to increase the overall efficiency of a motor circuit.

Digital multimeter. A battery-operated, hand-held digital multimeter is a must-have item for any technician doing electrical work (Fig. 2-20). Multimeters are a versatile instrument that can measure voltage and current, both ac and dc, as well as resistance. In addition, many are also designed to cover continuity checks and diode measurements.

Megger. A megger is a special high-voltage ohmmeter. It uses 500 volts or more to perform nondestructive testing, and it can measure very high resistance values. The megger can detect leakage that cannot be found using an ordinary ohmmeter. The megger is most commonly used to determine both the electrical condition hermetic compressors and their remaining electrical life.

Even though the internal wiring of a hermetic compressor is fully insulated, the normal infinite resistance reading between the windings and shell of the unit is not always present. This is because if the oil and refrigerant used to lubricate the compressor windings becomes contaminated, it can lower the winding resistance. This can create a significant electrical leakage path to the case or shell resulting in compressor failure.

Figure 2-20 A quality digital multimeter measures voltage, current, and resistance. *(Fluke)*

Test capacitors. You should own a set of nonpolarized motor start and run test capacitors to check compressor and fan motors. The test capacitors should be checked periodically for any malfunctions.

If the test capacitor indicates that a compressor or motor is defective, the best way to verify that the existing capacitor is giving accurate information is by substituting it with another, similar model.

A capacitor should include 4-microfarad, 15-microfarad, and 40- or 50-microfarad units, all rated at a minimum of 370 volts ac. If the system uses a 208/240 power line, the capacitor should be at least 440 volts.

A large, nonpolarized capacitor (100 to 200 microfarads) is often used to hard start a reluctant compressor. This capacitor should be at least 370 volts ac.

Capacitor tester. Capacitor testers are used to measure capacitor values and check for shorted, open, leaking, or good capacitors (Fig. 2-21). This type of tester plugs into a 115-Vac outlet. The tester alligator clips are attached to the capacitor terminals and the selector switch is set to the proper MFD ranges.

Hermetic (compressor) analyzer. Hermetic analyzers can measure current and voltage, as well as indicate the condition of the compressor. Some instruments can even create a "reverse running" mode that can free a stuck compressor.

Safety

The job of the refrigeration technician involves working with machinery and materials that can be dangerous if not handled properly and treated with respect. Refrigerant systems and refrigerant storage cylinders often contain highly pressurized gases with great potential energy if accidentally released. High-temperature vapors can burn while coming in contact with low-pressure liquid refrigerant and can instantly

Figure 2-21 A capacitor tester. *(Watsco Components Inc.)*

freeze your fingers and skin. Compressors, motors, and control systems are electrical equipment and pose electrical hazards if improperly serviced. Turning motors and belts can grab clothing or fingers.

Fire safety

Soldering and brazing torches present a potential fire hazard. Be aware of combustible materials in the working area. For example, grease build-up in restaurants and food-service locations is highly flammable. Use a metal shield to shield combustible material when using a torch to braze or solder connections.

A multipurpose dry-chemical fire extinguisher capable of handling ordinary combustibles, oil and grease, and electrical fires should be a standard item in your service truck or van. With a dry chemical extinguisher, the spray should be directed at the base of the fire to suffocate the flames.

Cold

Dress warm when working in walk-in coolers and freezers. You might need to stay inside a large low-temperature (0°F) freezer for more than an hour making a typical repair. Always have the proper coat, hat, and gloves when working in cold environments. If you allow your body to become cold, you might rush to finish the job and do sloppy or unsafe work.

Leaks on the low side of a refrigerant system can release refrigerant with the potential to quickly cool your skin down to −40°F, causing instant frostbite. Never try to stop a leak with your hands; locate the appropriate valve and shut off refrigerant flow to the damaged line. Always wear eye protection when handling refrigerants.

Electrical safety

Your body is an excellent conductor of electricity. Always remember this when working on the electrical system of refrigeration equipment. To prevent electrical shock, avoid becoming a conductor between two live wires or a live wire and ground. Use properly grounded power tools. Remove rings, watches, and metal jewelry when working around electricity. The following safety rules should be observed when working with electrical circuits:

- Always disconnect the electrical power and leave it off when working on electrical components of any refrigeration system.
- Perform lock-out tag-out on all circuit breakers in an open position so that no one inadvertently closes them during installation or servicing.

- Before touching any terminals, short any capacitor in the system using a resistor. A charged capacitor can release between 200 to 500 volts. Check the circuit with a voltmeter to ensure that it is not "live" before touching any wires, terminals, or electrical parts of the refrigeration system.
- Stand on a dry surface area such as wood or concrete. If the work area is wet or damp, put on rubber boots and gloves.
- Properly ground all electrical components of the refrigeration system to minimize the potential for shock, which could occur if the circuit was accidentally grounded.
- Always replace a fuse or circuit breaker with one of the same size and rating. Installing an oversized breaker or fuse overloads component and can cause fires.
- Solder or crimp solid metal terminals to the ends of stranded wire.
- Replace any wires that are cracked or have brittle insulation.

Pressure vessels

Move refrigerant vessels with the protective cap in place. The pressure inside a cylinder of refrigerant can reach 225 psi on a hot day. The total potential pressure release inside the vessel can be as high as 350,000 pounds of pressure.

Never heat a refrigerant storage tank with a torch when charging a system. This keeps the pressure from dropping in the vessel, but it is extremely dangerous. The only safe method of keeping the vessel pressure from dropping is to set the vessel in a drum of hot water, no higher than 90°F.

Material Safety Data Sheets and Right-to-Know laws

Refrigerants, compressor oils, cleaning solvents, water treatment chemicals, and other materials can pose certain hazards if they are not handled correctly. They might be flammable, emit harmful vapors, or be irritating to the eyes or skin.

All workers are protected by federal and state Right-to-Know laws concerning hazardous materials in the workplace. If you plan on working for a larger company, or eventually hiring workers of your own, these laws will be important to you. The laws specify that it is the responsibility of the employer to train all employees in the safe handling and disposal of all potentially hazardous chemicals in the workplace.

Important information about refrigerants, oils, and other chemicals used in the refrigeration industry is contained on Material Safety Data

Sheets (Fig. 2-22). The MSDS is issued by the manufacturer of the material in question and lists its chemical composition, ignitability, corrosiveness, reactivity, toxicity, and other distinctive characteristics. It often states specific handling instructions and safety precautions that must be followed. Emergency treatment for accidental inhalation, eye and skin contact, etc. are given. The Canadian equivalent to MSDS is known as Workplace Hazardous Materials Information Sheets.

It is the employer's responsibility to obtain all MSDS sheets for the materials used on the job and to make this information available to all

DuPont Chemicals

2187FR　　　　　　　　Revised 22-JUN-1995　　　　Printed 14-SEP-1995

"SUVA" 134a

CHEMICAL PRODUCT/COMPANY IDENTIFICATION

Material Identification
"SUVA" is a pending trademark of DuPont.

Corporate MSDS Number	DU000693
CAS Number	811-97-2
Formula	CH2FCF3
CAS Name	1,1,1,2-TETRAFLUOROETHANE

Tradenames and Synonyms
"SUVA" 134a
HFC 134a
VT1505

Company Identification
MANUFACTURER/DISTRIBUTOR
　　DuPont
　　1007 Market Street
　　Wilmington, DE 19898

PHONE NUMBERS
　Product Information　　1-800-441-7515
　Transport Emergency　　CHEMTREC: 1-800-424-9300
　Medical Emergency　　　1-800-441-3637

COMPOSITION/INFORMATION ON INGREDIENTS

Components Material	CAS Number	%
ETHANE, 1,1,1,2-TETRAFLUORO- (HFC-134a)	811-97-2	100

(Continued)

Figure 2-22 Material Safety Data Sheets contain important safety information on refrigerants, lubricants, and other chemicals used in the refrigeration industry. *(DuPont Fluoroproducts)*

employees. The employer must also provide formal training on the safe handling of all hazardous materials and update this training on a yearly basis. Containers storing potentially hazardous materials must be properly labeled with health, fire, reactivity, and handling hazard information. The simplest way to ensure this is to keep materials in their original containers. If a chemical is moved into another container, such as the transfer of refrigerant from a large to small storage vessel, the second container must be the proper type and it must be correctly labeled.

Lifting and carrying

Using the proper method to lift heavy objects is important. Even small or midsize components, such as a hermetic compressor or refrigerant cylinder can strain your back or pull a muscle if mishandled. When lifting any object, place your feet close to the load, spreading them slightly for better balance. Keep your back and elbows straight, bend at the knees until your hands reach the load, and grasp the object firmly. Keep the object close to your body and lift by straightening your legs. Turn your entire body when changing direction. When setting an object on the floor, lower it by bending your knees and keeping your back straight. Avoid bending forward, which can strain your back muscles.

Personal safety

Accidents can happen to you. To ensure your personal safety, think about what you are doing at all times and take common-sense steps to protect yourself.

Always wear safety glasses, goggles, or a safety face shield when using a brazing torch or halide leak detector. Eye protection must also be used whenever there is a possibility of dirt, metal chips, or refrigerant getting in your eyes. If you wear prescription glasses, use glasses with safety lenses and side shields. If you have long hair, tie it up or keep it under a hat so it doesn't get caught in the moving parts of compressors or motors.

Wear the proper work clothing, such as well fitted coveralls. Avoid wearing loose clothing or jewelry that can become caught in moving machinery. Wear heavy-duty work shoes or steel-toed safety shoes to protect your feet from sparks, falling objects, and even some acids and solvents. A protective hard hat is recommended when doing installation work in commercial and industrial sites.

Tool and equipment safety

Carefully follow all operating instructions and cautions covering the use of tools and equipment. Do not use any damaged tools or equipment. Misuse of hand tools is the leading source of accidents on the job. Know what a tool has been designed to do and use it properly. Select the correct size and type of hand tool for the job, and only use a hand tool for the job it was intended for.

Keep your hand tools and test equipment in good condition, and store them in a safe location when not in use. Replace cracked or worn wrenches. Pliers are made for holding, pinching, squeezing, and cutting, but not for turning. Do not substitute pliers for a wrench, because the pliers can slip and damage bolt heads and nuts.

Chapter

3

Basic Principles of Refrigeration

Refrigeration systems work because they are designed to take advantage of a number of scientific laws that are both predictable and unchangeable. These laws deal with heat and how it behaves, states of matter such as liquids and gases, and the relationship between pressure and temperature.

This is not a science book. It is a practical guide to understanding and servicing commercial and residential refrigeration systems. You do not need to be a scientist to service a refrigeration system. But if you understand the basics of thermodynamic law and you can apply them to a mechanical refrigeration system, you'll be well on your way to becoming a successful service technician.

Cold and Food Preservation

Food keeps longer when it is kept cold or frozen. Food spoils when bacteria grows on or in it. We have all seen mold on an old loaf of bread or a piece of cheese. Cooling or freezing food slows down or stops the activity and growth of bacteria.

The correct temperature range for refrigerating fresh food is between 35°F and 45°F. Most fresh food contains a considerable amount of water. If the refrigerating temperature drops to 32°F, this water freezes forming large ice crystals that damage the food tissue. The food's appearance and taste are ruined, and when defrosted, the food spoils very quickly.

However, we all know that many types of food are frozen to keep them usable for long periods of time. The difference is that in a freezer, temperatures are kept in the range of 0 to −15°F. This results in very quick freezing. The ice crystals formed are quite small. Damage to the food is minimal and taste and appearance are not compromised.

In ice-making machines, a temperature below 32°F is all that is required. However, lower temperatures are often specified by manufacturers to ensure that the ice is clear and hard.

What Is Refrigeration?

Most people associate refrigeration with cold or cooling. When a refrigerator or freezer does its job, cold temperatures are certainly the end result. But a refrigerated area is really one where heat has been removed, not one where cold has been added. So to be more precise, you should always think of refrigeration as dealing with the transfer of heat. Cold is the absence of heat, and if heat can be moved out of a given area, such as a walk-in cooler or a supermarket dairy case, that area and items stored in it will become cold (Fig. 3-1).

Mechanical refrigeration is a process that can transfer heat from one place to another efficiently, economically, and continuously. The closed-loop process is performed by flowing a refrigerant through a vapor compression cycle.

In its simplest form, the refrigeration system uses five components to perform the vapor compression cycle: an evaporator, a compressor, a condenser, a flow control, and interconnecting piping (Fig. 3-2). In a simplified vapor compression cycle, the flow control meters the correct amount of high pressure liquid refrigerant into the evaporator. As it enters the evaporator, the refrigerant turns from a liquid into a low-pressure vapor. As it turns from a liquid to a gas, the refrigerant absorbs heat from the surrounding area, so the evaporator becomes quite cold. The evaporator is located in or near the area to be kept cold. An electric fan is usually used to create air flow over the evaporator coils and throughout the refrigerated area.

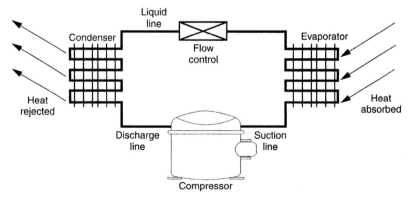

Figure 3-1 Refrigeration is the process of moving heat from one area to another.

Basic Principles of Refrigeration   41

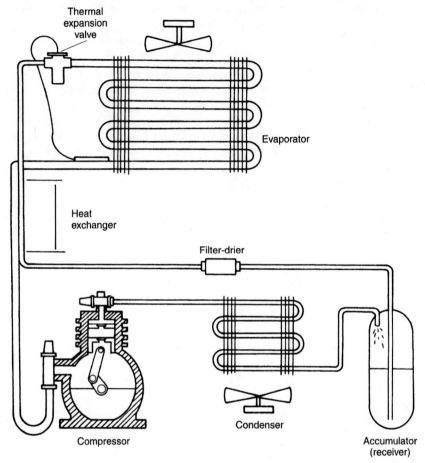

Figure 3-2 The major components of a mechanical refrigeration system.

The refrigerant has done its job by absorbing heat in the evaporator. But in a refrigeration system, the refrigerant must perform this job not once, but thousands and thousands of times. So, the vapor compression cycle must provide a method of removing the heat from the refrigerant and turning it back into a liquid. The way this is done is by increasing the pressure of the refrigerant so it will condense back to a liquid at a higher temperature. The system's compressor provides the energy and power needed to increase the pressure and temperature of the refrigerant vapor, and moves it into the condenser. The condenser is simply a cooling coil. As the high-pressure, high-temperature vapor cools down, it turns back into liquid. The high-pressure liquid refrigerant can once again be metered through the flow-control device into the evaporator, and the cycle can be repeated.

Refrigerant pressures play a very important role in refrigeration. Always think in terms of dealing with two pressures in the refrigeration cycle: the condensing or high pressure and the evaporating or low pressure (Fig. 3-3). All components of a system subjected to low pressures are said to be on the *low side*, and all components subjected to high pressure are commonly referred to as being the *high side*. High-side pressure is often called *head pressure*. The separating factor between the high and low pressure sides of a refrigeration system is the refrigerant flow control device, which is either a thermal expansion valve or a capillary tube.

Most of the refrigerant systems in use today use the basic vapor compression cycle while adding a number of auxiliary components, sensors, and controls that enhance its performance. These items include a receiver to store excess refrigerant, filter-driers to remove water and dirt from the refrigerant, timer-controlled evaporator defrost systems, heat exchangers, oil separators, and various temperature and pressure based control systems. System components are discussed in detail in Chapter 4.

Before discussing the vapor compression cycle in more detail, it is best to review a number of scientific principles that all refrigeration systems use. These principles deal with the transfer of heat, how heat behaves when a substance changes from a liquid to a gas, and how gases and liquids behave under different pressures.

Heat and Thermodynamics

The mechanical action of heat is a branch of science called *thermodynamics*. The first and most important law of thermodynamics states,

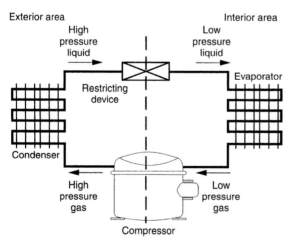

Figure 3-3 The refrigeration system can be divided into low-pressure and high-pressure sides.

Basic Principles of Refrigeration 43

"Energy can not be created nor destroyed." Heat is a form of energy. It cannot be created from anything; it must be converted from something that already exists. When a log is burned in a fireplace it releases energy in the form of heat. When a single atom is split, it releases tremendous amounts of energy. This energy is never lost, it simply disperses into the surrounding area or is converted into force and work. The compressor in a refrigeration system converts electrical power into mechanical force that it applies to the low temperature refrigerant vapor. The mechanical energy exerted on the gas is not lost. As the vapor pressure increases, the mechanical energy is converted to heat energy and the temperature of the vapor increases dramatically.

Heat is energy that is always moving, and the second law of thermodynamics states, "Heat always travels from a warm object to a cold one." For example, when you pour ice water into a glass, the glass loses its heat to the ice water and the glass becomes cold. If hot tea is poured into a cup, the tea quickly loses some of its heat to the cup. The cup becomes warm, and the tea cools down a bit. In each case, the energy is not destroyed, but simply moves from one area to another.

The size and mass of the objects do not affect the direction of heat flow. For example, consider an insulated box into which we place two metal objects, one on top of the other. The first object is a huge iron anvil weighing 500 pounds. The anvil's temperature is 200°F. The second object is a small steel bar weighing ten pounds. The bar's temperature is 210°F. Even though the massiveness of the anvil is able to contain much more heat than the much smaller steel bar, because of the temperature difference between the two objects, heat will travel from the warmer steel bar to the cooler iron anvil until the two objects are of the same temperature (Fig. 3-4).

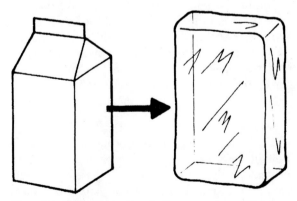

Figure 3-4 Heat always travels from a warmer object to a cooler object. The same is true of warm and cool areas.

The speed at which heat travels from the warm object to the cold one is determined by the difference of temperature between those two bodies. The greater the difference in temperature, the greater the heat flow.

Temperature versus heat

Temperature indicates the degree of warmth or how hot or cold a substance is. It does not by itself tell us how much heat is in the substance. The amount of heat in a substance equals the mass of the substance multiplied by its temperature. In the example given earlier, the 500 pound anvil contained much more heat than the 10 pound bar.

Mass is a term that is often interchanged with weight, but they are not the same in true scientific terms. *Weight* is the measurement of the gravitational pull on an object. An object with a given mass would have different weights on earth as compared to the moon or the other planets. But for all practical purposes, the terms *weight* and *mass* can be used interchangeably in refrigeration work.

Measuring temperature

Temperature is measured using a thermometer. Thermometers can use one of several scales including Fahrenheit (F), Celsius (C), Rankine (R), or Kelvin (K). The reference point for all temperature scales is absolute zero (Fig. 3-5).

To understand absolute zero, you must know that all materials are made up of *molecules*. Molecules are groups of atoms that are in constant motion. As heat is removed from a substance, its molecules move slower. The point at which all molecular motion stops completely is called *absolute zero*. On the Fahrenheit scale absolute zero is –460°. This corresponds to –273° on the Celsius scale. The Rankine and Kelvin temperature scales are absolute temperature scales. They have no negative readings, but begin at 0° absolute zero. The Rankine scale is the absolute scale for the Fahrenheit scale; 0°R equals –460°F. The Kelvin scale is the absolute scale for the Celsius scale; 0 K equals –273°C.

For the sake of simplicity, this book will use the U.S. Customary System of measurements. For temperature, this is degrees Fahrenheit.

Units of heat

The U.S. conventional unit of heat is the British thermal unit or *BTU*. It explains how much heat is contained in a given object. A BTU is defined as the amount of heat needed to raise the temperature of one pound of water one degree Fahrenheit. If we put one pound of water (roughly one pint) in a pot, measure its temperature as 50°F and then

Basic Principles of Refrigeration 45

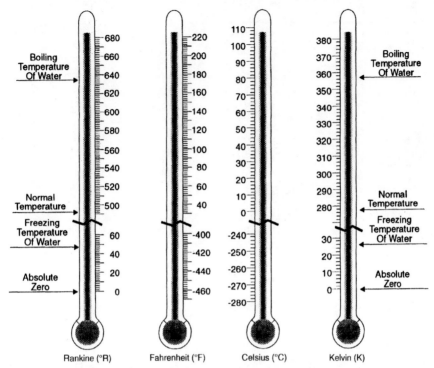

Figure 3-5 The four major temperature scales are the Fahrenheit, Celsius, Rankine, and Kelvin.

heat the pot on a stove until the thermometer in the pot reads 51°F, we will have added 1 BTU of heat to the water. The metric equivalent of the BTU is the *joule*.

Specific heat

Specific heat is the amount of heat it takes to increase the temperature of one pound of a substance one degree Fahrenheit. If you refer back to the definition of the BTU, you see that the specific heat of water is 1 BTU. But as Table 3-1 shows, the specific heat of other substances differs widely. The specific heat of ice and steam is also different from that of water. It only requires 0.5 BTU to raise the temperature of one pound of ice or steam one degree Fahrenheit.

States of matter

There are three states of matter: solids, liquids, and gases. Ice, water, and steam are a good example of the three states of matter. All pure substances can change their state if sufficient heat is added to or taken

TABLE 3-1

Substance	Specific Heat (BTU/lb./°F)
Water	1.000
Ice	0.504
Steam	0.500
Iron	0.129
Steel	0.115
Aluminum	0.224
Copper	0.095
Mercury	0.033
Brick	0.210
Concrete	0.155
Marble	0.200
Air	0.250
Salt Water	0.900
Glass	0.185
Wood	0.325
CFC-12	0.213
HCFC-22	0.260
R-502	0.255
HCFC-123	0.235
Refrigerant MP39	0.301
Refrigerant MP52	0.294
Refrigerant MP66	0.300

out of that substance. Two laws govern how substances change their state, and they are extremely important in understanding how refrigerants work. They are:

- The change of state points always happen at the same temperature and pressure combinations for a given substance.
- It takes the addition or removal of heat to produce these changes of state.

As heat is added to a substance, its molecules begin to move faster and farther apart. If enough heat is added, a solid becomes a liquid and a liquid becomes a gas. If heat is removed, the gas returns to its liquid state and, if enough heat is removed, the gas returns to its solid state.

Sensible heat

Sensible heat is heat absorbed by a solid, liquid, or gas, that results in an increase in temperature of that solid, liquid, or gas.

Latent heat

Latent heat is heat absorbed by a solid, liquid, or gas that does not result in an increase in temperature. Latent heat is also called *hidden heat*. When a substance is heated there are two points when the substance's temperature does not increase. The first is when it is changing from a solid to a liquid. The second is when it is changing from a liquid to a gas. At these points, the heat being added to the substance is needed to change the substance's state. Once the substance has changed from a solid to a liquid or a liquid to a gas, the heat again becomes sensible heat and the substances temperature increases.

This is an extremely important concept in understanding refrigeration systems. The operation of the vapor-compression cycle is based on sensible and latent heat. Before we apply the concept of sensible and latent heat to a refrigerant, let's review how they behave in a more familiar substance, water. The graph in Fig. 3-6 tracks the water temperature versus heat content as water is heated from its solid to gaseous states. At standard atmospheric pressure, water freezes at 32°F and boils at 212°F. Let's begin tracking a 10-pound block of ice when it is at 20°F, well below its melting point. Remember the specific heat of ice is 0.5 BTU, so for every 5 BTUs added to the 10-pound block, its temperature rises 1°F.

When 60 BTUs are added to the ice block, its temperature increases to 32°F. These 60 BTUs are sensible heat. The ice block is now at its melting point. Another way of thinking of the ice block at this point is that it is *saturated* with heat. Any additional heat added to the block at this time will not raise its temperature, it will cause the ice to melt.

To melt one pound of ice, 144 BTUs are needed. This heat is latent heat. To melt the entire 10-pound block, 1440 BTUs of latent heat are required.

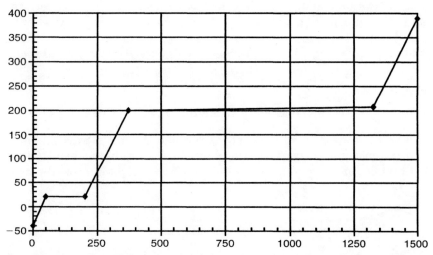

Figure 3-6 A graph tracking temperature versus heat content for water changing from a solid to a gas.

At this point the block of ice has become a 10-pound puddle of 32°F water. Any heat added to the water at this time is sensible heat and will raise the temperature of the water. If any heat were lost from the water, it would begin to turn back into solid ice. We already know that the specific heat of water is 1 BTU. So adding 10 BTUs of sensible heat will raise the temperature of the water 1°F. If we want to raise the temperature of the water to its boiling point of 212°F, we must add the following amount of heat:

$$10 \text{ BTU/°F} \times (212°F - 32°F) = 10 \text{ BTU/°F} \times 180°F = 1800 \text{ BTUs}$$

After 1800 BTUs of sensible heat is added to the 10 pounds of water, it has reached another saturation point. Adding additional heat will not increase the temperature, it will cause the water to boil and change from a liquid to a gas. At atmospheric pressure, it takes 970 BTUs of heat to change one pound of water into one pound of steam. So we require 9700 BTUs to boil off the water into steam.

Rather than allow the steam or vapor to escape into the atmosphere, let's capture the steam in a sealed container that keeps the steam at the same temperature (212°F) and pressure. If any heat were lost from the steam it would condense back into water. The same thing happens when you boil a pot of water on a cold day and moisture forms on your kitchen windows.

Heat can also be added to the steam to increase its temperature above 212°F. Remember 0.5 BTU of sensible heat will increase the temperature of one pound of steam 1°F. Raising the temperature of a vapor above its boiling point is called *superheating*. The sensible heat a vapor contains above its boiling point is called *superheat*. The terms *superheat*, and *saturation point* are very important in understanding how refrigerants behave. These terms will come up over and over again in discussing and servicing refrigeration equipment, so be sure you understand them.

Heat transfer

Heat can travel from one object to another in one of three ways:

- *Radiation.* Radiated heat moves through the air in waves similar to radio waves (Fig. 3-7). The sun's heat is radiated heat. Other examples of radiated heat are the heat from a light bulb, or the heat

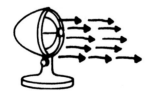

Figure 3-7 Radiated heat moves through the air, such as the heat from the sun or a lamp.

felt when standing close to a fire. All objects will absorb and give off radiated heat. Darker, rough-textured surfaces absorb more heat than light colored, polished surfaces. Both air and glass absorb very little heat by radiation. Radiated heat from light fixtures, large windows, and other sources must be taken into account when determining the cooling load of a particular refrigeration system.

- *Conduction.* Conduction is how heat moves through a single object or two objects touching one another. If a large cast iron frying pan is heated on a gas stove, heat will gradually move from the base of the pan out into the handle area (Fig. 3-8). This heat has traveled through the pan by conduction. Conduction also occurs when heat moves from one object to another through direct contact. In the example given earlier, the small metal bar gave up heat to the larger anvil through conduction. In a refrigeration system, you often have to think "in reverse" to understand how conduction is working. As air is passed over the coils of the cold evaporator, heat from the air moves through the evaporator coils and into the refrigerant via conduction. The warm air loses heat to the evaporator and becomes cool. As this cool air moves through the refrigerated area (a process know as *convection*) it contacts food and objects that are warmer than it is. Heat from the stored food is transferred to the cold air via conduction.

  Conduction can be improved by providing a large conducting surface made of good conducting materials. Copper, steel, and most metals are excellent conductors of heat, which is the reason why refrigeration lines, evaporators, and condensers are constructed of copper or steel tubing. Increasing the conduction area is done by tightly coiling the tubing or by installing metal fins to the tubing.

- *Convection.* Convection is when heat travels via an easily movable medium, which can be either a liquid or a gas. The heat is transferred to the medium at one source and the medium is moved to another location where it gives up this heat to the surrounding area. Convection heat transfer is used in a refrigeration system's con-

Figure 3-8 Heat moves through solid objects via conduction.

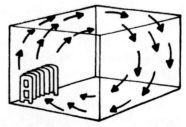

Figure 3-9 Convection moves heat by the flow of a gas or liquid.

denser (Fig. 3-9). Air passing over the condenser coils picks up heat from the coils that is vented to an outside area. Water can also be circulated around the condenser coils to pick up excess heat, and then pumped to another area where it gives up this heat.

Convection is also used to cool a refrigerated space. As mentioned earlier, heat is removed from the surrounding air by the evaporator and the air is then moved to the refrigerated space where it picks up heat from the items to be kept cold or frozen. Convection is improved if the flow of the moving medium is increased. This is why electric fans are normally used to blow air over the evaporator or condenser coils.

Pressure

Because a refrigeration system is a closed loop containing both liquid and vaporized refrigerant, it is a pressurized system. It is important to understand the relationship between pressure and temperature within a refrigeration system. Operation of the vapor-compression cycle depends mainly on pressure differences in the system.

Pressure is defined as force per unit area. Pressure can be measured in one of three ways: pounds per square inch (psi), inches of mercury (in. Hg), inches of water column. The refrigeration industry uses all three of these measuring systems, and understanding why and how each is used can be confusing at first.

Gauge versus absolute pressure

Although we are not aware of it, the air in the earth's atmosphere exerts a downward pressure on us and all things around us. At sea level this pressure is 14.696 pounds per square inch, commonly rounded off to 14.7 psi.

Pressure in a refrigeration system is measured using a manifold gauge. The gauge reads zero when it is not connected, but atmospheric pressure (14.7 psi) is always present at the gauge input. When the gauge is used to take a pressure reading, the pressure exerted on the gauge scale must overcome this 14.7 psi before the needle will move in the positive direction. So the absolute pressure existing in the system is actually 14.7 psi greater than the reading on the manifold gauge. To let the technician know that this is so, pressure readings taken by a manifold gauge are usually noted as the pounds-per-square-inch gauge (psig) difference between the actual reading and the absolute pressure.

In working out most pressure and volume problems, *absolute pressures* are used. Absolute pressure is simply gauge pressure plus atmospheric pressure. For example, if a manifold gauge connected to a refrigeration line reads 30 psig, the absolute pressure in the line is:

$$30 \text{ psig} + 14.7 \text{ psi} = 44.7 \text{ psia}$$

Because of the precise nature of pressure measurement in refrigeration work, most pressures are designated as psig or psia. If the pressure is simply listed as psi, it is commonly assumed that this is gauge pressure (psig).

Vacuum and inches of mercury

Any time pressure falls below the standard atmospheric pressure of 14.696 psia, it is said to be going to vacuum. On the absolute pressure scale, a perfect vacuum in which no pressure exists at all will register 0 psia. If the psig scale were used to read vacuum pressure, it would need to use negative numbers, with a perfect vacuum registering −14.7 psig.

Vacuum is commonly measured in inches of mercury (in. Hg). One psi equals 2.036 in. Hg. So a perfect vacuum would draw 29.92 in. Hg. Table 3-2 lists the most popular scales for measuring pressure and vacuum.

Because the refrigeration technician must often test both pressure and vacuum in the same system, manifold gauges are designed to measure both. These gauges are called *compound manifold gauges*. Pressures less than zero psig are measured on the manifold gauge with a scale calibrated in inches of mercury (Fig. 3-10).

Inches of water

When working with natural gas or propane regulators used in absorption refrigeration systems, extremely low temperature measurements

TABLE 3-2

	Pressure—pounds per square inch			
	Absolute (psia)	Gauge (psig)	Vacuum (in. Hg)	kPa
Positive				
pressures	100	85	—	690
	90	75	—	621
	75	50	—	518
	50	35	—	345
	40	25	—	276
	30	15	—	207
Atmospheric				
pressure	14.7	00.0	—	101.3
Vacuum				
pressures	10	−5	25.4	69
	5	−10	50.8	35
	0	−14.7	76	0

might be needed. In this case the mercury system is not precise enough. It is replaced with a 0 to 33 feet column of water scale. On the water scale, 27.7 inches of water column equals 1 psi.

The pressure–temperature relationship

The following law of physics is another important key to understanding refrigeration: The boiling point of a liquid can be changed and controlled by controlling the vapor pressure above the liquid.

To demonstrate this law, we will again look at water and how it behaves under different pressures. Water boils at 212°F at sea level where the atmospheric pressure pressing down on it is 14.7 psi. At higher altitudes, this is not true. For example, if we place a thermometer in a pan of boiling water located at the top of a 5000-foot mountain, it will read 203°F. The reason for this is that atmospheric pressure at 5000 feet above sea level is about 12.24 psi. With less pressure pushing down on the surface of the water, less heat is needed to move the water molecules apart and form steam. The water becomes saturated with heat at a lower temperature (203°F) and begins to boil. At this altitude or pressure steam will also condense back to a liquid at temperatures below 203°F.

Based on the same principle, the boiling point of water will increase if the pressure bearing down on the surface of the water increases. This

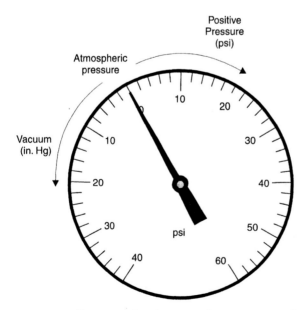

Figure 3-10 Various scales for measuring pressure and vacuum.

is the operating principle used in a pressure cooker. If the vapor pressure in the sealed pressure cooker is increased to 30 psi, any water in the cooker or the water content of food in the cooker will boil at approximately 250°F. The water does not become saturated with heat until it reaches 250°F. The food cooks faster because it can retain more heat.

Equally important, the water vapor above the liquid in the pressure cooker will condense back to a liquid at a temperature below 250°F. The fact that vapor under high pressure will condense back into a liquid at higher temperatures is extremely important in refrigeration. The compressor in the system raises the pressure of the vapor refrigerant to a point where it will condense back into a liquid at a relatively high temperature in the condenser.

Water as a refrigerant

It's difficult to think of water as a refrigerant because under normal conditions heat from an external source has to be added to water to raise its temperature and change its state from a liquid to a vapor. But if the pressure around the water were lowered sufficiently, the water would boil at a lower temperature. An external heat source would not be needed, and the water would absorb the heat energy needed to change its state from the air and the objects in the area around it.

For example, consider a pan of water in a sealed container. Both the water and the air inside the container are at room temperature 68°F. A vacuum pump is connected to a valve on the container wall and air is pumped out of the container. When the pressure inside the container is lowered to approximately 0.36 psia, the water will start to boil. As it boils, the water absorbs heat from the air remaining in the container and the container walls. The temperature inside the sealed container drops dramatically as its sensible heat is used as latent heat to convert water from liquid to vapor.

So at very low pressures, water would work just fine as a refrigerant. But operating at pressures lower than 0 psig are not practical in a refrigeration system. Such a system would be under vacuum. If a leak occurred, the refrigerant would not leak out. Instead, outside air would be sucked into the refrigerant lines.

For refrigeration systems operating in the 35° to 40°F range, operating pressures inside the evaporator are typically between 30 and 70 psig, depending on the refrigerant used.

Saturated vapor

Refrigeration systems have a precise relationship between pressure and temperature of liquids and gases within a fixed volume. In a sealed refrigeration system, where both liquid and gaseous refrigerants are present, the internal pressure is determined by its *saturated vapor pressure*. This is a condition of equilibrium in which the liquid refrigerant boils off, at the prevailing temperature, until the pressure in the system is at the point where any added pressure caused by boiling would force the vapor to condense back into a liquid.

The saturated vapor pressure of any liquid is directly proportional to its temperature. It is a known and predictable quantity, a fact that is important to remember when diagnosing or servicing a unit. At any given pressure reading, the accompanying temperature reading for that refrigerant is the boiling point of that refrigerant at that pressure level.

The pressure–temperature relationship in refrigerants is available to us in the form of a prepared chart on which several refrigerants are listed. Pocket-size versions of the chart are always included in the technician's tools and equipment because it is a very valuable device when used along with pressure gauges. Evaporating and condensing temperatures of a refrigeration system are easily determined when the pressures are known.

Critical temperature

The *critical temperature* of a substance is the highest temperature at which the substance will remain a liquid regardless of the pressure applied to it.

Critical pressure

The *critical pressure* is the minimum pressure needed to liquify a gas that is at its critical temperature. Less pressure will not liquify the substance.

Specific gravity

Specific gravity indicates the relative density of a liquid or solid substance as compared to water. Water has a specific gravity of one. Objects with a specific gravity of less than one will float on water. Those with specific gravities greater than one will sink.

Brines are mixtures of salt and water. Brines have specific gravities greater than one. They are useful in some refrigeration and air-conditioning applications because they freeze at temperatures below 32°F and boil at temperatures above 212°F at atmospheric pressure.

Specific volume

Specific volume is used to compare the densities of gases. Specific volume is the volume one pound of a given gas will occupy under the standard atmospheric conditions of 68°F and 14.7 psi. The relative density of gases is defined as the ratio of the mass of a given volume of a gas as compared to an equal volume of hydrogen gas at standard atmospheric conditions. For example, the volume of one pound of clean dry air under standard atmospheric conditions is 13.45 cubic feet. The same amount of hydrogen occupies 178.90 cubic feet.

Cooling capacity

The *cooling capacity* of a refrigeration system can be expressed using the term *tons of refrigeration*. This is an older term but it might still be used. One ton of refrigeration is the rate of cooling produced when one ton (2000 pounds) of ice melts during a period of one day (24 hrs.). The ice is assumed to be a solid at 32°F when cooling begins and a liquid at 32°F when the cooling period stops. The amount of heat energy absorbed by the ice is the total amount of latent heat used to change the solid ice to a liquid. As mentioned earlier in this chapter, it takes 144 BTUs to change one pound of ice into one pound of water. So a ton of refrigeration is 2000 pounds × 144 BTUs/pound or 288,000 BTUs/24 hours.

A refrigeration system is often rated on its ability to absorb heat over a 24 hour period. The heat absorbing ability (HA) is expressed as tons of refrigerating effect. For example, a walk-in freezer that can absorb 2,304,000 BTUs per 24 hours is said to have 2,304,000/288,0000 = 8 tons of refrigeration effect.

Enthalpy

Enthalpy measures the energy content of a given substance. *Enthalpy* is defined as all the heat in one pound of a substance from a given reference temperature. In refrigeration systems this reference temperature is −40°F. Both the temperature and pressure of the substance determine the amount of enthalpy.

Specific enthalpy is enthalpy per a given unit of mass. It is commonly stated as BTU per pound. The enthalpy of refrigerants is discussed in greater detail in later chapters of this book.

Relative humidity

Air absorbs moisture. The exact amount depends on the temperature and pressure of the air. Warm air absorbs more moisture than cold air. That's why we have hot and humid weather during the summer months, and cold and dry conditions during most of the winter. Relative humidity states the amount of moisture the air is holding as compared to the amount of moisture it could hold at that temperature and pressure. For example, a relative humidity of 60 percent means the air is holding 60 percent as much moisture as it will hold at the present temperature and pressure conditions.

Vapor compression cycle

As discussed earlier in this chapter, vapor compression refrigeration is based on the fact that as the refrigerant changes state from a liquid to a vapor, it absorbs heat. As the vapor changes back to a liquid, it gives up heat. A basic vapor compression refrigeration system is illustrated in Fig. 3-11. Now that we understand how heat and pressure behave in a closed system, it's time to study the vapor compression cycle in more detail.

The cycle begins as the liquid refrigerant flows through the flow control into the evaporator. Once in the evaporator, the liquid refrigerant absorbs heat from the surrounding air and begins to boil. The temperature at which boiling occurs depends on the pressure in the evaporator and the refrigerant in the system. For this example, HFC-134a is the refrigerant. A manifold gauge connected to the low side of the system reads 22.1 psig. The HFC-134a Saturated Vapor Temperature/Pressure data given in Table 3-3 tells us the refrigerant is boiling off at 25.5°F. In the first half of the evaporator, the refrigerant becomes a low-pressure liquid and vapor mixture. As the refrigerant flows through the evaporator, more and more liquid changes to a vapor. The heat being used is latent heat. The temperature of the liquid/gas mix remains a constant 25°F. At point A in Fig. 3-11, all of the liquid refrigerant has changed into a vapor at 25.5°F.

As the vapor flows through the remaining section of the evaporator, it picks up additional sensible heat and the temperature increases to

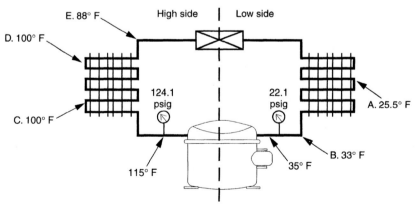

Figure 3-11 Reading positive pressure and vacuum on a manifold gauge.

TABLE 3-3

HFC-134a Saturated Vapor Temperature/Pressure Data			
Temperature °F	Pressure psig	Temperature °F	Pressure psig
−15.0	0.0	50.0	45.4
−10.0	2.0	55.0	51.2
−5.0	4.1	60.0	57.4
0.0	6.5	65.0	64.0
5.0	9.1	70.0	71.7
10.0	12.0	75.0	78.6
15.0	15.1	80.0	86.7
20.0	18.4	85.0	95.2
25.0	22.1	90.0	104.3
30.0	26.1	95.0	113.9
35.0	30.4	100.0	124.1
40.0	35.0	105.0	134.9
45.0	40.0	110.0	146.3

33°F, at point B. The refrigerant at point B is now superheated 8°F. Remember that superheat is the number of degrees a refrigerant is heated above its boiling point. In refrigerant work, it is safe to assume that any superheated refrigerant is in its vaporized state.

As it exits the evaporator, the low-pressure, low-temperature gas is pulled through the suction line to the compressor. As it passes through the suction line, the vapor picks up additional superheat and enters the compressor at a temperature of 35°F. The job of the compressor is to receive low-pressure refrigerant vapor from the evaporator and compress it into high-pressure refrigerant vapor. In our example, the compressor draws in vapor from the suction line at 35°F and 22.1 psig. The compressor raises the vapor pressure to 124.1 psig as it exits the compressor to the hot gas line. The heat of compression raises the vapor's temperature to 115°F in the hot gas line. Checking the temperature/pressure table for HFC-134a indicates that its saturation temperature at 124.1 psig is 100°F. But the actual temperature of the refrigerant measured in the hot gas line is 115°F. The vapor refrigerant is superheated 15°F. So before the vapor can begin to change back into a liquid, it must be cooled (desuperheated) 15°F to the condensing temperature of 100°F for 124.1 psig.

Slight cooling occurs in the hot gas line, but the majority of cooling takes place in approximately the first 5 percent of the condenser. At point C in Fig. 3-11, the refrigerant has cooled to 100°F and condensing begins. As the refrigerant flows toward point D, more and more of the vapor changes to liquid, still at 100°F. It takes about 90 percent of the condenser surface to condense all of the vapor back into a liquid. The refrigerant then flows through the rest of the condenser as a liquid.

Because all latent heat has been removed, any heat removed from the liquid refrigerant is sensible heat and results in a steep drop in temperature. At point E, the refrigerant temperature has dropped to 88°F, 12 degrees below the saturation (condensing) temperature of 100°F. The refrigerant has been subcooled 12 degrees. *Subcooling* is the number of degrees that the liquid refrigerant is cooled below its saturation temperature.

The high-pressure liquid refrigerant flows through the liquid line to the flow control device that meters it into the evaporator. Some subcooling occurs in the liquid line, so the refrigerant enters the flow control device at a temperature of 85°F. The refrigerant flow control is simply a restriction in the line through which refrigerant must pass. Thermal expansion valves (TEVs *or* TXVs) and capillary tubes are the two most widely used flow-control devices. Both are designed to allow only the proper amount of liquid refrigerant to pass through back to the low side of the system. For example, a thermal expansion valve can be set to allow a liquid flow of 12 pounds of HFC-134a per minute.

Bubbles in a liquid refrigerant stream are known as *flash gas*. Flash gas in the liquid line reduces the refrigerant flow through the control device. Vapor cannot exist in subcooled refrigerant. So if the system has sufficient condensing area, only subcooled liquid refrigerant will enter the liquid line and reach the flow-control device.

As refrigerant passes through the flow control, it moves from an area of high pressure (124.1 psig) to one of low pressure (22.1 psig). At the low-side pressure of 22.1 psig, the saturation temperature of HFC-134a is 25.5°F. But remember, the refrigerant temperature is still approximately 85°F. Because a liquid cannot exist at a temperature above its saturation point, a portion of the liquid refrigerant immediately boils or flashes off into a vapor. This boiling process quickly absorbs heat from the liquid refrigerant, cooling it to the low-side saturation temperature of 25.5°F. In a typical refrigeration system, one pound of refrigerant might flash off for every five pounds that passes through the flow-control device. So shortly after it clears the flow control, the refrigerant is about 20 percent vapor and 80 percent liquid. The 80 percent liquid is available to absorb heat in the evaporator and change to a vapor. The 20 percent of refrigerant lost to flash off is now vapor and cannot contribute to the system's cooling capacity. A certain percentage of flash gas leaving the refrigerant flow control is unavoidable. But any flash gas entering the flow control is undesirable and should be avoided.

Once the vapor/liquid mixture moves into the evaporator, the vapor compression cycle is complete and is set to continually repeat itself until the compressor is shut off or problems develop in the system.

Chapter 4

Refrigeration System Components

This chapter explains the construction and function of the components that make up a vapor-compression refrigeration system. Refrigeration systems can be either of the following:

- Self-contained units with the compressor, evaporator, condenser, and all piping housed within a single enclosure. Typically, self-contained systems are small and relatively simple (Fig. 4-1a).
- Remote or "split" systems where the compressor(s), evaporator(s), and condenser(s) are separate. In many cases, part of the system is located outdoors. Piping connects the various components to form the system (Fig. 4-1b). Remote systems can run from small and simple to very large and complex.

As explained in the previous chapter, refrigeration systems operate in a continuous cycle or loop. As shown in Fig. 4-2, this loop can contain many more than the basic components outlined in Chapter 3.

Low-temperature, low-pressure refrigerant vapor absorbs heat and is then compressed into high-temperature, high-pressure vapor that then releases this heat. A refrigerant metering or flow-control device separates the high-pressure (condenser) side of the system from the low-pressure (evaporator) side. The control device creates a restriction against which the compressor pumps. This creates high-side pressure. The control device also meters the correct amount of high-pressure liquid refrigerant into the evaporator where it then boils off into low-pressure/temperature vapor. Because the control device acts as a dividing line between the high and low sides of a refrigeration system, it is the logical point to begin a closer look at refrigeration components.

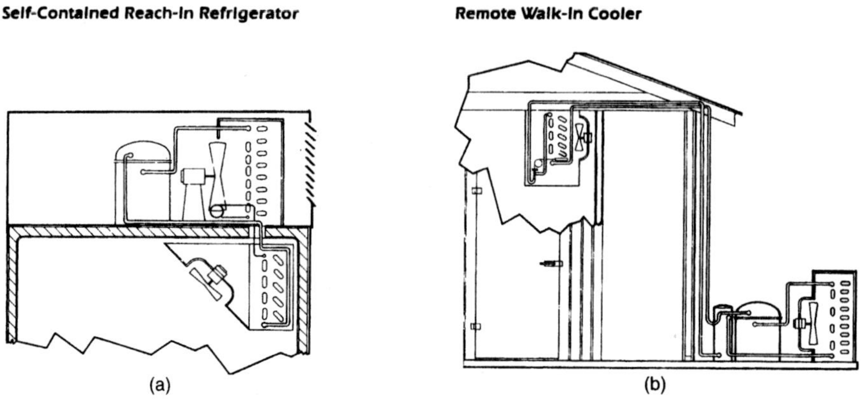

60   Chapter Four

Self-Contained Reach-In Refrigerator

Remote Walk-In Cooler

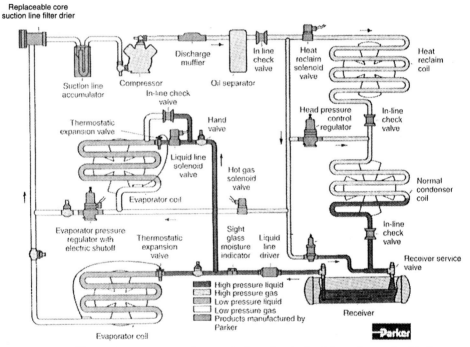

(a)   (b)

Figure 4-1 A self-contained, reach-in refrigerator (a), and a split-system, walk-in cooler (b). *(Ranco Incorporated)*

Figure 4-2 The major components in a typical mechanical refrigeration cycle. *(Parker Hannifin Corporation)*

Types of Control Devices

The selection of the control device is important to the operation of the refrigeration system. A control device that does not fit the application or is incorrectly sized can cause poor system performance. For example, if the control device is undersized, not enough refrigerant will flow into the evaporator and the system's cooling capability will decrease.

An oversized expansion device might allow too much refrigerant into the evaporator causing liquid refrigerant to flow back to the compressor. This condition is referred to as *floodback*. Both conditions will result in compressor damage if not promptly corrected.

Refrigerant control devices can be divided into four general types:

- Fixed opening restrictors, such as capillary tubes and plug orifices
- Automatic (constant pressure) expansion valves
- Thermostatic expansion valves
- Electronic expansion valves

Each type of control device has its advantages and disadvantages that make it best suited for certain types of applications.

Capillary tubes

Capillary tubes are often used as a means of controlling liquid refrigerant in self-contained refrigeration equipment, such as domestic refrigerators and small commercial applications. The capillary or "cap" tube is made of a length of small-diameter tubing (Fig. 4-3). Capillary tubes have inside diameters as small as 0.031 inch (1 millimeter). The inside diameter of the tube is constructed to very close tolerances.

The cap tube acts as a fixed size opening through which the proper amount of refrigerant flows into the evaporator. The capillary tube creates resistance to refrigerant flow by friction. A small amount of liquid moving through a small diameter tube moves slowly because there is more friction than in a larger diameter tube. Friction can also be increased by increasing the length of the cap tube. The diameter and length of the tube vary based on the BTU rating of the refrigeration unit. The BTU rating is generally specified by the unit's manufacturer.

The capillary tube opening is always open. When the system compressor shuts off, liquid refrigerant continues to flow into the evaporator until the high- and low-side pressures are equal.

Advantages of capillary tubes. The major advantages of the capillary tube are as follows:

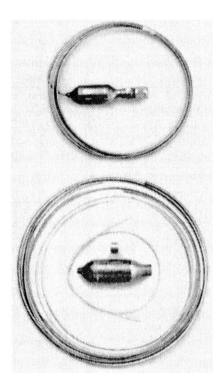

Figure 4-3 Capillary tubing with a mesh strainer. *(Watsco Components, Inc.)*

- It has no moving parts to wear out or stick. The capillary tube is a simple, trouble-free device if it is kept free of foreign material and moisture.
- It is relatively inexpensive when compared to a thermal expansion valve or automatic expansion valve.
- It allows the high-pressure and low-pressure sides of the system to equalize between operating cycles. This means that the unit might use a more economical, lower starting torque compressor.
- Because capillary tube systems are critically charged, no receiver is necessary for the storage of excess refrigerant.

Disadvantages of capillary tubes
- The capillary tube's ability to adjust to load fluctuations or varying ambient operating conditions is limited when compared to a thermal expansion valve.
- Repair and installation of capillary tubes are not as simple as those for thermostatic expansion valve systems.

Bubble point and metering control

Although a capillary tube system works at maximum efficiency at only one combination of high- and low-side pressures, it does have some ability to adjust refrigerant metering based on operating conditions.

The bubble point is the location in the capillary tube where liquid refrigerant begins to boil into vapor. Upstream of the bubble point, on the high side of the system, the refrigerant is all liquid. Downstream on the low side, the refrigerant is a mixture of vapor and liquid.

As the amount of vapor/liquid mix in the capillary tube increases, the resistance to flow also increases. This is an important fact to remember because the bubble point is not stationary. It moves based on operating conditions in the system, and as it moves the section of cap tube containing a vapor/liquid mix increases or decreases (Fig. 4-4).

For example, if large amounts of warm air pass over the evaporator coils, load on the system increases, and more and more refrigerant boils off in the evaporator. The compressor now has more vapor to pump. Its operating efficiency goes up, and the amount of high-temperature, high-pressure liquid refrigerant being pumped into the condenser increases. In turn, the pressure inside the condenser increases. This

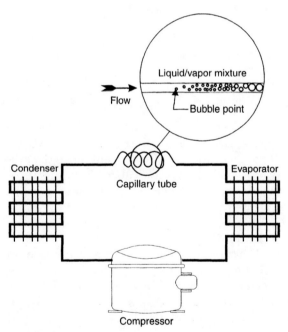

Figure 4-4 The location of the bubble point in the capillary tube affects the rate at which refrigerant is metered into the evaporator.

increased high-side pressure pushes against the bubble point in the capillary tube, moving it closer to the evaporator end of the tube. As the bubble point moves closer to the end of the cap tube connected to the evaporator, most of the cap tube is now filled with liquid refrigerant, not a vapor/liquid mix. Resistance to flow decreases and a greater volume of refrigerant enters the evaporator.

When system load decreases, the opposite occurs. Less refrigerant evaporates in the evaporator. The density of refrigerant vapor entering the compressor is lower, so compressor efficiency is lower. Less liquid refrigerant backs up in the condenser, so high-side pressure acting against the cap tube bubble point is decreased. The bubble moves closer to the condenser side end of the capillary tube. Most of the tube is now filled with a vapor/liquid mix. Resistance through the tube increases and a smaller volume of refrigerant enters the evaporator.

Suction accumulator

Under very low load conditions, the capillary tube might not be capable of decreasing flow low enough to prevent some liquid refrigerant from entering the evaporator. When liquid refrigerant in the evaporator might be a problem, a suction accumulator is installed in the system's suction line between the evaporator and compressor. (See Fig. 4-2.) The suction line accumulator provides a vessel where the liquid refrigerant can collect before it reaches the compressor. Always remember that liquid refrigerant cannot be compressed and must never reach the compressor.

Installation and service

It is very important that dirt or water does not enter a capillary tube during installation and service. Even the smallest amount of dirt can easily clog the tube's small passageway, and quickly shut down the system. A small pocket of water will freeze and also block the tube. Because dirt and water present such a great danger to smooth capillary tube operation, many capillary tubes are connected to a small filter-drier assembly. Some systems use a small screen mesh filter that has no moisture-removing capabilities.

If you suspect blockage in the capillary tube, service it as a freeze-up first. If this does not correct the problem, dirt is the likely cause, and the cap tube must be cleaned or replaced. While tools are available for cleaning the cap tube bore (see Chapter 2), it is best to replace a suspected cap tube, particularly when the relatively low cost is weighed against a return service call.

When changing out a bad capillary tube, the replacement tube must have the same inside diameter (bore size) and length as the original.

When cutting the replacement tube to length, take the time to make a clean smooth cut. Any burrs left on the inside of the tube act as unwanted restrictions in the line. The cap tube also must be installed in the same manner. Capillary tubes can be installed as a straight length of tubing, but more often they are coiled to reduce the amount of space required. Altering the size spacing of the coils can change the flow resistance inside the tube, so take the time to reproduce the original configuration as closely as possible. The tube is commonly strapped or soldered to the suction line, which helps subcool the liquid refrigerant inside the cap tube. Subcooling has a definite effect on the high-side pressure acting on the tube's bubble point, so reinstall the tube using the same method. Solder the replacement tube in place if the original was soldered. Strap the new tube down if the old one was held in this manner.

Temperature control

Themostats are always used for temperature control in capillary-tube controlled refrigeration systems. Because high- and low-side pressures in a capillary tube refrigeration system equalize when the compressor shuts off, temperature controls that operate based on system pressures will not work. The compressor would simply short cycle on and off and quickly fail.

Float valves

Float valves are another type of refrigerant flow control device. A float valve opens or closes its valve port as the liquid level in the float body changes. Both high-side and low-side float valves are available. Each operates in a slightly different way.

High-side float valves. A high-side float valve controls the rate liquid refrigerant flows into the evaporator so it equals the rate at which vaporized refrigerant exits the evaporator on its way to the compressor. Liquid refrigerant from the condenser enters the valve's float body. As the liquid level in the float body rises, the float ball moves upward, pulling the valve pin out of the valve port so liquid refrigerant can flow into the evaporator (Fig. 4-5).

Systems controlled by a high-side float valve are called *flooded systems* because most of the refrigerant charge is in the evaporator at any given time. It is also critical that a flooded system be charged with just the right amount of refrigerant. Too much refrigerant will cause floodback, while too little charge will reduce system capacity. Most systems using a high-side float valve as a refrigerant control device are single evaporator, compressor, and condenser systems.

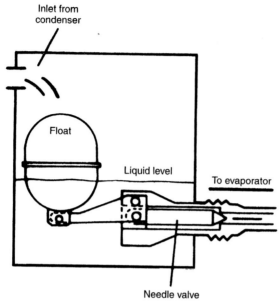

Figure 4-5 Details of a high-side float valve refrigerant flow control.

Low-side float valves. As its name suggests, a low-side float valve is connected to the low-pressure side of a refrigeration system. The float and valve mechanism are located inside the evaporator body, which is especially constructed to retain a constant level of liquid refrigerant under low-side pressure. When the liquid level in the evaporator drops, the float opens the valve to allow high-pressure liquid refrigerant from the liquid line to enter the evaporator (Fig. 4-6).

The compressor is normally controlled by a pressure switch on the suction line or evaporator. Refrigerant charge in a low-side float-controlled system is not critical, but the system does require a substantial charge to keep both the evaporator and high-side receiver supplied with refrigerant. Low-side float systems are often found on drinking fountains and other applications where a constant temperature is required. Because the liquid refrigerant in the evaporator will absorb heat during both the on and off cycles, low-side operating pressures will vary based on surrounding ambient temperatures.

Automatic (constant-pressure) expansion valves

Like the capillary tube and float valve, the automatic expansion valve (AEV or AXV) is best suited for refrigeration applications having a

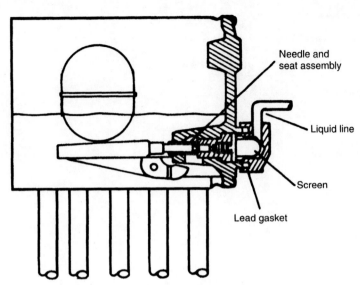

Figure 4-6 Details of a low-side float valve refrigerant flow control.

more or less constant evaporator loading, such as small commercial systems. The AEV (Fig. 4-7) is controlled by low-side pressure. When the evaporator is placed under increased heat load, a motor control sensor mounted near the outlet of the evaporator switches on the compressor. Once the compressor is running, the pressure inside the evaporator is lowered. This drop in pressure opens the valve a preset amount and liquid refrigerant is sprayed into the evaporator. The refrigerant vaporizes at a constant low pressure. The evaporator gradually cools. When the motor control sensor reads a sufficiently cool evaporator outlet temperature, it turns the compressor off. This immediately increases pressure inside the evaporator and the automatic expansion valve closes.

The opening pressure of the AEV is adjustable and must be set for the refrigerant being used. They also must be adjusted to atmospheric pressure, so operating such valves at high altitudes requires special adjustment.

The major drawback of automatic expansion valves is that they meter refrigerant into the evaporator at only one rate. As the heat load on the evaporator rises, the AEV decreases refrigerant flow to maintain evaporator pressure at the valve's setting. Conversely, the AEV increases refrigerant flow when the evaporator heat load decreases to maintain evaporator pressure at the valve's setting. As a result, the AEV starves the evaporator at high-load conditions, and overfeeds it at low-load conditions.

68   Chapter Four

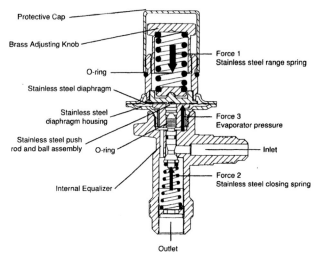

Figure 4-7 A constant-pressure or automatic expansion valve. *(Parker Hannifin Corporation)*

Thermostatic expansion valves

The thermostatic expansion valve (TEV or TXV) is the most popular device used today for controlling the flow of refrigerant into an evaporator.

A thermostatic expansion valve (Fig. 4-8) is designed to regulate the flow of liquid to the evaporator, at a rate equal to the evaporation of the liquid in the evaporator. This is done by maintaining a preset superheat at the evaporator outlet (suction line). As defined in Chapter 2, superheat is the temperature the refrigerant vapor is above its saturated vapor temperature. This superheat is normally about 10°F. In most thermal expansion valves, this superheat level might be adjusted by turning the superheat adjustment screw on the TEV. Thermostatic expansion valves are normally preset by the manufacturer and do not require adjustment in the field unless the valve spring weakens over time.

Maintaining a preset level of superheat at the outlet of the evaporator ensures that all liquid refrigerant vaporizes in the evaporator. No liquid refrigerant can reach the compressor.

Basic operation

The thermostatic expansion valve is installed in the liquid line at the evaporator inlet. Like all other control devices, it separates the high and low pressure side of the system. A thermostatic expansion valve is made of several major components (Fig. 4-9). A *sensing bulb* connects to the TEV by a length of capillary tubing. This tubing transmits bulb

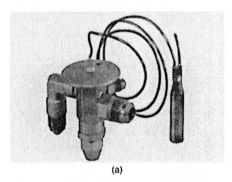

(a)

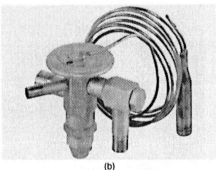

(b)

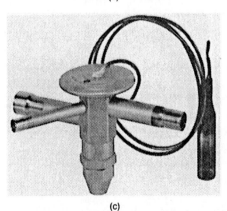

(c)

Figure 4-8 Thermostatic expansion valves: general valve for commercial applications (a), removable strainer design (b), and straight-through valve with extended sweat connections (c). *(Parker Hannifin Corporation)*

pressure to the top of the valve's *diaphragm*. The sensing bulb, capillary tubing, and diaphragm assembly is referred to as the *thermostatic element*. In most cases the thermostatic element is replaceable.

The sensing bulb is filled (charged) with refrigerant. The bulb might be *system charged* (charged with the same refrigerant used in the system) or *cross charged* (refrigerant different from that used in the system). Most thermostatic expansion valves are cross charged.

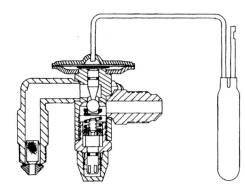

Figure 4-9 Major components of a thermostatic expansion valve. *(Parker Hannifin Corporation)*

The sensing bulb is securely attached to the suction line near the outlet of the evaporator. The refrigerant charge inside the bulb is sensitive to the temperature of the vapor at the evaporator outlet. The hotter the vapor, the more the charge inside the bulb expands. The expanding vapor charge increases pressure inside the bulb and capillary tube. More force is exerted onto the diaphragm of the TEV and the valve opens to increase refrigerant flow.

If the temperature of the vapor exiting the evaporator decreases, the bulb charge contracts and less pressure is exerted on the valve diaphragm. As a result, the valve closes and the amount of liquid refrigerant fed into the evaporator decreases. The net result of this type of regulation is that the evaporator remains as active as possible under all type of load conditions.

The movable diaphragm actuates members of the valve. Its motion is transmitted to the *pin* and *pin carrier* assembly by one or two *pushrods*. The pin(s) move in and out of the valve *port*. The *superheat spring* is located under the pin carrier, and a *spring guide* sets it in place. On externally adjustable valves, an external *valve adjustment* permits the spring pressure to be altered.

There are three basic pressures acting on the valve diaphragm that affect its movement: sensing bulb pressure (P1), equalizer pressure (P2), and spring pressure (P3). (See Fig. 4-10.) As explained in the previous paragraphs, the sensing bulb pressure is a function of the temperature of the sensing bulb charge. This pressure acts on the top of the valve diaphragm causing the valve to move to a more open position. The equalizer and spring pressures act together underneath the diaphragm causing the valve to move to a more closed position. During normal valve operation, the sensing bulb pressure must equal the equalizer pressure plus the spring pressure. The TEV functions by controlling the difference between bulb and equalizer pressures by the amount of the spring pressure.

Ideally, the bulb temperature will match the refrigerant vapor temperature. As the bulb temperature increases, bulb pressure also

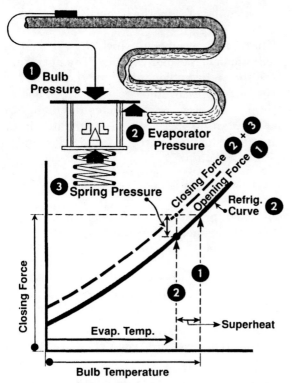

Figure 4-10 Forces acting on a thermostatic expansion valve. *(Sporlan Valve Company)*

increases causing the valve pin to move away from the valve port. This opens the valve, allowing more refrigerant to flow into the evaporator. The valve continues to open until the equalizer pressure increases to the point where the sum of the equalizer and spring pressures balance with the bulb pressure. In the same way, as the bulb temperature decreases, the bulb pressure decreases causing the valve pin to move toward the valve port, closing the valve and allowing less refrigerant to flow into the evaporator. The valve continues to close until the equalizer pressure decreases to the point where the sum of the equalizer and spring pressures balance with the bulb pressure.

Superheat

When discussing TEVs, superheat can be further defined as follows:

Static superheat. The amount of superheat needed to overcome the superheat spring force biased in a closed position. Any additional superheat (force) would open the valve.

Opening superheat. The amount of superheat needed to open a valve to its rated capacity.

Operating superheat. The superheat level at which the valve operates at normal running conditions or normal capacity. The operating superheat is the sum of both the static and opening superheat.

Figure 4-11 illustrates the three types of superheat described above. Their reserve capacity is important because it provides the ability to compensate for occasional substantial increases in evaporator load, intermittent flash gas, reduction in high-side pressure due to low ambient conditions, or shortage of refrigerant.

Calculating superheat

Working with and troubleshooting thermostatic expansion valves requires calculating the amount of superheat at the evaporator outlet (Fig. 4-12). This is done in the following manner.

1. Determine suction pressure at the evaporator outlet with a gauge. On close-coupled installations, suction pressure can be read at the compressor suction connection.

2. Use a pressure-temperature chart to determine the saturation temperature at the observed suction pressure. For example, with an R-22 system: 52.0 psig = 28°F.

3. Measure the temperature of the suction gas at the expansion valve's remote bulb location. For example: 37°F.

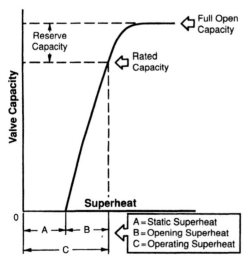

Figure 4-11 A graph of three different types of superheat. *(Sporlan Valve Company)*

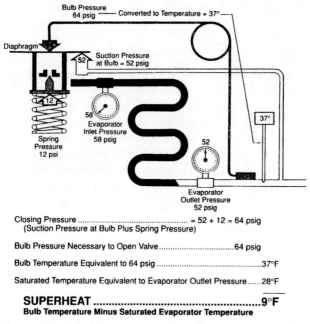

Figure 4-12 Calculating superheat at the evaporator outlet. *(Sporlan Valve Company)*

4. Subtract the saturation temperature of 28°F (step 2) from the suction gas temperature of 37°F (step 3). The difference, 9°F, is the superheat of the suction gas.

A change in refrigerant vapor temperature at the outlet of the evaporator can be the result of one of two things: changing the spring pressure by adjusting the valve, or a change in the heat load on the evaporator.

Spring pressure adjustment

The thermostatic expansion valve superheat is adjusted by means of an adjusting screw on the TEV. In most cases the valve is factory set at the static condition described above. Turning the adjusting screw clockwise increases the static superheat, while turning the adjusting screw counterclockwise decreases the superheat. When a system is operating, any adjustments to the TEV will change the operating superheat. The static superheat range of adjustment is 3° to 18°F. One full turn clockwise will typically increase superheat 2° to 4°.

Note: Refer to the valve manufacturer's instructions for specific directions on superheat adjustment.

Spring pressure determines the superheat at which the valve controls. Increasing spring pressure increases superheat, while decreasing spring pressure decreases superheat.

Spring pressure is increased by turning the valve adjustment clockwise. This reduces refrigerant flow into the evaporator. Vapor temperature at the evaporator outlet increases, because the point where the refrigerant completely vaporizes moves further back within the evaporator, leaving more evaporator surface area to heat the refrigerant in its vapor form. The actual refrigerant vapor and bulb temperature will be controlled at the point where bulb pressure balances with the sum of the equalizer and spring pressures. Conversely, decreasing spring pressure by turning the valve adjustment counterclockwise increases refrigerant flow into the evaporator and decreases refrigerant vapor and bulb temperature.

TEV off-cycle pressure equalization

As explained in the earlier sections on capillary tube flow controls, a refrigeration system using low starting torque single phase compressor motors (e.g., a permanent split capacitor motor) requires some means of pressure equalization during system off-cycle. Pressure equalization is necessary because low starting torque compressors are not capable of restarting against a large pressure differential. Typical applications requiring pressure equalization are small systems that frequently cycle on and off in response to a thermostat.

To provide for this pressure equalization, thermostatic expansion valves can be equipped with a *bleed port*. Standard bleed port sizes are: 5 percent, 10 percent, 15 percent, 20 percent, 30 percent, and 40 percent. Bleed ports are designated by the percentage they increase nominal valve capacity at 40°F evaporator temperature. For example, a two-ton TEV with a 30 percent bleed will have the following capacity: $2 \times 1.3 = 2.6$ tons.

The subject of pressure equalization during system off-cycle should not be confused with the external equalizer of the TEV. System pressure equalization is accomplished by allowing a certain amount of refrigerant to leak through a machined notch or hole in the valve seat during system off-cycle. The external equalizer of the TEV, however, simply allows the valve to sense evaporator pressure. The external equalizer does not provide pressure equalization during system off-cycle.

Effects of heat load on operation

As explained earlier, an increase in the heat load on the evaporator causes refrigerant to evaporate at a faster rate. As a result, the point of complete vaporization of the refrigerant flow is moved further back within the evaporator. Refrigerant vapor and bulb temperature increase, causing bulb pressure to rise and the valve to move in the opening direction until the three pressures are in balance. Conversely, a reduction in the heat load on the evaporator will cause the vapor and bulb temperature to fall and the valve to move in a closed direction until the three pressures are in balance. Unlike a change in the spring pressure due to valve adjustment, a change in the heat load on the evaporator does not appreciably affect the superheat at which the thermostatic expansion valve controls. This is due to the fact that the TEV is designed to maintain an essentially constant difference between the bulb and equalizer pressures, thus controlling superheat regardless of the heat load.

Keep in mind that a thermostatic expansion valve monitors evaporator coil temperature alone, and does not monitor any other system temperatures. The valve restricts refrigerant flow as the coils get colder, which lowers the evaporator pressure. If the flow of warm evaporator inlet air is restricted by a clogged air filter, the valve could close, resulting in freezing of the coils.

Noncritical refrigerant charge

The thermostatic expansion valve provides an additional benefit when charging the system with refrigerant. When a TEV is used, the system refrigerant charge is usually not as critical as with a capillary tube or AEV.

Electronic (solid-state) expansion valves

An electronic-controlled expansion valve (Fig. 4-13) uses a thermistor to sense the temperature of the refrigerant in the suction line. Based on input from the thermistor, a low-voltage transformer provides the power to open or close the valve port and increase or decrease refrigerant flow into the evaporator. Additional information on electronically controlled refrigeration systems is detailed in Chapter 5.

Evaporators

An evaporator is found in all vapor compression refrigeration systems. Evaporators vary considerably in design, but the most widely used is the forced-air unit cooler or blower coil (Fig. 4-14). Its basic construction is copper or aluminum tubing held together by sheet metal fins. The fins increase the contact area for air flow generated by an electric motor fan. A controlled amount of refrigerant is metered into the evaporator by the capillary tube or thermal expansion valve. As the liquid

Figure 4-13 Electronic controlled temperature-control valve blower coil-type evaporator. *(Sporlan Valve Company)*

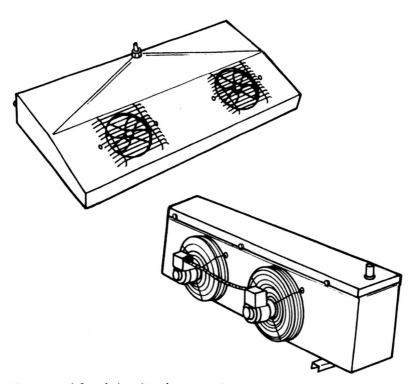

Figure 4-14 A forced-air unit cooler evaporator.

enters the evaporator tubing, the ambient low pressure causes it to evaporate or turn into vapor.

A forced-air unit cooler consists of a direct expansion coil and a motor-driven fan that are both contained in a housing. In small unit coolers, such as those found in reach-in coolers, the coil normally contains only one continuous tubing circuit. As the size of the unit cooler increases, it becomes necessary to divide the coil into two or more circuits to reduce the pressure drop that increases as the length of the circuit increases. The individual circuits are then fed through a specially designed distributor that ensures an equal feed to each circuit. All circuits empty into a common suction header.

Alternate evaporator designs might include flat plate surfaces for heat transfer, bare tubes, or, in the case of many small domestic refrigerators, an ice cube making compartment that acts as the heat-transfer device for a larger compartment. In these instances, natural convection air flow is used.

Evaporators in which refrigerant is fed directly into the cooling coil through the flow-control device are called *direct-expansion evaporators*. In direct expansion evaporators, the heat is absorbed by the evaporator directly from the area to be cooled. This differs from other types of systems where water might be chilled by a refrigeration system. The chilled water is then circulated through the area to be cooled. In this case, the water acts as a refrigerant.

As the refrigerant liquid changes to a gas inside the evaporator, it absorbs heat from surrounding metal surfaces. The warm, humid air passing over the fins and coils gives up heat to the metal areas. This cool air recirculates back to areas that must be refrigerated.

As discussed in Chapter 3, the boiling temperature of the liquid is directly related to the evaporator or suction pressure found in the system. This temperature is normally lower than the dew point of the unconditioned air. This is why water vapor condenses and freezes on the outside of evaporator coils and fins, forcing most system to rely on a defrost system.

Dehumidification of the air will absorb BTU energy in the form of latent heat. The amount of BTUs absorbed will depend on coil temperature and air velocity as well. BTUs that are used up removing moisture from the air cannot be used to lower the temperature. The amount of moisture removed from the air, and the inlet/outlet air differential temperatures for a given unit will depend on the manufacturer's specifications for that unit.

A pressure drop of 1 to 2 pounds in medium- or high-temperature evaporators is acceptable, while ½ to 1 pound is desired in low-temperature evaporators. Excessive pressure drop occurring in the evaporator tends to create a loss of system capacity due to a reduction in suction pressure at the outlet of the coil. If the evaporator tubing is increased too much in an effort to reduce pressure drop, the refrigerant gas velocity might be reduced to a point where oil will accumulate in the tubing and it will not return to the compressor. Also, the heat transfer ability of the coil will be significantly reduced if the gas velocity is not sufficient to literally "scrub" the interior tubing wall, keeping it free of an oil film. So, a compromise is really necessary in the final design of an efficient evaporator.

Temperature difference-humidity relationship

In the selection of a forced-air unit cooler for a given installation, one controllable variable is the *temperature difference*, or *TD*. TD is the difference in temperature between the evaporating refrigerant and the area being cooled. Temperature differences of 5 to 20 degrees are widely used, but as a general rule, the TD should be kept as low as possible to permit efficient operation of the compressor that occurs at higher suction pressures.

The amount of moisture condensed out of the air is directly related to the temperature of the evaporator coil. Careful selection of the unit cooler will provide the desirable condition of humidity in the refrigerated space. If the TD is too large (the evaporating temperature of the coil is much colder than that of the refrigerated space), low humidity and dehydration will be a problem. In the case of perishable fruits and vegetables, a very high relative humidity is desirable. To achieve this condition, a unit cooler should be selected that will provide a TD of 8 to 12 degrees. The rule is: The larger the coil, the higher the humidity.

Multicircuited evaporators

Multicircuited evaporators use distributors to equalize the refrigerant feed to each circuit. They normally require a thermostatic expansion valve with an external equalizer connection attached to an external equalizer tube that originates near the outlet of the evaporator. This fitting routes evaporator outlet pressure under the thermostatic expansion valve diaphragm instead of inlet pressure. This allows the valve to operate without the influence of the pressure drop.

Suction Line

The passageway from the evaporator to the compressor is commonly called the *suction line*. Check valves that allow fluid flow in one direction only are often installed in the suction line to prevent vapor, compressor oil, and liquid refrigerant from backing up into the evaporator or other devices where it could condense and cause problems.

Suction line filter-drier

A filter-drier is often installed in the suction line just ahead of the compressor (see Fig. 4-2). Such an installation is frequently used to clean up a new system or a system that has experienced a hermetic compressor motor burnout. A filter-drier directly ahead of the compressor offers maximum protection to the compressor from all contaminants, even those that are in the low side of the system.

Compressor low-side or suction line service valve

Service valves are plumbed into the refrigerant line at various locations to allow for the connection of manifold gauges to check pressures and to add or take out refrigerant and refrigerant oil (Fig. 4-15). The compressor low-side suction line service valve is connected to the compressor at the compressor inlet union. The suction line from the evaporator connects to the valve's low-side inlet.

If the valve stem is turned all the way in, the connection from the compressor to the suction line is closed. With the stem turned all the way in, the compressor inlet union to valve connection can be opened and the suction line will remain sealed. If the valve stem is turned all the way out, it closes off the charging and gauge connection port. The pressure gauge or charging/evacuating line can now be connected to this port without opening the system. Once the gauge or charge line is in place, the valve stem can be turned in so internal pressure reaches the gauge

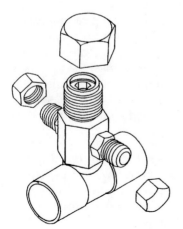

Figure 4-15 A typical refrigerant line service valve.

or the system is open for charging or evacuation. Both the valve stem and gauge/charge port are protected by removable sealing caps.

Compressors

The heart of any refrigeration system is its compressor. The compressor supplies the energy needed to create refrigerant flow and raise the temperature and pressure of the refrigerant vapor entering from the low-side suction line. System operating pressures and temperatures are set by the refrigeration equipment manufacturer. The size of the compressor determines how much refrigerant will circulate through the system in a specified amount of time. This is the basis of the BTU rating for the equipment.

All compressors are powered or driven by an electric motor. There are several major types of compressors used in residential and commercial refrigeration systems. These include:

- Reciprocating compressors
- Rotary compressors
- Centrifugal compressors
- Screw compressors
- Scroll compressors

In addition to these various model types, compressors are classified in a number of other ways, such as open-type or hermetic (sealed), or air or water cooled.

Open (external drive) compressors

An open or externally driven compressor uses a drive motor that is separate from the compression element. Large reciprocating compressors, such as the one shown in Fig. 4-16, make up the largest percentage of external drive installations. The external motor can connect to the compressor via a direct drive shaft, or the compressor can be belt driven off the motor. An external drive compressor has inlet and exhaust ports on its cylinder head that can be fitted with service valves. Open-type compressors offer the following advantages over hermetically sealed units.

- If a motor burns out it can be quickly replaced without opening the refrigeration system. A spare motor can be kept on hand to reduce downtime in critical applications. Replacing the drive motor is less expensive than replacing an entire hermetic compressor unit.

Figure 4-16 A large-horsepower reciprocating compressor used in open or external drive installations.

- An open compressor does not rely on suction gas cooling for efficient operation. This means a refrigerant leak or a bad TEV will not result in compressor failure as it can with a hermetic unit.
- Open systems have compiled an impressive history or reliability.
- For those willing to learn the skills needed, the open compressor and drive motor can be rebuilt. Hermetic units are virtually always replaced.

Hermetic (sealed) compressors

The dome-shaped hermetic compressor contains the compressor element along with the drive motor (Fig. 4-17). The motor is connected directly to the compressor, which can be any one of the designs outlined earlier in the chapter. The entire unit is sealed in a welded steel casing that leaves only the connecting tubing and electrical connections exposed. A hermetic compressor, when installed properly in the refrigeration system, forms a perfectly hermetic seal.

Internal components are inaccessible so field repairs are possible. The compressor motor is lubricated by the oil carried in the vapor refrigerant. The motors used in hermetic units do not have brushes or open points. These could cause arcing that would pollute the refrigerant and oil, and eventually lead to burnout.

Most small hermetic compressors, such as those used in residential applications, are not equipped with service valves. To install a pressure gauge to these systems, a tapping or piercing valve must be installed in the refrigerant line. Some larger hermetic compressors can have service valves on their inlet and exhaust lines, but this is not normally the case.

The hermetic compressor has gone through several design changes over the years. The first units had four pole induction motors that oper-

Figure 4-17 A scroll-type hermetic compressor that is completely sealed inside a protective dome. *(Copeland Corporation)*

ated at about 1800 rpm. They were quiet, durable machines that lasted for years. Newer designs had two pole, 3600 rpm motors. This meant they were smaller, but the higher speeds increased noise levels. Many hermetic units still in use are two-speed.

Hermetic compressors are universally used on residential refrigerators and freezers. They are also extremely popular in small to midsize commercial applications.

Because hermetic units are simply changed out if there is a compressor-related problem, the system can be repaired in a relatively short time by a technician possessing basic mechanical skills. However, the cost of changing out the entire unit often excesses the cost of repairing an open-type compressor.

Accessible hermetic motor-compressor

Like a fully hermetic unit, this design employs an electric motor coupled directly to the compressor crankshaft inside a hermetically sealed housing. However, the head, stator covers, bottom plates, and housing covers are all removable, making field repairs to the unit possible.

Because an adequate supply of oil must be maintained in the crankshaft of all refrigeration compressors, accessible hermetics are normally fitted with a sight glass to monitor oil level. While still found in some older installations, accessible hermetic compressors are gradually disappearing from modern refrigeration.

Reciprocating compressors

The reciprocating compressor (see Fig. 4-16) is basically a positive displacement pump. In many ways, a reciprocating compressor resembles an automobile engine. Many of the compressor components are called by the same names and function in the same manner as their automotive counterparts. However, in an internal combustion engine, the energy released when a gas/air mix is ignited in the combustion chamber forces the piston down in the cylinder. The flywheel stores some of this energy so it can move the piston back up to its original position. The crankshaft changes this up-and-down motion into rotary motion that is transferred to the transmission and drive shaft.

In a reciprocating compressor, the flow of energy is reversed. An electric motor provides the power to turn a drive shaft connected to the compressors crankshaft. The crankshaft changes the rotary motion of the driveshaft into the up-and-down motion of the pistons within their cylinders. Connecting rods connect the crankshaft to the pistons. As the compressor pistons move up and down, they alternately draw in refrigerant vapor on the downstroke and compress it on the upstroke. Each revolution of the crankshaft provides both a suction stroke and a compression stroke.

A reciprocating compressor can have one or more cylinders. Multiple cylinder compressors can have their cylinders arranged in-line or in a V-configuration. The crankshaft is housed in a crankcase, which also holds the lubricating oil needed to operate the compressor. The crankcase and cylinder bodies are normally cast in a single-cast iron housing. The outside walls of the cylinders are often designed with cooling fins to help dissipate heat. In the case of an open compressor, the crankshaft protrudes from the crankcase so it can be connected to an external drive motor. The crankshaft seal is a critical component in this type of unit.

Like an automobile engine, the compressor uses a cylinder head to seal the top of the cylinders and house the valves needed to seal the intake and outlet ports of each cylinder. It also provides the passageways for refrigerant vapor to move in and out of the compressor. The valve assembly typically is made of a valve plate and intake and exhaust valves, and valve retainers (Fig. 4-18). Pressure differences acting on each side of the valve cause it to open or close. For example, when the piston is moving downward on the suction stroke, pressure on top of the intake valve is greater than pressure on the underside of the valve.

84 Chapter Four

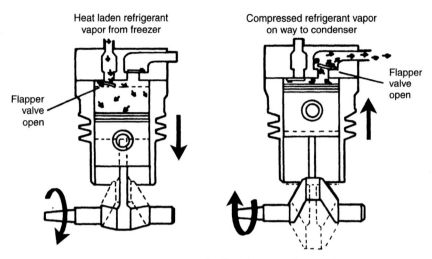

Figure 4-18 Reciprocating compressor valve-head parts and operation.

This forces the valve downward. It opens, and refrigerant vapor enters the cylinder.

When the piston reaches the bottom of its stroke, the cylinder is filled with vapor. Pressure difference between the vapor in the cylinder and vapor in the low side of the system is very slight, so the intake valve closes. As the piston begins to move upward, it compresses the vapor in the cylinder chamber. When pressure reaches a certain level, the exhaust valve is pushed upward and open, and the high-pressure vapor is expelled into the discharge line to the condenser. The high pressure does not open the intake valve as this valve can only open in the downward direction. Pressures inside the compressor can run as high as 300 psi.

Rotary compressors

Rotary compressors have replaced many smaller reciprocating compressors. A rotary unit is much smaller and lighter weight than a comparable reciprocating unit. A typical two-blade rotary compressor operates as follows (Fig. 4-19):

1. As the first blade passes the suction port, low-pressure vapor is drawn into the compressor.

2. When the second blade passes the suction port, it seals the refrigerant between both blades, then starts a new cycle behind as it moves along. The gas is trapped and compressed as the blades revolve.

3. As the first blade reaches the exhaust port, it pushes the compressed gas into the exhaust port. This entire cycle repeats itself as the blades rotate between intake and exhaust ports of the compressor.

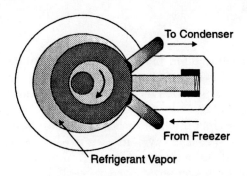

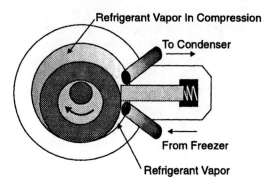

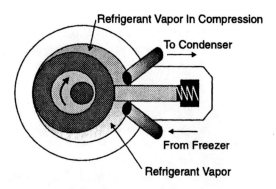

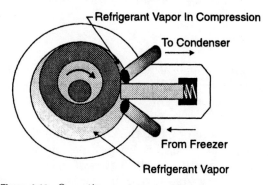

Figure 4-19 Operating a rotary compressor.

Rotary compressors can have up to eight blades. Some rotary units have a stationary-blade system. This system has an eccentric shaft in the cavity that compresses the gas as it travels between intake and exhaust ports.

A rotary compressor is often used as the first or booster compressor in a cascade refrigeration system. A cascade system uses two separate refrigeration systems to produce ultra-low temperatures around −50°F. (−46°C). The evaporator in the first system is used to cool the condenser of the second system. The evaporator of the second system is used to generate the ultra-low temperatures required for multiple compressors to produce ultra-low temperatures.

Rotary compressors also provide a large size opening into the suction line, so substantial amounts of vapor can be drawn in on each stroke, maximizing compressor efficiency. Small clearances used in rotary compressor construction ensure that all low-pressure vapor drawn into the unit is compressed and pushed out on the discharge stroke.

When a rotary compressor is used, a check valve is normally installed in the suction line to prevent high-pressure vapor and compressor oil from flowing back into the evaporator.

Screw compressors

Screw compresses use a set of precision-matched helical rotors to trap and compress refrigerant vapor as they turn inside a precisely machined compression cylinder. The drive or male rotor is driven by the compressor motor. This rotor has four lobes that fit into the six interlobe spaces on the female rotor. As the rotors turn, low-pressure vapor is drawn into the intake side of the compressor, filling the interlobe spaces (Fig. 4-20). As the screws turn, the vapor is trapped, compressed, and then discharged from the outlet or exhaust port. Operation is continuous and extremely smooth and vibration free. Oil injection is used on many screw compressors to help seal clearances between the rotors and the compression cylinder. Screw compressors are highly efficient.

Scroll compressors

Scroll compressors are quiet, smooth-operating units with the highest efficiency ratio of all compressor types. Scroll compressors are typically available in 1½ to 5 horsepower hermetic units.

Two offset spiral disks are used to compress refrigerant vapor (Fig. 4-21). The upper disk remains stationary while the lower disk is driven by the unit's motor. As the lower disk rotates, it draws vapor in through the inlet ports at the perimeter of the scroll. The vapor is drawn through and compressed in spiral passageways that decrease in width

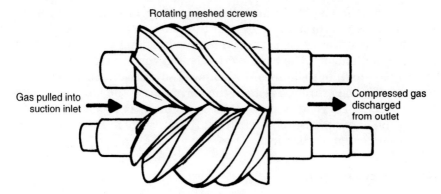

Figure 4-20 Operating a screw-type compressor.

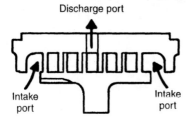

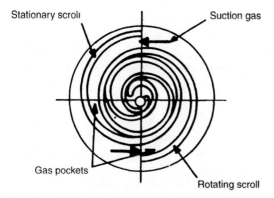

Figure 4-21 Operating a scroll compressor.

as they reach the center of the disk. The high-pressure vapor discharges through a port at the center of the disk.

Centrifugal compressors

A centrifugal compressor resembles a long horizontal cylinder. A drive shaft is mounted inside the cylinder and a series of impellers are mounted to the shaft. The impellers spin very fast. The refrigerant vapor is drawn in through vanes near the center of the impeller. As the impeller turns, the vapor is thrown outward against the outer wall of the impeller housing. The force exerted on the vapor increases its pressure by a small, but significant amount. The pressurized vapor exits the impeller body through slots in the perimeter of the body. It is then picked up by a second impeller and the process is repeated, further increasing vapor pressure. Centrifugal compressors are used for some large commercial applications.

Air-cooled compressors

Air-cooled compressors require an adequate supply of cooling air to prevent overheating. When the compressor is part of an air-cooled condensing unit, air from the condenser fan is directed over the compressing body.

Water-cooled compressors

Water-cooled compressors normally use a copper water coil wrapped around the compressor body. Water is circulated through the coil when the motor-compressor operates.

Suction-cooled compressors

Compressor motors can also be cooled by routing the cold suction line over the motor coils. Another cooling method is to circulate the compressor crankcase oil through it's own air- or water-cooled coil. Removing heat from the lubricating oil helps reduce overall compressor temperature.

Low and high starting torque compressors

When the compressor is started, it must have the power to overcome the high-side pressure in the system before it can begin moving vapor and liquid through the refrigerant lines.

On systems using a capillary tube as the control device, system pressure equalizes to a rather low pressure when the compressor is turned off. When the compressor's electric drive motor is switched back on, it requires very little starting torque, so a low starting torque compressor can be used.

On systems using thermostatic or automatic expansion valves, the system pressure does not equalize on compressor shut-down. High-side or head pressure remains at its operating level. To overcome this higher pressure, the compressor must be equipped with a starting capacitor that helps the unit develop a high start-up torque.

Some compressors are equipped with an unloader, which is a device that temporarily reduces high-side pressure so the compressor can start up without the use of a start capacitor. Unloaders can be operated by a solenoid valve, hydraulically or mechanically.

Mufflers

Mufflers are often installed on both the intake and exhaust ports of hermetic and external drive compressors to minimize operating noise. Most compressor mufflers are little more than cylinders with baffles mounted inside.

Oil separator

Oil is required for lubrication of the compressor. A small amount of special oil is placed inside the compressor's crankcase or housing. As the compressor operates, this oil circulates to the various parts of the compressor requiring lubrication. In a hermetically sealed compressor housing, this oil also lubricates the motor bearings. Oil belongs in the compressor crankcase or housing, but as the compressor runs, some oil is pumped out of the compressor along with the compressed refrigerant vapor. In small amounts, this oil does not pose a problem to other system components such as the control device, evaporator, or condenser.

Any time the compressor is running, some oil is circulating throughout the refrigeration system along with the refrigerant. Normally, the velocity at which the refrigerant moves through the system keeps the oil entrained in the vapor so it can return to the compressor. Although most well-designed refrigeration systems do an effective job of returning oil to the compressor, some systems require assistance in this area. Such systems include ultra-low temperature systems, large commercial systems, or systems where effective oil return is a problem because of inherent design factors. On ultra-low temperature systems, the evaporator pressure is often so low that an effective job of "scrubbing" the interior tubing walls of oil film is not possible.

Oil separators are installed between the compressor exhaust and the condenser. The oil separator intercepts the oil entrapped in the high-pressure vapor exiting the compressor and returns it immediately to the compressor crankcase or housing, rather than allowing it to circulate with the refrigerant.

The oil separator is simply a chamber with baffles on its inlet and outlet connections (Fig. 4-22). These baffles reduce refrigerant vapor velocity so that most of the oil separates out and collects at the bottom of the separator. When the oil in the chamber reaches a certain level, a float valve opens a passage to the compressor's crankcase. The pres-

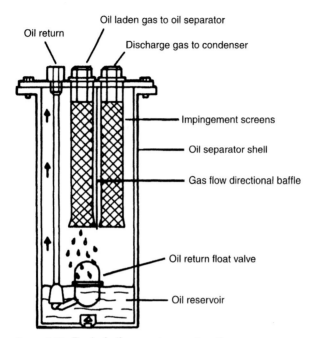

Figure 4-22 Typical oil separator construction.

sure differential between the crankcase and the oil separator quickly transfers the oil to the crankcase. This process is completely automatic.

Electric heater elements are often attached to oil separators to prevent migration of the refrigerant to the chamber where unwanted condensation of the vapor would occur. For all installations, the oil separator must be insulated over its exterior surface.

Compressor high-side service valve

The high-side service valve allows refrigerant flow to be shut off between the compressor and the condenser. It also provides a connection for taking high-side pressure readings with a pressure or manifold gauge.

Hot gas/discharge line

Refrigerant flows from the compressor to the condenser through this line.

Condensers

The condenser is the device used to dissipate the heat that was absorbed by the refrigerant during the evaporation process, plus the heat of compression created within the compressor when energy was used to compress the gas.

The condenser is similar to the evaporator in design—both have a series of coils or tubing. However, they perform opposite jobs. An evaporator allows a refrigerant to change from a liquid to a gas. The condenser, however, provides a large surface area to the high-pressure, high-temperature gas, causing it to release heat and condense back into a liquid.

When a refrigerant gas leaves the compressor, it is superheated. Because the high-pressure gas enters the inlet side of the condenser at such a high temperature, heat energy in the gas is released into the atmosphere. As explained in Chapter 3, heat always travels from an area of higher temperature to a lower one.

The refrigerant in the condenser gives up its excess heat to the atmosphere, and as the temperature falls it condenses to a liquid. It reaches the condenser outlet in a 100 percent liquid state. The temperature should be near ambient temperature at this point.

Condenser construction

The most easily recognized and most widely used condenser is of the fin-and-tube type, employing a motor-driven fan to force large quantities of air through the fin and tube assembly (Fig. 4-23). Except for very small refrigerators and freezers, which often use condensers that depend on gravity for air circulation, the air-cooled condenser is most often chosen because no water is required, it is easy to maintain, and it does not freeze in cold weather.

For maximum efficiency, the fins of the air-cooled condensers should be kept free of accumulations of dust, dirt, and foreign material. Wire-and-tube condensers are frequently used as well, and although some sacrifice must be expected in heat-transfer capabilities due to the absence of fins, the tendency to remain relatively obstruction-free over extended periods of time is often desired.

Figure 4-23 Air-cooled remote condenser. *(Heatcraft Inc.)*

Water cooling

Where an adequate supply of low-cost water is available, water-cooled condensers are often used because of water's excellent heat transfer characteristics. Well water, for example, is often much colder than air temperatures and lower condensing pressures result. Generally, a pressure-controlled water valve is used to modulate the flow of water through the condenser thereby maintaining the condensing pressures within a desired range.

Condensing units

A condensing unit consists of a compressor and condenser mounted on a common base (Fig. 4-24). Condensing units used on thermostatic expansion valve-controlled systems have a third component mounted to the base—the refrigerant receiver. Remember, capillary-tube-controlled systems do not require a receiver. Condensing units are available in a number of different applications.

Figure 4-24 A typical condensing unit. *(Heatcraft Inc.)*

Air-cooled condensing unit. A single-unit compressor (½ to about 30 horsepower) with air-cooled condenser systems can be mounted in racks up to three high to save space in supermarket and convenience store applications. The air-cooled condensing unit can be fitted with a heated compressor crankcase, insulated receiver, and controls suited for exterior installations. Typically, air-cooled condensing units have condensers sized to create a temperature differential (TD) in the 10 to 25°F range. Larger condensers are often used to achieve lower TDs and higher Energy Efficiency Ratios (EER).

Water-cooled condensing unit. Water-cooled units range in size from 0.5 to 30 horsepower and are best for hot, dry climates. The city water-cooled condensing unit is usually no longer economical due to the high cost of water and sewer fees. Cooling towers, in which one tower cools the water for all compressors, have been used instead. Closed-water systems are also used in areas that can use an evaporative water cooler.

Remote condenser–compressor unit. Remote units operate efficiently with minimum condenser maintenance. Evaporative condensers are also used in some areas. Sizes range from 0.5 to 30 horsepower.

Recommended sizing for remote air-cooled condensers is 10°F TD for low-temperature application and 15°F TD for medium-temperature application. Remote water-cooled condensers are often used in areas with abundant water.

Single-compressor control. Single-compressor units make up half of the supermarket compressor equipment currently used. A solid-state pressure control for single units can help control excess capacity when the ambient temperature drops. The control senses the pressure and adjusts the cut-out point to eliminate short cycling, which ruins many compressors in low load conditions. This control also saves energy by maintaining a higher suction pressure than would otherwise be possible and by reducing overall running time.

Liquid line

Refrigerant from the condenser flows through this line to the refrigerant flow control.

Receiver

The receiver is a storage tank that holds the system's excess refrigerant. Receivers are used on vapor compression refrigeration systems that use thermal expansion valves as the refrigerant metering control device.

The receiver also provides storage space for the refrigerant when maintenance is being performed on the system. When sizing a receiver for a particular system, keep in mind that it must have the capacity to store the entire refrigerant charge. A service valve at the outlet of the receiver allows all refrigerant in the system to be pumped into the receiver. This procedure is commonly known as *pumping down* the system.

To be absolutely certain liquid refrigerant is always available to the high side of the system, the outlet must be positioned, or some provision made, to prevent vapor from entering the liquid line. When the outlet is at or near the top, where it often is, a dip tube is employed that extends to about ½" from the bottom of the receiver.

The high pressure in the receiver forces high-pressure liquid through the pick-up tube and on into the liquid line and through the filter-drier.

Filter-driers

Even though many precautions are taken during assembly of a refrigeration system, certain undesirable substances, such as water, dirt, sludge, and acids are introduced into the system during manufacture or service, or build up in the system over time. These substances can lead to freeze-ups, contamination, or restrictions. Foreign matter destroys the close working tolerances of moving parts and water always leads to corrosion in varying amounts.

To remove and trap harmful substances, a filter-drier is plumbed into the recirculation line. Its function is to keep the circulating liquid free of contaminants.

The selection of the proper filter-drier, generally, involves the following considerations:

- Water capacity and refrigerant flow
- Filtration
- Acid removal

Several of the most common filter drier types are illustrated in Fig. 4-25. Manufacturers of filter-driers establish ratings for their own products, but the final selection is a matter of selecting a size that is adequate in all respects for the refrigeration system in which it is to be installed.

A filter-drier normally contains a desiccant or drying agent plus fine mesh filters at the inlet and outlet end of the unit.

Filter-drier location

The filter-drier is usually located in the liquid line, immediately ahead of other liquid line controls, such as the thermostatic expansion valve,

Refrigeration System Components   95

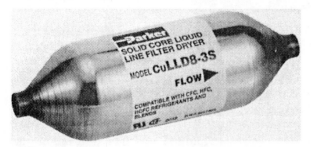

Figure 4-25 Typical receiver construction. *(Parker Hannifin Corporation)*

solenoid valve, and the moisture/liquid indicator. When located in this position, the filter-drier provides maximum protection for the expansion valve and solenoid valve from contaminants in the system. If the lines contain significant amounts of moisture, a filter-drier in this location also protects the expansion valve from freeze-up. If possible, locate the filter-drier in a cold area.

Suction line filter-driers

A filter-drier is often installed in the suction line just ahead of the compressor. Such an installation is frequently used to clean up a new system or a system that has experienced a hermetic compressor motor burnout. A filter-drier directly ahead of the compressor offers maximum protection to the compressor from all contaminants, even those that in the low side of the system.

The main disadvantage of the suction line location is the need for a larger, more expensive filter-drier. This is because the refrigerant velocity in the suction line is about six times the velocity in the liquid line. The larger filter size is needed to maintain a sufficiently low pressure drop across the unit.

The water capacity of the filter-drier in the suction line is equal to or slightly greater than the liquid line water capacity. Filtration and acid removal in the suction line are comparable to the liquid line. Filter-driers are not recommended for use in the discharge line because the high operating temperature needed would cause a reduction in the water capacity. The acid removal ability of the filter-drier is the same regardless of whether it is installed in the liquid line or suction line.

A filter-drier can be installed in either the top or bottom feed position. It's a good idea to mount replaceable core models horizontally so that dirt and water will not fall into the outlet fitting when the cores are changed. Always note the flow direction marked on the unit, and take care not to install the filter-drier in a reverse-flow position.

Capillary tube filter-driers

Some filter-driers are designed specifically for use in capillary tube controlled systems. The ideal filter-drier location is immediately ahead of the capillary tube. The unit typically consists of a filter-drier with ¼-inch copper tubes brazed into each end. Capillary tubes of any size can be inserted into the ¼-inch filter-drier tube. The ends of the tubing are then pinched down and soldered. Proper installation will maximize the contaminant removal properties of this filter-drier, which is commonly used in domestic refrigerators and freezers. The unit might also have a service valve for refrigerant charging.

Bypass installations

For maximum protection, install the filter-drier in the main liquid line. When the filter-drier is located in a bypass, dirt or foreign material might pass into the system through the unprotected main line. However, when a bypass installation is needed (Fig. 4-26), a hand throttling valve is recommended. This throttling valve allows a certain percentage of refrigerant to pass through the filter-drier. Two additional hand valves are needed if the filter-drier is replaced without pumping down from the receiver. In this case, pump out the section of the line containing the filter-drier by closing hand valves A and B (noting direction of the flow). Wait until the isolated section is pumped out, close valve C, then change the filter-drier.

Warning: Dangerous hydraulic pressures can develop if hand valves B and C are closed when the filter-drier is full of liquid. If there is a chance that inexperienced personnel could change the valves without pumping down, a pressure relief device should be installed as an added safety device.

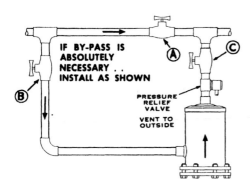

Figure 4-26 Solid-core liquid line filter drier. *(Parker Hannifin Corporation)*

Brazing and soldering

The fittings on most sealed model filter-driers are constructed of heavily copper plated steel, suitable for all types of brazing and soldering alloys. The fittings on the replaceable core model filter-driers are copper and can be used with all the same alloys. Proper brazing technique involves using the proper torch tip for rapid heating, and also directing the flame away from the filter-drier shell. This technique will minimize any damage to the paint.

To remove seals from either sweat or flare connections, gently cut them away with a knife. With flare connections, exercise caution to avoid damaging the flare surface. The seals cannot be removed and replaced without tearing them.

Chapter 5

Refrigeration Control Systems

The job of any refrigeration system, be it a cooler or freezer, is to maintain a desired temperature range that maximizes operating efficiency. To do this, the system must be capable of starting and stopping when needed, providing for safe, continuous operation when running, and immediately shutting itself down if operating conditions become unsafe. Modern refrigeration systems use a variety of control switches and control valves to perform these important tasks.

Control Switches

A control switch is a device that operates one or more sets of electrical contacts. When the contacts are closed, electrical power flows through the switch to open or close solenoid valves in refrigerant or cooling water lines, to engage or disengage the clutch coils or relay of a compressor motor, to operate an evaporator or condenser fan, or activate defrost timers or thermostats.

Note: If you are unfamiliar with the principles of electricity and electrical components, review Chapter 9 on understanding electrical systems used in refrigeration.

The operation is similar to working a wall-mounted light switch, except there is no one available to throw the switch or to decide when the switch should be thrown. To perform these tasks the control switch must be able to:

- Respond to changes in the physical operating conditions of the system, such as changes in temperature, pressure, liquid level, flow velocity, or proximity.
- Generate a mechanical movement or force that can open and close switch contacts.

The two most popular types of control switches used in commercial and residential refrigeration systems are thermostats and pressure switches.

As explained in Chapter 4, systems using a capillary tube as their refrigerant flow control are commonly used on smaller, self-contained single evaporator units. Typically, a capillary tube system uses only thermostat control for operating the compressor.

By cycling the compressor, the temperature in the refrigerated space is maintained at the desired level. During the off cycle, the pressure within the system equalizes, preventing the use of a low-pressure control for temperature control.

Refrigeration systems that use thermostatic expansion valves as the flow-control device can use either thermostats or pressure switches.

Pressure switches

A pressure switch closes or opens a set of electrical contacts when a given amount of pressure is exerted on the switch diaphragm. A pressure switch can be adjustable or nonadjustable, and high-, low-, and dual-pressure models are available (Fig. 5-1).

The pressure switch closes at a pressure level called the *cut-in pressure* and opens at a pressure known as the *cut-out pressure*. The cut-out pressure determines the lower temperature limit. To set the uppermost temperature limit, the cut-in pressure must also be determined. The difference between the cut-in and cut-out pressures of the switch is called the *differential*. The *range* is the pressure range over which the switch can accurately operate. Pressure switches can be used as a refrigeration system's primary operating control, as a safety devise, or to operate a certain part of the total system.

High-pressure switches. High-pressure switches are commonly connected to the discharge line between the compressor and condenser to sense discharge pressure. They normally read in a 150 to 450 psig range with 40 to 150 psig differentials. If compressor discharge pressure rises above a certain level considered unsafe, the pressure switch opens to shut down the compressor. High-pressure switches are also used to control condenser fan operation.

A high-pressure switch most commonly employs an SPST switch to cut off the compressor. An SPDT switch can be used to shut off the compressor and simultaneously activate an alarm or indicator light. Many pressure switches are equipped with a manual reset button that locks in the open position. Pressure switches might also have an adjustable limit stop to prevent control adjustment above a preset pressure.

There are also high-pressure switches designed to close when exposed to a rise in pressure. These pressure sensors are used to oper-

Refrigeration Control Systems   101

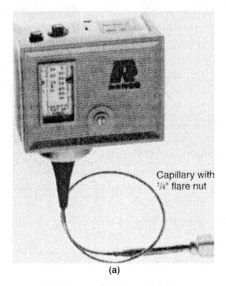

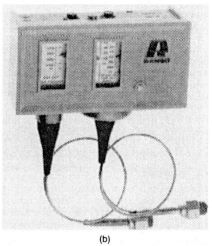

Figure 5-1 Single-function (a) and dual-function (b) pressure controls. *(Ranco Incorporated)*

ate another device that controls discharge pressure. The pressure setting of a high-pressure switch is dependent of the type of refrigerant used in the system.

Low-pressure switches. Low-pressure switches are installed in the suction line to sense suction pressure into the compressor. They can be used for temperature control, pumpdown control, capacity control, low-pressure limit control, and alarm control. Low-pressure switches for low-pressure cycling and pumpdown applications on low- and medium-temperature refrigeration units read in a 12-in. Hg vacuum to 50 psig range with a 5 to 35 psi adjustable differential. To cover all possible applications, a 10-in. Hg vacuum to 100 psig with a 10 to 40 psi adjustable differential is available. A low-pressure switch can be an open on pressure rise or close on pressure rise depending on the application. SPST, SPDT, and DPST switch configurations are all used depending on the application.

Troubleshooting. The major problems encountered with pressure switches is pitting, wear, or mechanical failure of the contacts. Check that the contacts are in the correct position for the set pressure. The contacts should show continuity and there should be low resistance (no more than three ohms) through the contacts. Leaks or kinks in the pressure lines to the switch will hamper or completely stop switch operation. Check for leaks with a refrigerant leak detector and replace any faulty lines.

Most pressure controls have a scale graduated in psig; however, these indicators are not always reliable. To ensure that the cut-out on the pressure control is correct, it is best to use a pressure gauge. In addition, it is wise to use a thermometer to measure unit temperature, because the unit temperature/refrigerant temperature will vary from one unit to another. Once the control is set to cut-out at the desired temperature, the corresponding pressure should be written down for quick reference in case the unit needs to be serviced.

Thermostats

Thermostats control compressor and motor operation based on the temperature the thermostat is sensing. The sensing element of a thermostat can be a sensing bulb, a bimetal switch, or a solid-state electronic device (thermistor).

Sensing bulb. The sensing bulb thermostat is the most popular thermostat used in refrigeration work (Fig. 5-2). Its operation is quite similar to the thermostatic expansion valve discussed in Chapter 4. The sensing bulb is filled with a volatile liquid that vaporizes at a low temperature. The sensing bulb is mounted in contact with the evaporator coil. The coil surface temperature, not the air temperature, is actually measured. As the temperature of the evaporator coil increases, the vapor pressure inside the bulb also increases. Vapor pressure passes out of the bulb through a capillary tube to a bellows diaphragm. Pressure on the diaphragm produces mechanical movement that is used to open or close the thermostat's electrical contacts. Opening the contacts shuts down the compressor or motor or removes operating voltage from a solenoid valve.

Bimetal switch thermostats. This type of thermostat is simply a set of electrical contacts connected by a strip or arm made of two different types of metal, usually copper and steel. As the surrounding temperature rises, the two different metals begin to expand until they move into position to close contacts and complete the circuit.

In some bimetal thermostats, the movement of a metal coil is used to operate a mercury switch. A mercury switch is a small bulb partially filled with mercury, which is a good conductor of electricity. Tilting the bulb to one side allows the mercury to cover both switch contacts and the circuit is complete. Tilting the bulb in the opposite direction uncovers one or both contacts and the circuit is opened.

Thermistor. A thermistor is a type of solid-state variable resistor. It is designed to change its resistance to current flow as its temperature changes. A thermistor with a positive temperature coefficient (PTC) will increase in resistance as temperature increases. A thermistor with a negative temperature coefficient (NTC) will decrease in resistance as temperature increases.

The thermistor is used as the sensing element for an electronic control unit. The control unit calculates temperature based on the resistance of the thermistor. It will then open or close relays, contactors, or solenoid valves based on the temperature readings.

As described earlier in this chapter, there is a definite relationship between air temperature and refrigerant temperature. So measuring the refrigerant temperature is as reliable as measuring the pressure.

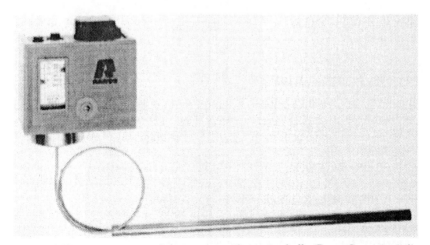

Figure 5-2 Temperature control thermostat with sensing bulb. *(Ranco Incorporated)*

When thermostats are used to regulate temperature, the cut-in temperature is fixed. The cut-in is set to allow the unit to self-defrost. The cut-out temperature is typically adjusted to create a higher or lower unit temperature.

Humidistats

In certain refrigeration applications, it is necessary to control air humidity (moisture content). A humidistat contains a moisture-sensitive element that moves in response to changes in humidity. The movement operates a mechanical linkage that controls the open and closed positions of electrical contacts.

Time-delay relays

The purpose of a time-delay relay is to delay the starting of a load for a specific period of time. The relay typically consists of a heating element and a bimetal element. As the bimetal element is heated, it moves. This motion closes the contacts of an electrical circuit. The time delay period is based on the amount of time that it takes the heating element to close the bimetal element. Relay contacts are rated "pilot duty," while the coil or heater can be rated either 24, 110, or 240 volts.

Time-delay relays can be used to prevent simultaneous start up of two heavy loads in the same system or of two units connected in parallel. This is done by placing the contacts of the time-delay relay in series with the load or controlling element of the unit.

When the refrigeration system uses a part winding motor, a time-delay relay is used to energize the windings at different times. The relay's time delay varies. It is typically 15 to 30 seconds, although it can be longer.

Time clocks (mechanical linkage)

Time clocks are used to control the operation of large or small installations based on preset periods of time (Fig. 5-3). One of the most common time clock applications is to start and stop the defrost cycle in a refrigeration system.

Time clocks have a mechanical linkage that opens and closes a set of contacts at a predetermined time. The mechanical linkage is positioned between the contacts and the clock. There are two basic types: the one-day and the seven-day. The one-day clock follows a 24-hour time period regardless of the day of the week. A seven-day clock might operate on an hourly or daily basis during a week.

Refrigeration Control Systems 105

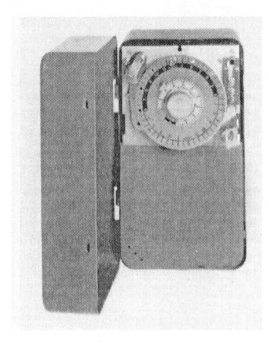

Figure 5-3 Mechanical linkage time clock set by attaching clips to the clock wheel or dial. *(Paragon Electric Company, Inc.)*

A time clock is set by attaching two clips to the clock wheel or dial. When the dial moves, the first clip closes the contacts at the point in time where it is attached to the wheel. The second clip opens the contacts at the point in time where it is set on the dial.

Flow switches

Flow switches are used to detect the movement of liquid refrigerant or cooling water in lines or cooling coils. The switch uses a set of spring-loaded vanes positioned perpendicular to the flow of liquid. When the moving liquid strikes the vanes, they deflect and make contact in an electrical circuit. When flow slows or stops, the vanes move back to their original position and the control circuit is opened. Liquid flow switches are often used to monitor water or refrigerant flow through condensers and chillers.

Differential control switches

These switches measure the difference in pressure or temperature between two separate locations, such as two pipes or spaces. Bellows are commonly used as the power elements that move in response to pressure differences to close or open a control circuit. The most common use for a differential control is the oil pressure safety switch used with reciprocating compressors. Details on this type of switch are given later in this chapter.

Float switches

A float switch reacts or moves with changes in liquid level. This movement can open or close circuit contacts that operate a pump, sound an alarm, or perform other jobs in the refrigeration system.

Pressure Regulators

A pressure regulator is a two-port device that maintains a preset pressure level at one port regardless of the pressure at the other. Hold-back pressure regulators keep a constant outlet pressure even when the input pressure exceeds this level. Back-pressure regulators keep the pressure entering the regulator at a constant rate. Pressure regulators are used in a number of important refrigerant system applications.

Evaporator pressure regulator

A back-pressure regulator installed in the evaporator outlet or section line is shown in Fig. 5-4. It keeps pressure entering the regulator and thus evaporator pressure at a constant level. Evaporator pressure regulators are used when low-limit control of the evaporator pressure or temperature is required.

The inlet pressure pushes against the bottom of the seating disk, and is opposed by adjusting spring pressure. The outlet pressure acts on the top of the seating disk and the underside of the adjusting spring bellows. The spring is adjusted to set the desired inlet (evaporator) pressure. If the evaporate pressure increases above this setting, the inlet port seat opens venting the excess pressure through the outlet port to the suction line. Once the inlet pressure drops to the preset level, the inlet port seat closes.

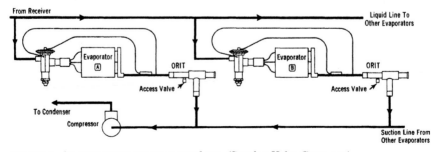

Figure 5-4 An evaporator pressure regulator. *(Sporlan Valve Company)*

The pressure change needed to operate the valve is called the *pressure differential*. The pressure drop is the difference between the pressure at the inlet and outlet. Some evaporator pressure valves are pilot-operated. The pilot pressure can be line pressure.

Evaporator pressure regulators are often installed in the suction line when evaporator frosting must be prevented. They are also used when a constant evaporator pressure must be maintained at a level higher than suction line pressure to prevent dehumidification. Evaporator pressure regulators are also used to provide individual evaporator pressure control on multievaporator systems. The pressure regulators are used to maintain the required evaporator temperature in the warmer units, while the pressure switch or thermostat-controlled compressor continues to operate for the coldest unit. In some cases a solenoid valve is installed in the regulator's external pilot line to allow the regulator to act as a stop valve as well.

Suction pressure regulator

A hold-back pressure regulator is used to regulate compressor suction pressure (regulator outlet pressure) to a preset safe level. Installation of a compressor suction pressure regulator prevents excessive start-up loads, high-suction pressure after a defrost cycle, prolonged compressor operation at high-suction pressures, and low-voltage/high-suction pressure conditions.

Condenser pressure regulators

These regulators are used on air-cooled condensers to maintain sufficient pressure to allow for proper cold-weather operation.

Check valves

A check valve allows liquid or vapor flow in one direction only. Once past the check valve, the liquid or vapor cannot back up through the valve. Check valves are often used to prevent high-pressure vapor from backing up into the evaporator during the refrigeration system's off cycle.

Solenoid Valves

Solenoid valves (Fig. 5-5) are used to automatically control the flow of liquid or gas in refrigerant liquid or suction lines, hot gas defrost circuits, water-cooling coils, and similar components. They are one of the most widely used of all refrigeration control devices, equally suitable for many other less common forms of refrigerant control.

Figure 5-5 Typical solenoid valve. *(Parker Hannifin Corporation)*

Principles of operation

Solenoid valve operation is based on the principles of electromagnetism. The coil of the solenoid valve coil generates a magnetic field when electrical current is passed through it (Fig. 5-6). If a piece of magnetic metal, such as steel, is exposed to the magnetic field, the pull of the field will move the metal and center it in the hollow core of the coil. If a stem is attached to the steel plunger, the magnetic force can then be used to open the port of a valve. When the electrical circuit to the coil is broken, the magnetic field collapses and the stem and plunger return to their original position, sometimes with the help of a spring.

A solenoid valve can use a hammer-blow effect to help overcome unbalanced pressures across the port. When the coil of such a valve is energized, the plunger starts upward before the stem. The plunger then picks up the stem by making contact with a collar at the top. The added momentum of the plunger assists in opening the valve.

Normally closed and normally open valves. There are two basic types of solenoid valves. The most common is the normally closed (NC) valve. In this type of solenoid, the valve opens when the coil is energized and closes when the coil is de-energized. The other type is the normally open valve that opens when the coil is de-energized and closes when the coil is energized.

Direct-acting solenoid valves. On a direct-acting solenoid valve, the stem and plunger assembly opens the port of the valve directly (Fig. 5-7). This type of construction is normally used on smaller valves with port sizes of less than ¼ inch.

Refrigeration Control Systems 109

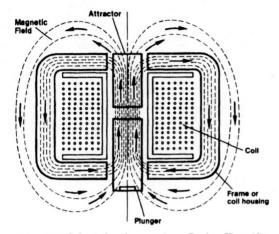

Figure 5-6 Soleniod coil operation. *(Parker Hannifin Corporation)*

Pilot-operated solenoid valves. Normally closed solenoid valves are often pilot operated. When the valve is energized, the stem and plunger assembly opens a pilot port. Opening this small port relieves the pressure acting on the top of the disc, piston, or diaphragm that is holding the valve port closed. The disc, piston, or diaphragm can now move upward, opening the main valve port. Figure 5-8 illustrates the four phases of the operating cycle of a typical normally closed pilot-operated valve.

Normally open solenoid valves operate very similar to the normally closed type. The system pressure is used to open and close these valves.

The major difference in the normally open construction is that with the coil de-energized, a spring is used to push the stem and plunger assembly upward holding the pilot port open. This then allows the disc to rise because of the pressure difference between the bottom and top of the disc and permits flow to take place.

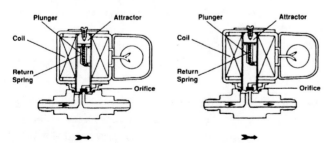

Figure 5-7 Operating a direct-acting solenoid valve. *(Parker Hannifin Corporation)*

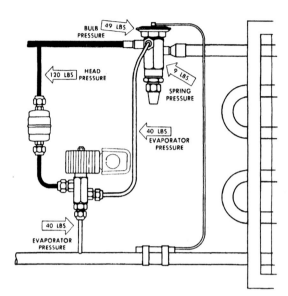

Figure 5-8 Operating a normally closed pilot-operated solenoid valve. *(Sporlan Valve Company)*

When the coil is energized, the stem and plunger assembly is pulled down, closing the pilot port. The pressure on top of the disc then equalizes with the incoming pressure and the disc moves down, closing the main port. The valve will remain closed as long as the coil is energized.

Direct-connected piston assemblies, with the stem and plunger mechanically connected to the piston, are used when the valve must be sized for very small pressure drop, such as on suction lines. The pilot-operated principle is used to open the valve, but the magnetic pull of the coil is used to hold the piston open to prevent any pulsations.

Direct-connected solenoid valves are somewhat sensitive to dirt, which can prevent free piston movement. The plunger might not be able to center itself in the coil, which will cause the coil to overheat and possibly burn out. Some valves employ a "floating" piston or disc that permits independent operation of the plunger. This ensures a complete magnetic circuit regardless of the piston's position and eliminates the possibility of coil burn-out due to restricted piston or disc movement.

Maximum operating pressure

Maximum operating pressure differential, generally abbreviated MOPD, is the maximum pressure differential against which a solenoid valve can open. Safe working pressure should not be confused with the MOPD rating of a valve. The working pressure of a solenoid valve is a design specification indicating the maximum pressure under which the valve must be able to withstand five times its safe working pressure to qualify for listing by Underwriters' Laboratories.

When it is necessary for the solenoid valve to operate independently of the electrical power, a manual lift stem might be used.

Suction line solenoid valves

On certain suction line applications, pressure drops in excess of two psi cannot be tolerated and only valves capable of opening at very low pressure drops can be used.

High-temperature applications

There are some high-temperature applications (180 to 300°F) where a high temperature coil construction must be used.

Protection against dirt

Because pilot-operated solenoid valves operate with rather close tolerances, conventional liquid line strainers do not offer adequate protection against dirt and other system contaminants. Strainers will pass particles large enough to lodge between the piston or disc and the valve body, locking up the valve.

To safeguard against this problem installing a filter-drier ahead of every solenoid valve on the refrigeration system is recommended by many valve manufacturers.

Controlling Refrigeration System Temperature

While the method used to control the product or refrigerated space temperature might vary from system to system, most methods of temperature control focus on the system's evaporator(s) and/or compressor(s).

Single-evaporator systems

In refrigeration systems using only one evaporator, the operating control cycles the compressor. Three different control schemes can be employed:

- Temperature (thermostat) control to cycle the compressor on and off (used on capillary tube or TEV systems).
- Low-pressure control sensing suction pressure to cycle the compressor directly (used on TEV systems only).
- Temperature control operating a solenoid valve with a low-pressure switch cycling the compressor (normally TEV systems only).

Temperature control. When using a temperature control, the sensing point can produce temperature, evaporator temperature, or space or evaporator return air temperature (Fig. 5-9).

Pressure control. A low-pressure control might be used as the operating control to cycle the compressor (Fig. 5-10). By sensing the suction pressure, the control can regulate the evaporator temperature. On "bleed" type TEV systems that equalize system pressure during the off cycle, low-pressure controls cannot be used as an operating/cycling control.

Temperature control of flow with pumpdown. On some systems, a temperature control operates a refrigerant solenoid valve, rather than the compressor (Fig. 5-11). A low-pressure control sensing suction pressure cycles the compressor (pumpdown system).

Multiple evaporator, single-compressor system (TEV only)

On systems with more than one evaporator, several control schemes can be employed.

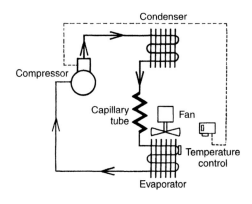

Figure 5-9 Methods of controlling refrigerated space temperature: (a) sensing evaporator temperature, and (b) sensing refrigerated space air temperature.

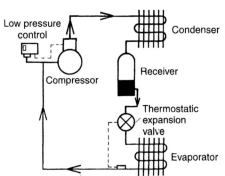

Figure 5-10 The compressor can be cycled using a low-pressure control that senses suction line pressure.

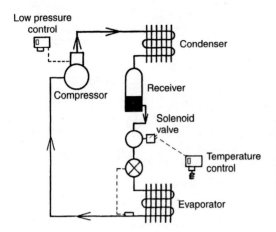

Figure 5-11 In this control scheme, refrigerant flow is controlled by a temperature controller and solenoid valve. The compressor is cycled using a low-pressure control in the suction line.

Parallel evaporator/low-pressure control. In a parallel evaporator system with a single low-pressure control switch, the evaporators are connected in parallel. System control is identical to the single evaporator system. By sensing the suction pressure, the low-pressure control switch regulates the evaporator temperature (Fig. 5-12a).

Thermostat/low-pressure control. This control system has a temperature control that operates a solenoid valve instead of the compressor. A low-pressure control sensing suction pressure cycles the compressor. Each evaporator in the system has its own thermostat to cycle its solenoid (Fig. 5-12b).

Evaporator pressure regulator/low-pressure control. On some multiple evaporator systems, evaporator pressure regulators (EPRs) are used with TEV valves and a low-pressure control to cycle the compressor. The EPR senses the evaporator pressure and regulates the flow of refrigerant exiting the evaporator to meet the evaporator's load requirements (Fig. 5-12c).

When multiple evaporators are set at different operating temperatures, the low-pressure control is set at a value below the lowest EPR setting. When EPRs are used, thermostats are typically not used at the evaporator.

While both the temperature control/solenoid valve system and EPR/TEV system allow each evaporator to operate at a different temperature, the temperature control/solenoid valve system is often preferred over the EPR/TEV system because the temperature control can directly sense product or return air temperature.

Due to their size, multiple evaporator systems are normally equipped with high-pressure protection. A dual-pressure control or individual high-pressure and low-pressure controls might be used.

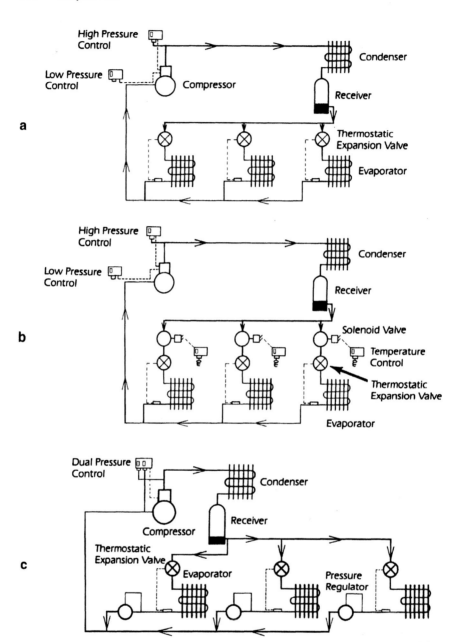

Figure 5-12 Methods of controlling multiple evaporator systems: (a) Low-pressure control in suction line, (b) temperature control of solenoid valve in each liquid line, and (c) evaporator pressure regulators for temperature control.

Multiple compressor systems (TEV)

Multiple compressor refrigeration systems are used to handle wide-load fluctuations or intermittent extremely high heat loads. They are also used where the complete loss of cooling could lead to substantial loss of product. With multiple compressors, if one compressor is lost due to breakdown, the system can still operate at a slightly lower capacity. Supermarkets are a prime example.

The multiple compressors are connected in parallel. When the heat load is greater than the first compressor's capacity, the second compressor is started. If the demand still exceeds this combined capacity, the next compressor is started. Compressors are brought on-line until either the demand levels off or lowers, or all the compressors are operating.

As the heat load decreases, the compressors will shut down one by one until the demand is equal to the capacity of the compressors remaining on line.

The temperature controls or EPR valves at the evaporator operate similarly to a single compressor system. The cut-in settings of the low-pressure switches for each compressor are typically set about two pounds or less apart to provide the proper compressor start-up and shut-down sequence.

Multiple compressor systems are normally equipped with a high-pressure control to provide for system shutdown if high-side pressures reach dangerous or damaging levels. This can be a separate pressure control, or a dual high/low pressure controller can be used. Typical control schemes for multiple compressor systems are shown in Fig. 5-13.

Control Applications

The following sections describe many of the typical refrigeration control systems in use today that employ pressure, temperature, and humidity controls.

Temperature control using evaporator temperature

Evaporator temperature sensing for system temperature control is used on many types of self-contained medium and low-temperature refrigeration systems. It is especially common on capillary tube controlled systems.

Evaporator temperature is sensed using a capillary tube, or a capillary and bulb. The sensing element can be mounted in a variety of ways, as shown in Fig. 5-14.

If the capillary tube is inserted into a tube well secured to the evaporator coil, apply sealant to prevent moisture from entering the tube

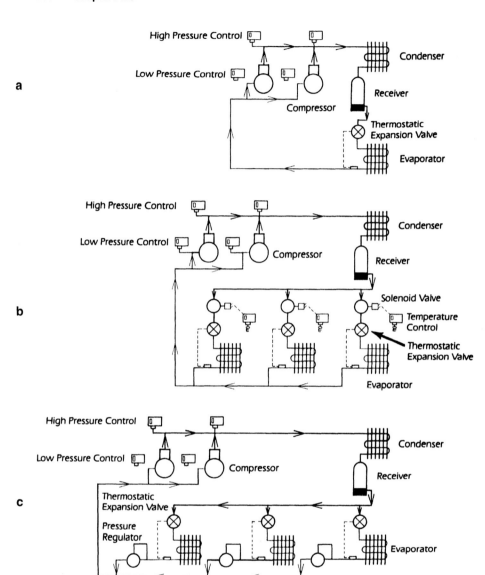

Figure 5-13 Methods of controlling multiple compressor systems: (a) Low-pressure control in suction line, (b) temperature control of solenoid valve in each liquid line, and (c) evaporator pressure regulators for temperature control. Note the use of high-side pressure control for safty shutdown.

well. A minimum of six inches of capillary should be in direct control with the evaporator or tube well to ensure proper control.

The control cut-in setting is set at the maximum desired space or product temperature. The differential setting is determined by adding the difference between the maximum and minimum space or product

Refrigeration Control Systems   117

Bulb Clamped To Evaporator End Turn

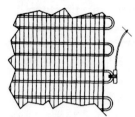

Capillary Wound on Evaporator

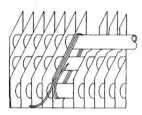

Capillary Mounted In Bulb Well On Evaporator End Turn

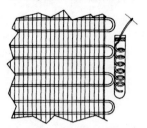

Capillary Mounted On Plate Type Evaporator

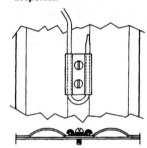

Figure 5-14   Typical mounting of a temperature-sensing bulb or element. *(Ranco Incorporated)*

temperature, plus the evaporator temperature differential. The following example is for a medium-temperature, reach-in commercial refrigerator:

Specifications:

Maximum desired temperature 40°F

Minimum desired temperature 35°F

Evaporator temperature differential 25°F

The cut-in is set at: 40°F

The differential is set at: (40° − 35°) + 25 = 5° + 25° = 30°F

The cut-out is: 40° − 30° = 10°F

Evaporator sensing systems use wide differential temperature controls. Proper setting of the differential prevents short cycling of the compressor. An added advantage of evaporator temperature control on medium temperature applications is guaranteed evaporator defrosting during the off cycle. This is because the evaporator coil will not reach the cut-in setting until the coil is defrosted.

Temperature control using product temperature sensing

In some applications, such as bulk milk and wine tanks, and slush ice, soft ice cream, and yogurt machines, precise temperature control is achieved by directly sensing the product temperature. A narrow differential control, combined with direct product temperature sensing is used in these cases.

Bulk milk and wine tanks. In these applications the control sensing element is inserted into a well surrounded by the product. The sensor is securely attached to the vessel and insulated from outside air.

Soft ice cream, yogurt, and slush drink machines. For these applications, the evaporator coils are wrapped around the drum in which the product is mixed and stored. On some designs, a tube is welded to the drum to accept the temperature control capillary. When a tube is not used, the capillary is securely wrapped around the product drum.

Temperature control using space/return air temperature sensing

The main advantage of this widely used control scheme is the close control you have of the refrigerated space temperature. As the space temperature changes, the control senses the changes and cycles the refrigeration system.

Space temperature sensing controls equipped with air coils are often used in such applications as walk-in coolers or freezers. Mount the control sensor in a protected area where it will not be affected by air circulation from door openings or evaporator discharge (Fig. 5-15). Make certain the refrigerated product does not block air circulation around the sensor.

To monitor return air temperature, the temperature control's capillary or bulb is positioned in the return air stream. The sensor must not contact the evaporator. As shown in Fig. 5-16, the control unit can be located outside of the refrigerated space.

Temperature control using ice thickness sensing

Ice banks are used to provide reserve thermal capacity on medium-temperature refrigeration equipment where cooling demand fluctuates widely. These systems build a reserve of ice during off-peak periods that the system then uses during peak operating, compressors then would be needed normally to meet the peak demand. Ice bank storage

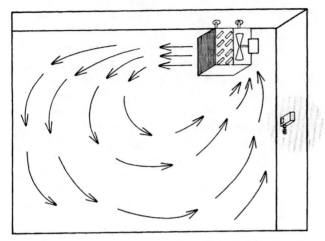

Figure 5-15 Space temperture sensing of a walk-in cooler. *(Ranco Incorporated)*

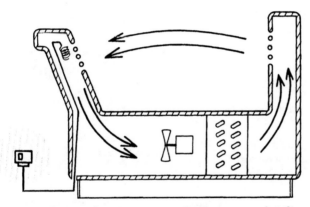

Figure 5-16 Return air temperture sensing in supermarket display case. *(Ranco Incorporated)*

is often used for applications such as drink dispensers where ice is built during evening or low-activity hours, and is then used for cooling beverages during high-activity periods such as lunch hour.

A unique ice bank control is used to cycle the system. This control uses a bulb filled with water. When the ice grows and surrounds the bulb, the water in the bulb freezes and expands. This expansion is transmitted to the control mechanism through a liquid-filled capillary to open the control switch. The switch might cycle the compressor directly, or on larger systems, close the refrigerant solenoid valve to start the pumpdown cycle.

Temperature control using suction pressure sensing

On TEV systems, a low-pressure control might be used for temperature control (see Fig. 5-10). By sensing suction pressure, the control cycles the compressor in response to the evaporator temperature.

As discussed in Chapter 3, a given refrigerant vapor pressure corresponds to a specific vapor temperature. So by properly setting a low-pressure switch to keep vapor pressure in an evaporator at a certain level, we can control the temperature of the air passing over the evaporator coils and thus the temperature of the cooled space.

There is a fixed, reproducible relationship between the temperature of the refrigerant inside the evaporator and the temperature of the air passing over it. Evaporator refrigerant to air temperature differences are typically from 12°F to 15°F. For example, when the refrigerant temperature is 21°F, the air temperature over the evaporator might be 34°F. Different refrigeration units will have different temperature spreads, but there is always a predictable relationship between the two temperatures.

The control cut-in is set at the equivalent pressure for the maximum desired space or product temperature. The differential setting is determined by adding the difference between the maximum and minimum temperature (converted to the equivalent pressure), plus the evaporator temperature differential.

Example: Reach-in refrigerator (37° average temperature)

Specifications:

Maximum desired temperature 40°F

Minimum desired temperature 35°F

Evaporator temperature differential 20°F

Equivalent pressure 37 psi

Settings:

The cut-in is set at 40°F

The differential temperature is (40° − 35°) + 20° = 25°

The cut-out is: 40° − [(40° − 35°) + 20°] = 40° − (5° + 20°) = 40° − 25° = 15°F (18 psig)

The differential pressure is: 37 psi − 18 psi = 19 psi

A narrow differential maintains close control, but can cause short cycling of the compressor. A wide differential gives longer running time, but it can cause wide temperature fluctuations.

In establishing low-pressure control settings, the pressure drop between the evaporator and the low-pressure control must be taken into consideration. The calculated pressure drop must be subtracted from the calculated cut-out setting.

Note: When low-pressure controls are used for temperature control, the condensing unit and system piping must be located in an ambient temperature warmer than the controlled temperature.

In the above example, the 15°F refrigerant vapor in the evaporator is at a pressure of 18.0 psig. If the pressure switch is set to cut out at 18.0 psig, then whenever the temperature over the evaporator drops to 35°F, the pressure control will open, shutting off the compressor. If the cut-out pressure is reduced to a lower value, a colder temperature can be reached. If the cut-out pressure is increased, a warmer temperature can be maintained.

There are two things to remember when setting the cut-in temperature/pressure. The pressure must be low enough so that the unit does not get too warm before the compressor starts. The pressure must also be high enough to allow defrosting of the evaporator coils.

Ice typically forms on the evaporator coils as the refrigerant temperature drops into 25° to 15°F range. As the ice builds up, it acts as an insulator, preventing good heat transfer between the air and the refrigerant. It might also begin to obstruct air flow, further decreasing cooling capacity. This is why it is important that the control shuts off long enough to allow the ice to melt from the evaporator.

While the cut-out pressure varies from one unit to another depending on design, the cut-in pressure is determined by the defrosting needs of the unit. Remember to set the cut-in pressure using a pressure or manifold gauge because the scale on the pressure control switch might not give an accurate enough reading.

Figure 5-17 shows a typical scale found on a pressure control. The previous example gives a cut-in pressure of 37 psig and a cut-out pressure of 18 psig. The differential is 37 minus 18, or 19 psi. The control should be set for a cut-in of 37 with a differential of 19. Making the final adjustment with a pressure gauge allows you to view the cut-in and cut-out pressures on the gage.

Pumpdown control using suction pressure sensing

Pumpdown removes refrigerant from the low side of the system and isolates it in the condenser and receiver when the compressor is not in operation. Pumpdown prevents migration of refrigerant to the compressor, which eliminates liquid slugging and possible loss of oil from the compressor during startup.

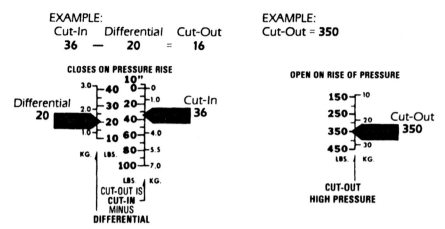

Figure 5-17 Scale settings on a typical pressure control.

Pumpdown control is typically used where the evaporator is remote from the condensing unit. It is particularly popular where the condensing unit is located outdoors in a northern, cooler climate. In this type of situation, the condenser is in a low ambient position, while the evaporator is working at an above-ambient temperature.

Operation. In pumpdown systems, the operating temperature control cycles the liquid line solenoid valve. When cooling is required, the temperature control opens the normally closed solenoid valve allowing refrigerant to flow into the evaporator. The suction pressure rises and a low-pressure control starts the compressor (see Fig. 5-11). When the temperature control is satisfied, it closes the valve and refrigerant flow to the evaporator ceases. The compressor will continue to pump the refrigerant from the evaporator and suction line until the suction pressure reaches the low-pressure control cut-out setting. At this cut-out setting the compressor shuts off. This is the pumpdown cycle.

If, during the on and off cycle, refrigerant leaks past the solenoid valve, the low-side pressure will rise to the cut-in setting of the low-pressure control. The compressor will recycle and run for a brief period until the low-side pressure drops to the cut-out point. These brief, occasional cycles keep refrigerant out of the compressor crankcase.

Note: Frequent recycling might indicate excessive leakage at the liquid line valve or the compressor valves.

Pressure control settings for pumpdown are very important. Because the cut-in setting determines how high the pressure in the low side can go, it is the cut-in setting that needs to be carefully selected.

For condensing units located outdoors, the pumpdown control cut-in setting is determined by selecting either the coldest unit operating or the coldest anticipated outdoor ambient, whichever is the coldest value.

For compressors located indoors, determine the lowest operating temperature of the unit. Subtract three to five degrees from the lowest operating temperature, select this temperature from a refrigerant pressure-temperature chart, then set the pumpdown control cut-in at that value. The cut-out setting should be a reasonable number of psig lower than the cut-in setting, but never so low that the machine will have difficulty reaching the cut-out setting during warm weather operating conditions.

Alarm control using suction pressure sensing

Low-pressure switches are often installed on the suction line of systems with remote evaporators to act as alarm controls (Fig. 5-18). In cases of extreme cooling demand, evaporator temperature goes up and suction line pressure increases to the level where the low-pressure switch closes the contacts in the alarm circuit. The control is typically located in the same room as the compressor so wiring to the refrigerated space is not needed, as would be the case if a thermostat was used as the alarm sensor.

In most cases the low-pressure control is used along with an automatic reset time-delay device. The alarm control energizes the time delay. If the suction pressure does not return to a normal level within the time-delay setting, then the timer contacts close, activating the alarm.

The time-delay circuit prevents nuisance alarms caused by temporary transient conditions. On systems with a defrost cycle, a time-delay setting is chosen to include the duration of the defrost cycle plus sufficient time for the unit to pull back down to normal temperature after defrost. Time delays of 45 to 60 minutes are normally used.

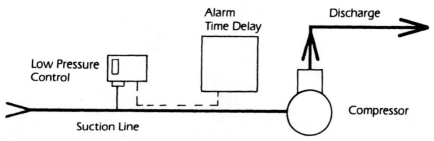

Figure 5-18 Low-pressure control sensing suction line pressure for high suction level alarm control.

Low-pressure limit control using suction pressure sensing

On non-pumpdown systems where the compressor is cycled by a thermostat, a low-pressure control can be used to protect the system from loss of charge or abnormally low evaporator temperatures (Fig. 5-19). Low-pressure limit controls protect against evaporator freeze up and will shut off the compressor if the suction pressure (temperature) drops below the freezing point.

Reset protection. When a low-pressure limit control trips, it typically indicates a problem that needs immediate repair or action. Because continued operation under these conditions might damage the compressor, the low-pressure controls used for low-pressure limit protection are normally equipped with manual reset buttons. An automatic reset control can be installed if the risk of product loss is high and/or the low-pressure limit control might trip due to a temporary condition. A manual reset time delay can also be used in conjunction with an automatic reset low-pressure limit control. This arrangement minimizes tripping due to transient conditions while providing the protection of a manual reset device. In this control scheme, the time delay is energized when the low-pressure switch contacts close. If the suction pressure does not increase to a normal level within the time-delay setting (typically 2 to 5 minutes), then the system shuts down.

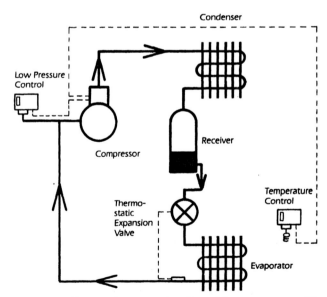

Figure 5-19 Temperature control cycling of the compressor with a low-pressure limit control. *(Ranco Incorporated)*

High-limit control using high-pressure sensing

As mentioned earlier in this chapter, high-pressure controls are often used to sense the compressor discharge pressure and stop the compressor in case of excessively high head pressure (see Fig. 5-13). High pressure limit controls protect the system from operating at excessive head pressure due to loss of fan motor or a blocked condenser in air cooled systems, or loss of water, inadequate water flow, or clogged condenser tubes in water-cooled systems.

An open high/close low switch is used in this control scheme. Pressure control settings vary with different refrigerants and system designs. The high pressure control cut-out settings recommended by the condensing unit manufacturer must be followed.

If an SPDT high-pressure control is used, the back contacts of the switch can be wired to an alarm circuit to visually or audibly indicate that the switch has opened. High-pressure controls are available with manual or automatic reset. Manual reset controls are frequently recommended by compressor manufacturers to provide fail-safe compressor protection.

Automatic reset controls are often preferred on systems refrigerating high-cost, perishable products in cases where occasional transient conditions would cause the high-pressure control to cut out.

Condenser fan control using high-pressure sensing

On air-cooled condenser systems operating at ambient temperatures below 50°F, head pressure controls are often provided to ensure proper refrigerant feed and system operation.

A high-pressure switch is used to sense the condensing pressure (head pressure). It's input is used to cycle the condenser fan on and off (Fig. 5-20). This application requires a close high/open low switch action. (The fan circuit closes when the head pressure reaches the high setting and opens when the head pressure falls below the control's low setting.)

Condensing temperatures commonly range from 80° to 95°F. The correct adjustment of the pressure switch ON/OFF control differential is very important. Always look up the refrigerant's pressure equivalent for the exact condensing temperature. A narrow differential can cause excessive fan motor cycling with a resultant short fan motor life. A wide differential can contribute to uneven TEV operation.

Condensing units with multiple fans can use pressure or temperature controls to maintain head pressure. When pressure controls are used, they are set to cut in approximately 10 psi apart for fan staging.

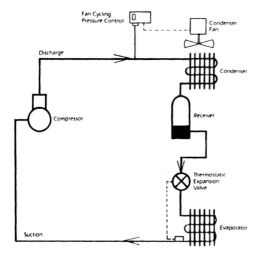

Figure 5-20 High-pressure control of the condenser fan. *(Ranco Incorporated)*

When temperature controls are used, they are set to cut out at various outdoor ambient temperatures.

The system might also use a high-pressure control to cycle one of the fans in addition to the temperature controls. On multiple circuit condensing units (condensers served by two or more independent compressors), temperature controls are typically used for fan cycling because pressure controls can only sense one circuit at a time.

Temperature controls can also be used along with head pressure control valves to reduce the amount of refrigerant needed to flood the condenser during low ambient temperature operation. Such a setup allows for a reduced refrigerant charge and a smaller, less expensive receiver.

Temperature control used for defrost termination/fan delay

Thermostats are often used for terminating defrost on commercial refrigeration systems. Typically, a timer or clock is used to initiate defrost at a preselected time. Defrosting is accomplished by the use of electric heat, hot gas, or reverse cycle.

Electric heat defrost. The timer initiates the defrost cycle by opening a circuit to stop refrigeration, and closing a circuit to energize electric heaters in the evaporator coils. These heaters melt the accumulated ice. A thermostat with a single pole, double throw switch senses the evaporator temperature and terminates defrost by energizing the timer release solenoid, starting refrigeration. The circuit to the evaporator fan(s) remains open until the evaporator temperature reaches the control setpoint, re-energizing the evaporator fans.

Hot gas defrost. The defrost timer starts defrost by closing the circuit (normally open) to a hot gas bypass solenoid valve and opening the circuit (normally closed) to the evaporator fan(s). An additional contact in the timer closes to bypass the temperature control circuit. The flow of hot gas is diverted from the compressor to the evaporator where it melts the frost on the evaporator.

When the evaporator temperature reaches the defrost termination/fan delay control setpoint, the switch closes, allowing the timer release solenoid to de-energize the hot gas defrost bypass valve. The circuit to the evaporator fan(s) remains open until the evaporator temperature reaches the control setpoint. Once the control setpoint temperature is reached the evaporator fans circuits are re-energized.

In both electric heat and hot gas defrost systems, the fan-delay feature prevents warm, moist air from being circulated in the refrigerated space. On multiple evaporator systems, stagger the time clock settings so all evaporators do not defrost at the same time. A low-pressure control also can be used to terminate defrost. The pressure control senses the suction pressure increase that occurs during defrost and closes a switch that energizes the timer-release solenoid to terminate defrost and start refrigeration.

Oil pressure sensing for lube oil protection

The compressor's lube oil protection control is a differential pressure switch. This pressure differential control uses two opposing pressure sensing elements to monitor the effective or "net" oil pressure, and will shut down the compressor if a lubrication failure occurs (Fig. 5-21).

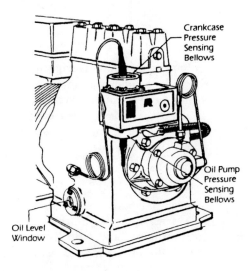

Figure 5-21 Oil pressure sensor installation on a compressor. *(Ranco Incorporated)*

The net pressure is the pressure difference between the crankcase pressure and the oil pump discharge pressure. Because the compressor crankcase frequently operates at pressures other than atmosphere, it is necessary to measure this net pressure to determine if proper lubrication is occurring. The pressure-sensing elements are connected to a normally closed switch that opens on an increase of pressure difference.

Pressure differential setpoints of lube oil pressure controls vary according to compressor manufacturer's specifications. Some controls offer adjustable pressure setting of 8 to 60 psi; other settings are factory fixed. Many manufacturers use a fixed-setting type to avoid the possibility of improper field adjustment.

Time delay. The normally closed control switch is connected to a built-in time delay switch. The time delay allows the oil pressure to build up to its operating pressure at compressor start-up, and also prevents nuisance shut-off of the compressor due to momentary drops in oil pressure during the running cycle.

The time-delay switch circuit is controlled by a thermal resistor heater that is powered by the control switch when the control senses a low-pressure differential. If the low-pressure condition is not corrected within the allowed time, the resistor heater becomes warm enough to open a bimetal switch in the compressor control circuit. The time delay also incorporates an ambient compensating device to provide accurate timing under varying ambient temperature conditions. The time-delay switch has a manual reset button to restart the compressor after shutdown from a loss of oil pressure. Time delay settings vary by compressor manufacturer, but are typically 60, 90, or 120 seconds.

Control Systems for Capacity Reduction

During periods of low cooling demand, many commercial refrigeration systems use some method to temporarily reduce their cooling capacity. As we has seen, the simplest way to reduce system capacity reduction is to cycle the compressor using a thermostat or pressure control switch. Other methods of capacity reduction include using compressors equipped with cylinder unloaders or hot gas bypass systems.

System design factors

Predicting TEV performance at reduced system capacities can be difficult. Many factors, such as TEV sizing, refrigerant distribution, TEV setting, evaporator coil design, suction line piping, and bulb location, can all affect operation. However, by closely following manufacturer's recommendations, a conventional TEV will operate quite well down to approximately 35 percent of its rated capacity. Balanced port thermostatic expansion valves, can operate even lower—down to about 25 percent of rated capacities.

Valve size. The TEV should be sized as close as possible to the system's maximum designed heat load condition. A valve with a capacity rating up to 10 percent below the full-load conditions can be selected if the system is to operate at reduced loads for long periods of time, and if slightly higher-than-normal superheats can be tolerated at full-load conditions.

Refrigerant distributor sizing. The function of the refrigerant distributor is to evenly distribute refrigerant to a multicircuited evaporator. The proper sizing of the distributor is extremely important for systems using methods of capacity reduction. If the distributor cannot perform properly at all load conditions, erratic TEV operation can occur. For the pressure drop type distributor, the distributor nozzle and tubes must be checked for proper sizing at both minimum and maximum load conditions.

Superheat adjustment. The superheat setting of the TEV should be set at the highest possible superheat tolerable at full-load conditions. A high superheat setting reduces problems associated with mild TEV hunting at low-load conditions.

Evaporator coil design. When the evaporator is circuited to provide counterflow of the refrigerant relative to the direction of the air flow, superheat will normally have the least effect on evaporator capacity and suction pressure fluctuations will be minimized.

The velocity of the refrigerant moving through the evaporator must be high enough to prevent excessive trapping of liquid refrigerant and oil, which might cause TEV hunting. Each evaporator circuit should be exposed to the same heat load. Air flow across the coil must be evenly distributed.

As explained earlier in this chapter, large capacity evaporators are often split into multiple sections so that one or more of these sections can be shut off for capacity control during part-load operation. A separate TEV is required to feed each of these sections.

Suction line piping. Approved methods of suction line piping, including recommended bulb locations and use of traps, must be followed.

Sensing bulb location. The TEV's sensing bulb should be located on a horizontal section of suction line near the evaporator outlet and, in the case of an externally equalized valve, upstream of the equalizer connection on the suction line.

Vapor-free liquid refrigerant. Another important aspect in ensuring proper TEV operation is providing vapor-free liquid refrigerant to the inlet of the TEV. Vapor in the liquid line can severely reduce the capacity of the TEV, hindering proper refrigerant flow to the evaporator. An adequately sized liquid-to-suction heat exchanger will help ensure vapor-free liquid by providing some amount of subcooling to the liquid. In addition, the heat exchanger provides an added advantage to the system by vaporizing small quantities of liquid refrigerant in the suction line before the liquid reaches the compressor. A sigh glass installed near the TEV inlet offers a visual check for vapor-free refrigerant.

Compressor unloader systems

Large reciprocating compressors are often equipped with internal unloaders. Unloaders reduce a compressor's pumping capacity by holding open the intake valves on one or more cylinders. The unloader is operated by compressor oil pressure, which acts on the intake valve holding it closed. During periods of low demand, a solenoid valve in the oil line is closed removing oil pressure to the unloader. With oil pressure removed, a spring supplies the force needed to hold open the intake valve. A low-pressure control is commonly used to cycle the solenoid valve. As the cooling load decreases, the suction pressure also decreases. At the pressure control cut-in pressure, the unloader is switched on to reduce the system's capacity. One or more pressure switches can be used, depending on the number of unloaders on the compressor. The unloader valve pressure control setting must be higher than the low-pressure control used to cycle the compressor on and off for system temperature control.

System design for part-load conditions. On systems where the compressor has an unloader that can reduce output to one half of its rated capacity, care must be exercised when selecting expansion valves and refrigerant distributors. If the compressor can unload below one-third of its rated capacity, special design considerations might be necessary to ensure proper TEV operation. Figures 5-22, 5-23, and 5-24 illustrate piping schematics for three possible methods of balancing the capacity of the TEV and distributor with the compressor during low-load operation.

Figure 5-22 illustrates two parallel evaporators, each controlled by a separate TEV and refrigerant distributor. Each evaporator shares half of the total common load. The liquid line solenoid valve ahead of each TEV is electrically connected to the compressor capacity modulating system. When the compressor capacity is reduced to 50 percent, one of the two solenoid valves closes, stopping refrigerant flow to one TEV. The TEV remaining in operation will then have a rated capacity approximately equal to the compressor capacity operating 50 percent unloaded.

Refrigeration Control Systems   131

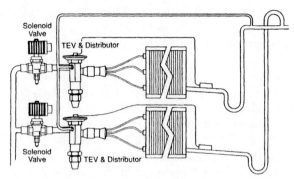

Figure 5-22 Two or more evaporator sections handling the same load. *(Sporlan Valve Company)*

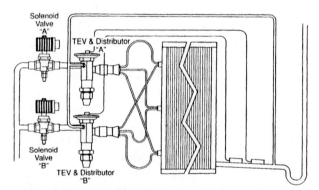

Figure 5-23 Capacity reduction using a single evaporator controlled by two TEV and two solenoid valves. *(Sporlan Valve Company)*

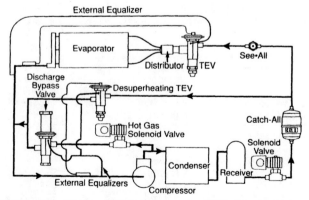

Figure 5-24 System with desuperheating TEV. *(Sporlan Valve Company)*

This technique can be carried further by using additional evaporator sections, each controlled by a separate TEV and refrigerant distributor. Using multiple evaporator sections will allow highly reduced loads to be properly controlled.

Figure 5-23 illustrates the use of two TEVs and two distributors feeding a single evaporator. Each evaporator circuit is fed by two distributor circuits, one from each distributor. The solenoid valves are connected to the compressor capacity modulating system, as mentioned before. Using this configuration, TEV and distributor capacities can be reduced in three stages. As an example, assume that TEV and distributor combination A are sized to handle 67 percent of the load, and combination B, 33 percent of the load. The three stages of valve and distributor capacity reduction result from opening or closing the solenoid valves according to the information in Table 5-1.

Another variation of this technique is to have each evaporator circuit fed by a single distributor circuit and size the TEVs and distributors on the expected load of the total number of circuits fed by each TEV. Reducing evaporator capacity is accomplished by closing a solenoid valve, which deactivates the circuits being fed by the TEV and distributor downstream of the solenoid valve. This method of capacity control, however, requires a degree of care, because the heat load on the evaporator circuits will be affected in the manner in which circuits are deactivated.

Balanced port TEVs. One of the factors limiting a TEV's ability to operate at part-load conditions is a variation in pressure drop across the TEV during normal system operation due to changes in head pressure. A pressure drop across the TEV influences valve operation. To counteract the effects of this force, balanced port valves have been developed.

The refrigerant flow entering a balance port valve is divided between two ports, so the force of the refrigerant flow is transmitted to the midsection of the piston. The force of the flow heading to the lower port is largely canceled out by the force of the flow heading to the upper port due to the design of the piston. A semibalanced valve is achieved,

TABLE 5-1 Valve and Distributor Capacity per Solenoid Valve Position

Percentage of compressor capacity	Position of solenoid valve "A"	Position of solenoid valve "B"	Total valve and distributing loading (percentage of rated capacity)
100	Open	Open	100
83	Open	Open	83
67	Open	Closed	100
50	Open	Closed	75
33	Closed	Open	100
16	Closed	Open	50

allowing the valve to operate at a lower percentage of its rated capacity than a conventionally designed valve.

Satisfactory operation down to 25 percent or lower of rated capacity can be expected with the type O valve, provided that these design recommendations are followed.

Recent efforts by system manufacturers to reduce operating costs of refrigeration systems by allowing condenser pressures to fall or float with lower ambient temperatures has created a need for small-capacity TEV with a balanced port design and superior modulating characteristics. This effort is particularly apparent with supermarket applications.

Hot gas bypass and desuperheating TEVs

If the compressor is not equipped with an unloader, low-pressure controls might be used to energize hot gas bypass valves to reduce system capacity. Bypassing a controlled amount of hot gas to the suction side of the system provides a practical solution to part-load operation. Bypassing hot gas is accomplished with a modulating control valve known as a *discharge bypass valve*.

For close-coupled systems, the preferred method of hot gas bypass is bypassing the inlet of the evaporator. When this method is used, the TEV responds to the increased superheat of the vapor leaving the evaporator and provides the liquid required for desuperheating. The evaporator also serves as an excellent mixing chamber for the bypasses hot gas and the liquid vapor mixture from the TEV. Finally, oil return from the evaporator is improved because the refrigerant velocity in the evaporator is kept high by the hot gas.

For multievaporator or remote systems, hot gas can be bypassed directly into the suction line (Fig. 5-24). In addition to the discharge bypass valve, an auxiliary TEV known as a *desuperheating* TEV is used to supply the needed liquid refrigerant to cool the discharge gas entering the suction line. Compressor manufacturers generally rate their compressors for a specific return gas temperature. Many refrigeration and low-temperature compressors require low suction gas temperatures to prevent discharge gas temperatures from rising too high and damaging compressor parts and carbonizing oil. Consult the compressor manufacturer if the maximum permissible suction gas temperature for a compressor is not known.

Special desuperheating thermostatic charges are used in this type of valve. Sizing a desuperheating valve involves determining the amount of refrigerant liquid necessary to reduce the suction gas temperature to the proper level. For hot gas bypass applications, a properly sized desuperheating valve can be used.

An externally equalized TEV is recommended for most desuperheating applications. If the piping of the desuperheating TEV is close coupled, an internally equalized valve can be used. Figure 5-24 illustrates the use of an externally equalized desuperheating TEV.

When piping the discharge bypass valve and the desuperheating TEV, remember that good mixing of the discharge gas and liquid must be obtained before the mixture reaches the sensing bulb of the desuperheating TEV. Improper mixing can produce unstable system operation causing the desuperheating TEV to hunt. Proper mixing can be accomplished by installing a suction line accumulator downstream of both valve outlet connections with the desuperheating TEV bulb downstream of the accumulator. An alternate method is to tee the liquid vapor mixture from the desuperheating TEV and the hot gas from the bypass valve together before connecting a common line to the suction line (Fig. 5-24).

Remember, the bypass valve pressure control setting must be higher than the low-pressure control used to cycle the compressor on and off for system temperature control.

Electronic Temperature Control Systems

Electronic temperature control systems are becoming increasingly widespread throughout the industry, and any technician entering the field must be capable of troubleshooting these units. Fortunately, these systems are designed with built-in diagnostic boards for easier troubleshooting. Advantages of electronic temperature control systems include:

- Compact, reliable, solid state electronics.
- Solenoid valve tight seating, which can be used as a suction stop valve for defrost.
- Close temperature control with a range of ± (plus or minus) 1°F.
- Can operate at less than ½ psi pressure drop.
- Can be installed without disassembly using phosphorous-copper—no expensive silver solder is needed.

An electronic temperature sensitive suction throttling valve (Fig. 5-25) is designed for more precise and energy-efficient control of temperature in refrigerators. Proper temperature is maintained by regulating refrigerant flow in the suction line allowing the evaporator capacity to exactly match the existing load.

An electronic temperature control system replaces evaporator pressure sensing regulators, conventional thermostat and solenoid valves, or compressor cycling devices. The system consists of a electronic temperature sensitive valve and components that are easily wired without the need for polarization.

Refrigeration Control Systems 135

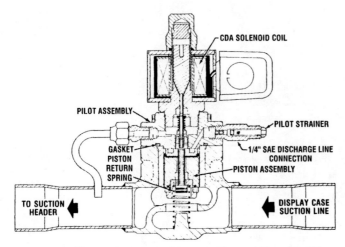

Figure 5-25 Electronic temperature-sensitive suction throttling CDA valve. *(Sporlan Valve Company)*

The valve is normally open, pilot operated, requiring high-side pressure to move it to a modulated or full-closed position. When in the full-closed position the tight seating allows the valve to be used as a suction stop valve on systems required for defrost.

Valve components are a solenoid, pilot assembly, piston, piston return spring, and a body assembly that allows for complete serviceability without removing the valve from the line. The valve modulates by the change in magnetic pull of the dc-operated solenoid pilot. The main port of the valve is modulated by bleeding a controlled amount of discharge gas to the chamber above the valve piston. The discharge gas enters the piston chamber through a fixed orifice and then flows through the modulating pilot valve to the suction line downstream of the valve; thus, as the pilot valve opens, the pressure decreases in the piston chamber and the piston return spring pushes the piston upward, opening the valve.

The valve can be used as a suction stop valve by providing full power to the solenoid coil. The pilot closes tight and high pressure builds in the piston chamber, closing the valve.

The electronic components that control the valve consist of the panelboard, a plug-in thermostat, an air temperature sensor, and an optional diagnostic board.

Panelboard. The thermostat plugs into a panelboard (Fig. 5-26), and each panelboard is equipped with two or four slots to accept thermostats for control of independent electronic throttling valves. The panelboard has wiring connections for each thermostat; two wires to the valve solenoid, and two wires to the temperature sensor in the refrigerated space. There are also terminals for wiring from the defrost

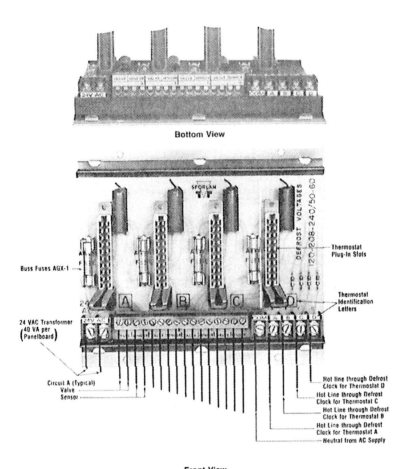

Figure 5-26 Electronic control panelboard. *(Sporlan Valve Company)*

clock and for the 24-volt ac power supply from the transformer. The panelboard is equipped with a solid state relay that will cause the valve to close when the defrost clock applies 120, 208, or 240 volts to the panelboard connections.

Plug-in thermostat. The plug-in thermostat is the brain of the electronic control system (Fig. 5-27). Some thermostats have the capability for temperature readout when used in conjunction with a digital voltmeter. The set point or sensor location temperature is obtained by setting a voltmeter on the 2 or 20-Vdc scale and connecting the leads to the appropriate poles on the thermostat. Move the decimal point two places to the right and the readout indicates temperature. The temperature range is −25°F to +60°F.

The plug-in thermostat operates in the following manner. The 24-Vac power supply from the transformer is rectified to a dc voltage. When

Refrigeration Control Systems   137

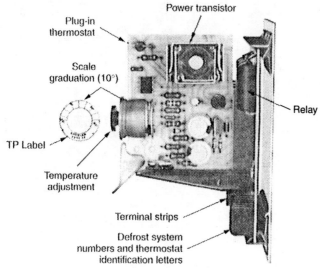

Figure 5-27 Plug-in thermostat. *(Sporlan Valve Company)*

the temperature sensor's resistance is increased by a drop in temperature, a voltage difference occurs across the bridge circuit. Voltage difference across the bridge is measured and amplified 50 times by the operational amplifier. The operational amplifier transmits the amplified signal to the base of the transistor.

The current flow through the power transistor increases to a value equal to 100 times the base current. This current flows through the

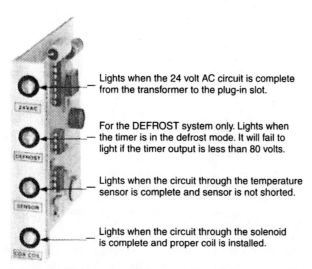

Figure 5-28 Light-activated diagnostic board of electronically controlled system. *(Sporlan Valve Company)*

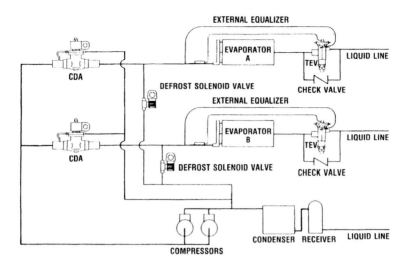

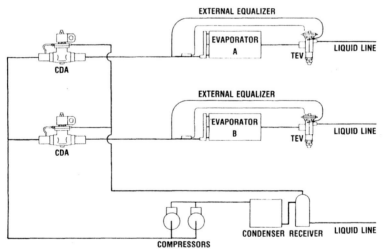

Figure 5-29 Applications for CDA valve, electronically controlled systems. *(Sporlan Valve Company)*

solenoid coil to reposition the valve piston and thereby increase or decrease refrigerant flow from the evaporator to regulate air temperature.

Air temperature sensor. The temperature sensor consists of a thermistor with a protective shield and a neoprene covered lead wire.

Diagnostic board. The diagnostic board plugs into the panelboard slot and is used to check operation during start up and service. The diagnostic board is equipped with four indicator lights, which confirm correct wiring for the following functions: 24-Vac supply; defrost clock wiring and operation; valve solenoid wiring for open or short; and temperature sensor wiring for open or short. Figure 5-28 shows a front view of a diagnostic board.

Application Considerations

The system's temperature sensor should be located in a typical sensing location similar to that for a standard thermostat sensing bulb. Stray air currents and heat from antisweat heaters will affect the set point and should be avoided.

In a single evaporator system, care must be taken to ensure the suction pressure does not pull down to an undesirable low level when the electronic solenoid valve throttles to maintain air temperature. Discharge gas bypass can be successfully applied to the suction line in conjunction with the electronically controlled temperature system to provide system unloading.

On multievaporator systems (Fig. 5-29), the evaporators can operate at different temperatures. An electronically controlled throttling valve might be required on one or more of the evaporators to maintain pressures higher than that of the common suction line. For example, if evaporator A in Fig. 5-29 is designed for –10°F (22 psig, refrigerant 502), and evaporator B for –20°F (15.3 psig), the valve will throttle to allow the evaporator's temperature difference to match the existing load.

It is very important to follow all manufacturer directions and specifications when installing and servicing electronic control systems.

Chapter 6

Working with Tubing

Tubing is used to carry liquid refrigerant and refrigerant vapor from component to component in the refrigeration system loop. It is the circulation system of any refrigeration unit, and all service technicians must be proficient in working with tubing.

Types of Refrigeration Tubing

While steel, aluminum, plastic, and flexible tubing is sometimes used, copper tubing is by far the most common material used for refrigeration system lines.

Copper tubing

All copper tubing used in refrigeration work must be classified as Air Conditioning and Refrigeration (ARC) tubing. Table 6-1 lists the common outside diameter sizes for ARC tubing. Refrigeration tubing is measured by its outside diameter, not inside diameter as in the plumbing industry.

TABLE 6-1 Common Outside Diameter Sizes for ARC Tubing

Tubing outside diameter	Tubing wall thickness
¼"	0.030"
⅜"	0.032"
½"	0.032"
⅝"	0.035"
¾"	0.035"
⅞"	0.045"
1⅛"	0.050"
1⅜"	0.055"

ARC tubing is manufactured to high cleanliness standards. The manufacturer normally charges the tubing with nitrogen gas and seals the ends to ensure it stays clean and dry until it is installed. The nitrogen gas prevents oxidation inside the tubing. Never remove the tubing seals until just before you are ready to use the tubing. Braze the end of the tubing shut after cutting a piece from the coil to keep dirt and moisture out. Copper tubing is further classified as soft or hard.

Soft copper tubing. Soft copper tubing is annealed (heated and cooled) to make it flexible and easy to cut, flare, and bend. Soft copper tubing is normally available in 25- to 100-foot rolls. It is most commonly joined using soft solder connections or flared connections.

Hard copper tubing. Hard copper tubing is completely rigid and is installed in straight vertical and horizontal runs connected by elbows and union fittings. It is sold in 20-foot lengths. All hard copper connections must be brazed connections.

Steel tubing

Tubing made of mild steel (SAE 1008) or stainless-steel (No. 304) can be used. Stainless-steel tubing is often specified for food service and dairy applications. It is also required for ammonia (R-717) refrigeration systems because copper will corrode when exposed to ammonia. Steel tubing sizes are essentially the same as those for copper tubing.

Joints for mild steel can be double lap brazed or butt welded. Stainless-steel tubing is joined with flare fitting or brazed joints.

Flexible refrigerant hose

Flexible tubing or hose manufactured specifically for refrigeration applications is often used in the liquid and suction lines of commercial refrigeration systems. This flexible hose is constructed of a nylon tube center covered with yarn reinforcement and a durable polyethylene cover. Nylon or brass fittings, often fitted with synthetic rubber O-rings, are used to make connections to refrigerant lines.

Plastic tubing

Polyethylene tubing is often used for the cold water lines of water-cooled condensers. It is easy to cut, bend, and join using simple fittings supplied by the tubing manufacturer. Plastic tubing is rarely, if ever, used in the actual refrigeration system to carry refrigerant liquid or vapor.

Cutting and Cleaning Tubing

Tubing is cut to length using a hacksaw or tube cutter. Hacksaws with wave-set teeth at 32 teeth per inch are a good choice for cutting hard copper tubing. The tube cutter is the preferred tool for cutting soft copper.

If a hacksaw is used, a sawing fixture should also be used to ensure square cuts. Remove all inside and outside burrs with a reamer, file, or other sharp-edged scraping tool. If tube is out of round it should be brought to true dimension and roundness with a sizing tool.

The joint surface areas should be clean and free from oil, grease, or oxide contamination prior to brazing or soldering. Clean the tubing end by brushing with a stainless-steel wire brush, or by rubbing the end with an emery cloth. If oil or grease is present on the tubing, clean with a commercial solvent. Remember to remove small foreign particles, such as emery dust, by wiping with a clean, dry cloth. The joint surfaces must be clean prior to soldering or brazing.

Bending Tubing

When replacing soft copper refrigeration lines, it is often necessary to bend the replacement tubing to duplicate the shape of the old one. Although large diameter copper tubing can be bent by hand to obtain a gentle curve, any attempt to hand bend tubing into a tight curve will usually kink the tubing (Fig. 6-1). A kink in a refrigeration line weakens the line at that point. Never use a kinked line. To avoid kinking, always use a bending tool. There are several types of tube bending tools available.

For small diameter copper tubing, a bending coil spring is the most popular choice. This coil spring is slipped over the outside of the tubing. The coil spring prevents the tubing from kinking as you slowly bend it by hand. Bend the tubing slightly further than required and back off to the desired angle. This releases spring tension in the bender so it can be easily removed.

Spring coil benders can also be inserted into a larger tube and the tube bent with the spring supporting it from inside. Spring-type tubing benders come in many sizes, and are often sold in sets.

On larger diameter tubing, or where more precise bends are needed, a lever-type or gear-type bender should be used. Slip the bender over the tubing at the exact point the bend is required.

If flare fittings are being used, assemble the flare nuts on the tubing after bending it to shape but before flaring the tube ends. Once the ends are flared, the flare nuts will not fit over the end of the tubing.

When bending a tube, always remember that the bend radius must be equal to or greater than the outside diameter of the tubing. A bend

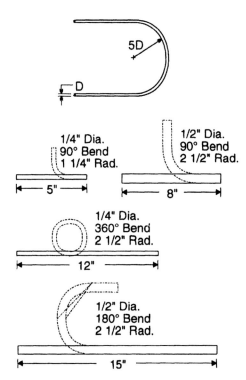

Figure 6-1 The minimum radius a tube should be bent should equal five times the tube's outside diameter.

radius that is smaller than the outside diameter of the tubing can result in stress or collapse of the metal.

Joining Tubing

A number of methods are used to join tubing to tubing and tubing to connector fittings. These include: flared connections, brazed or soldered connections, and compression and O-ring fittings.

Flared connections

Flared connections are made using special flare connectors. They are used on soft copper and steel tubing. The connectors consist of two parts, the fitting and the flare nuts (Fig. 6-2). The flare nuts are slipped onto the tubing pieces to be joined. The ends of the tubing are then carefully flared to increase their end diameters. Flares are typically made at a 45° angle to the tube. Steel tubing does not flare easily, however, and should have only a 37° angle.

Once the tube end is flared out to the proper angle, the flare nuts are then threaded onto the fitting. This pulls the ends of the tubing tightly into the fitting forming a vapor-tight joint. Flared fittings are available

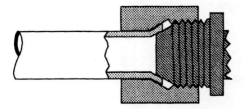

Figure 6-2 Properly made flared connection.

for straight, tee, and elbow type joints. Flared joints for refrigeration work can be formed using single- or double-flared tube ends.

Single-thickness flare. Using a flaring block and flaring tool, make a properly sized flare as follows:

1. After the end of the tube has been cut, use a smooth mill file to square off the tube end and remove any inside burrs with a reaming tool (Fig. 6-3). Take care that no filings enter the tube.
2. Place the flare nut on the tubing with the open end toward the end of the tubing. Insert the tube into the flaring tool so that it extends above the surface of the block (Fig. 6-4a). Proper positioning will leave enough metal exposed to form a full flare.
3. If too much tubing extends above the block, the resulting flare will be too large for the nut to fit over it. If the tube does not extend far enough, the flare could end up small enough to be squeezed out when the fitting is tightened.
4. When forming the flare, place a drop of refrigerant oil on the flaring tool spinner at the point where it contacts the tubing. Tighten the

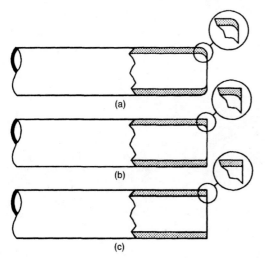

Figure 6-3 Three stages of preparing a tube end: (a) tube after cutting, (b) tube after squaring with file, and (c) tube after reaming.

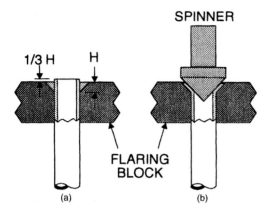

Figure 6-4 Steps in making a single-thickness flare: (a) install the tubing in the block and (b) turn down the spinner into the tubing the correct depth.

spinner ½ turn against the tube end, then reverse it ¼ turn. Continue, advancing the spinner ¾ turn and reversing it ¼ turn repeatedly until the flare is formed (Fig. 6-4b).

Never overtighten the spinning tool because you could thin or weaken the wall of the tubing at the flare. In addition, always set the flare nut in its proper position on the tube before the flare is made. Once the tube has been flared, it is very difficult to reposition the nut.

Double-thickness flare. Special tools are used to form a double thickness flare. Figure 6-5 shows a cross section of a simple block and punch type tool used in double flaring. The correct final shape of the double flare is also shown in the illustration. Some flaring tools have adapters that allow them to form either a single or double flare (Fig. 6-6).

Double thickness flares are typically recommended for larger tubing, $5/16$-inch minimum. Smaller tubing is not a good candidate for the

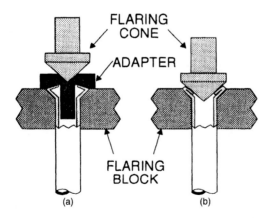

Figure 6-5 Steps in forming a double flare with a block and punch flaring-tool set. (a) Female punch bends tubing inward. (b) Male punch folds the end of the tube over to form a final double flare.

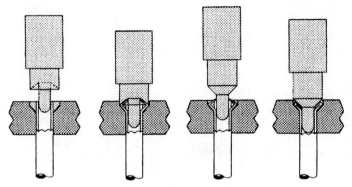

Figure 6-6 Alternate method of forming a double flare.

double-flaring technique. When used, double flaring creates a stronger joint than the single flare.

1. Cut the tubing to the required length using a tube cutter, and clean and square the ends as described earlier.
2. Place the tube nut on the tubing in the correct direction.
3. Place the tubing in the flaring bar so the end protrudes slightly above the face of the bar.
4. Firmly clamp the tube in the bar so the force generated during flaring does not push the tubing down through the bar.
5. Fold over the end of the tubing by placing the adapter or anvil in place over the tube opening and tightening down the flaring clamp.
6. Loosen the flaring clamp and inspect it to confirm that the end of the tubing is properly flared outward at a 45° angle for soft copper, and slightly less for steel tubing.
7. Install the cone onto the tube opening and retighten the flaring clamp.
8. The cone completes the double flare by folding the tubing back on itself. This doubles its thickness and creates two sealing surfaces.

Flared tube fittings

The standard type of flared tube fitting used in refrigeration is a forged fitting, using either pipe thread or Society of Automotive Engineers (SAE) National Fine Thread (Fig. 6-7).

Fittings are typically made of drop-forged brass. Each fitting consists of accurately machined National Fine (NF) or National Pipe threads with a hexagonal shape for a flare nut wrench attachment and a 45°

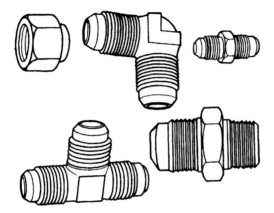

Figure 6-7 Common flare-type fittings.

flare that fits against the tubing flare. Always handle threaded fittings carefully in order to minimize the possibility of damage.

Fitting sizes correspond to tubing sizes. A ¼-inch flare nut will connect ¼-inch tubing to a flared fitting, even though the fitting has ⁷⁄₁₆-inch NF internal threads and needs a ¾-inch wrench to turn it.

Manufacturer's listings of tube fittings generally use code numbers to indicate size. For example, a code number of 3 means that the fitting was made for ³⁄₁₆-inch tubing, while a number 4 fitting corresponds to ¼-inch (⁴⁄₁₆) tubing. Code number 8 fits ½-inch (⁸⁄₁₆) tubing, and so on.

For fittings that have pipe threads on one end, keep in mind that the pipe threads taper ¹⁄₁₆-inch in diameter for every inch in length.

Special fittings have been designed for use with plastic tubing. Common fitting materials are brass, aluminum, and polyethylene. Fittings on plastic tubing are not flared but are a compression type.

Metric sizing. Metric-sized fittings must be used on metric-sized tubing. The metric fittings are comparable in terms of materials, shapes, and styles to U.S. conventional fittings. Care should be taken when working with these two types of fittings not to mix one type with the other.

Fittings for brazed joints

Brazed sleeve fittings can be used to connect tubing of both equal and unequal diameters. The fittings are available in many sizes. Two tubes of unequal diameter can be joined with a sleeve fitting that is brazed to the tubing. The two tubes can also be fitted using a *constrictor*, a tool that restricts the diameter of the larger tube to fit over the smaller one (Fig. 6-8). Some tube-cutting tools have one wheel for cutting and a second wheel for constricting. A constrictor is a good tool to use before brazing because the silver brazing alloys can allow for larger gaps between the parts.

Figure 6-8 Joining tubes of dissimilar diameters using a constricted fitting. *(Adapted from* Practical Air Conditioning Equipment Repair *by Anthony J. Caristi, McGraw-Hill, copyright 1991.)*

Swaging. Swaging is the best method available for preparing the ends of equal-diameter tubing for brazing. It requires no additional fittings. One end of the first tube is expanded so that its inner diameter is the same as the outer diameter of the second tube. The two ends are slip fitted together. It's more reliable than using the sleeve fitting method because with swaging there is only one connection to be brazed (Fig. 6-9).

One way to prepare a joint for swaging is to use a specially designed punch that is hammered into the tubing to expand it to the proper size. Another method involves a lever-type tool that stretches the diameter of the tubing when the lever is squeezed. The tubing can be done either way. Both tools can handle a variety of tubing sizes.

When one tube can be slipped into another, such as when brazing a capillary tube to a larger soft copper tube, a "pinched" joint, might be

Figure 6-9 Swaged joint prior to brazing operation. *(Adapted from* Practical Air Conditioning Equipment Repair *by Anthony J. Caristi, McGraw-Hill, copyright 1991.)*

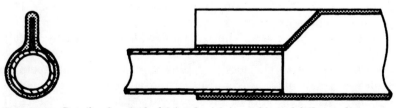

Figure 6-10 Details of a pinched joint for joining tubing of differing diameters. *(Adapted from* Practical Air Conditioning Equipment Repair *by Anthony J. Caristi, McGraw-Hill, copyright 1991.)*

acceptable (Fig. 6-10). To prepare the tubes for this procedure, thoroughly clean them. Insert the smaller tube at least ½ inch (1.2 cm) into the larger one. With the capillary tubing, ¾ inch (2 cm) is preferred to minimize the chance of brazing material entering the capillary tube.

Next, using the Vise grip or arc-joint-type pliers, squeeze the larger tube so it fits snugly against the smaller one. Any excess from the larger tube can be pinched tightly into a U shape. After making sure that the joint is mechanically sound, silver braze it to ensure that it will not leak.

Compression fittings

Although compression fittings are not commonly used as refrigerant fittings, they are frequently used for water line connections. A related type of fitting, the O-ring, is used quite often with automotive refrigeration systems.

A compression fitting is a threaded assembly with a part called a *ferrule* that is placed on the tubing before it is assembled into the fitting. When the nut is tightened, the ferrule presses against the tube for a leak-proof seal. Most compression fittings will need one ferrule for each section of tubing.

O-ring fittings

O-ring fittings have a round-shaped rubber ring that seals the parts together. This seal does not need the high torque that is used with flared fittings. This type of seal is often used on the connecting hoses in an automotive refrigeration system.

Before assembling an O-ring fitting, make sure that the ring is new (not used) and lubricated with clean refrigeration oil. As the parts are joined, the O-ring should seal the connection with relatively little compression. Use a backup wrench to prevent rotation of the parts when tightening the connection. Tighten only to manufacturers torque specifications.

Process tube adapter

A process tube adapter is used to make connections to a hermetic system in the same way that system charging is done at the factory. The process tube adapter permits access to the refrigerant system while providing a more reliable seal. There is no need to permanently install an access valve.

A process tube is a small piece of tubing that normally connects to the low-pressure side of the refrigerant system (it could also connect to the high side). When the copper tube and valve assembly is attached to the low-pressure process tube, it allows evacuation and recharge of the system.

The process tube can be either capped or pinched off and left attached to the system once recharging is complete. A pinch-off tool can be used to temporarily close the process tube and reseal it. The valve assembly is removed, and the open end of the process tube is brazed shut for a permanent seal. At this point the pinch-off tool is removed.

Brazed and Soldered Connections

In refrigeration work, the majority of tubing and fitting connections made are brazed or soldered connections. The method of making brazed and soldered connections is basically the same. A base metal is heated to the point where the filler is melted and drawn by capillary action through the joint. The joint, once cooled, consists of a strong bond between the filler and the two base metals. Brazed and soldered joints differ in a number of important ways.

Brazing produces a stronger joint than soldering. For this reason, brazed joints are used for all refrigeration tubing and pipe connections. Soldered joints are only used for water pipes and drain lines. Soldered joints are never used in refrigerant lines.

Soldered joints draw molten solder into the area between the tubing and fitting. This molten solder might be a 50/50 tin/lead mix that is suited for moderate pressures and temperatures. This half tin/half lead solder has a melting point of 360°F (182°C) and becomes fluid at 415°F (213°C). For higher pressures or where greater joint strength is needed, a 95/5 tin-antimony solder can be used. This 95 percent tin solder is a more durable mixture that melts at 450°F (232°C) and turns liquid at 465°F (241°C).

Brazing is a metal joining process utilizing a filler metal that melts above 840°F and below the melting point of the base metals. The filler metal is then drawn into the joint by capillary attraction, producing a sound, leak-proof connection. Two advantages of the brazed joint are its high strength and the low amount of heat needed to create it. If the bond is well constructed, a brazed joint will actually be stronger than the metals used to form the joint. The temperatures needed to create the joint will be much lower than the melting temperatures of the metals as well.

Brazing alloys

Selecting the proper brazing alloy for the job is extremely important. When brazing copper to copper, use an alloy containing phosphorus that is self-fluxing on copper. When brazing brass or bronze fittings, white flux is required with these alloys. When brazing steel tubing, use a silver alloy with white brazing flux. Do not use phosphorus-bearing alloys as the joint might be brittle.

Fluxing

Proper fluxing is important because the flux absorbs oxides formed during heating and promotes the flow of filler metal. Flux is commonly brushed on. To prevent excess flux residue inside refrigeration lines, apply a thin layer of flux to only the male tubing. Insert the tube into the fitting and, if possible, revolve the fitting once or twice on the tube to ensure uniform coverage.

Most manufacturers of brazing/soldering alloys and fluxes offer handy guides on the selection and application of their products. To guarantee good results, it is best to rely on manufacturer's recommendations for a particular alloy or flux.

Cleanliness and purging

Pay attention to the condition of the materials. Use only clean, dry, sealed refrigeration grade tubing. During brazing, the piping should be purged with dry nitrogen or carbon dioxide flowing through the pipes at low pressure. Purging with nitrogen or carbon dioxide prevents the formation of copper oxide and scale, which can easily clog the small ports on the pilot and other valves in the system.

Caution: Pressure regulators must always be used with nitrogen or carbon dioxide.

Brazing safety tips

In brazing, there are two main hazards for the brazing operator. One is chemical fumes, and the other is the heat and rays of the torch flame itself. Take the following precautions when safeguarding against these hazards:

Fumes. Ventilate confined areas, using fans and exhaust hoods, as well as respirators for operators, as needed. Clean all base metals to remove surface contaminants that might create fumes when the metals are heated. Use flux (where needed) in sufficient amounts to prevent oxidation and fuming during the heating cycle. Heat broadly, heating only the base metals, not the filler metal. Remove any toxic coatings (such as cadmium, lead, or zinc) before heating.

Torch heat and rays. Operators should wear gloves to protect hands against heat. Shaded goggles or fixed glass shields protect operators against eye fatigue and vision damage.

Making the joint

Maintain support during this operation to ensure the proper alignment until the brazing alloy solidifies. This might require holding it for just a few seconds, or more, depending on the size of the joint area. Follow these general guidelines when brazing a tube into a fitting or into a larger tube:

1. Insert the fluxed tube end into the fitting.
2. Adjust the oxygen-acetylene torch so that it emits a neutral to slightly reducing flame, which is a flame that contains more fuel gas than oxygen (Fig. 6-11). The neutral flame has a well-defined inner cone. Avoid an oxidizing flame (more oxygen than fuel gas). An oxidizing flame can form excess contaminating oxides. The flame should also be soft, and large enough in diameter to encircle both tube and fitting.
3. Always keep the torch flame moving in short back and forth strokes. Start heating the tube first, applying flame at a point about ½ to 1 inch just adjacent to the fitting. Work the flame alternately around the tube and fitting until both reach brazing temperature before applying the brazing filler metal (Fig. 6-12). As you move the torch slowly back and forth from tube to fitting, hold the torch a bit longer over the slower-to-heat fitting. Do not allow the heat to directly contact the fitting face, however, because this small area could become overheated. When using flux, continue to heat the tube until the flux passes the "bubbling" temperature range and becomes quiet, completely fluid and transparent. It should have the appearance of clear water.

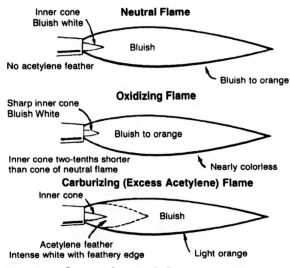

Figure 6-11 Oxy-acetylene torch flame types. *(J.W. Harris Company, Inc.)*

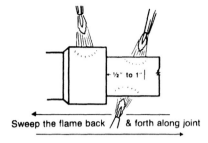

Figure 6-12 Proper method of heating the joint. *(J.W. Harris Company, Inc.)*

4. Pull the flame back and apply filler firmly against the tube at the point where tube and fitting meet (Fig. 6-13). If using a phosphorous type filler metal, which tends to flow slowly, lay it on the joint and draw it around. Make sure that the filler metal penetrates and fills the joint completely, always flowing toward the hottest area. Always keep both the fitting and the tube heated by playing the flame over the tube and the fitting as the brazing alloy is drawn into the joint. The brazing alloy will diffuse into and completely fill all joint areas. Do not continue feeding brazing alloy after the joint area is filled. Excess fillers do not improve the quality or the dependability of the braze and are a waste of material.

5. Once the joint is done, make a final pass at the joint, twisting the joint if possible. This will maximize wetting by the filler metal as any trapped gas or flux is released.

Brazing techniques can vary, depending on the type of joint being made. Joints commonly used in heat exchanger units are the vertical down, the vertical up, and the horizontal.

Vertical down joints. Bring the entire joint area to brazing temperature quickly and uniformly, heating the tube first, then the fitting. When the joint area has reached brazing temperature (as indicated by

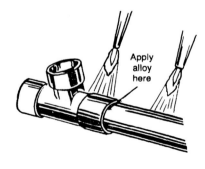

Figure 6-13 Location for applying alloy to the heated joint. *(J.W. Harris Company, Inc.)*

the flux appearance), apply extra heat to the fitting. This will encourage the flux to flow toward the fitting area.

Vertical up joints. Begin by heating the tube to a temperature of about 800°F (425°C). Slowly transfer the heat to the fitting, sweeping back and forth from fitting to tube, all around the joint area. Be careful not to overheat the tube below the fitting, as this would encourage the filler metal to run down the tube out of the joint. When brazing temperature is reached, touch the filler metal to the joint with the flame, on the wall of the fitting. This heating pattern will draw the filler metal up completely through the joint area.

Horizontal joints. Preheat both tubing and fitting quickly and evenly. When brazing temperature is reached, apply the filler metal to the top of the joint. The combination of gravity and capillary action will draw the filler metal completely around the tube to its bottom. (Applying slightly more heat to the bottom of the fitting will allow the filler metal to completely penetrate the joint.) Check the joint face to be sure that the filler metal is visible all around it. Also check that there is filler metal showing at the top of the joint. If filler is not visible, keep applying it until the filler does show.

When heating the assemblies for brazing, it is important to heat the joint area as quickly and uniformly as possible. When applying heat to metals of unequal thickness, the thicker area will need more heat because it heats more slowly. If the metals are different, they will not have the same conductivity. With steel and copper, for example, you would need to apply more heat to the copper because copper quickly conducts heat away to cooler areas.

No metal, in any case, should be heated to the melting point in a brazing operation.

Cleaning a joint

All flux residues must be removed for inspection and pressure testing. Immediately after the brazing alloy has set, quench or apply a wet brush or swab to crack and remove the flux residues. Use emery cloth or a wire brush if necessary.

Taking a brazed joint apart

To break a brazed joint, first flux the visible alloy and all adjacent areas of the tube and fitting. Then heat the joint (tube and fitting) evenly, especially the flange of the fitting. When the brazing alloy becomes fluid throughout the joint area, the tube can be easily removed. To rebraze the joint, clean the tube end and the inside of the fitting and proceed as directed for making a new brazed joint.

Troubleshooting

The art of brazing is relatively simple. But there are times when things can go wrong and the brazing process fails to work as it should.

Poor flowing filler. If brazing alloy does not flow into the joint, even though it melts and forms a filler, the following problems might exist.

- The outside of the joint might be hot, but the inside is not up to brazing temperature. Follow the correct heating procedures outlined earlier in this chapter. Remember to heat the tube first to conduct heat inside the fitting.
- The flux might be breaking down due to excessive heat. If overheated, the flux can become saturated with oxides and the brazing alloy will not flow. Try using a "softer" flame and/or applying a heavier coating of flux. On heavy sections where heating is prolonged or on stainless-steel, special black flux is recommended.

If the brazing alloy doesn't wet surfaces but "balls up" instead of running into the joint, check the following:

- Review your heating technique. The base metals might not be up to brazing temperature while the alloy has been melted by the torch flame. The joint might have been overheated and the flux is no longer active.
- Base metals might not be properly cleaned.

If brazing alloy flows away from instead of into the joint, make sure the fitting is up to temperature and the flame is directed towards the fitting.

Cracked joints. If the filler metal cracks after it solidifies consider the following points:

- When brazing dissimilar metals, the different coefficients of expansion might put the filler metal in tension just below liquid temperature during cooling. This sometimes occurs in a copper-into-steel joint. The copper expands and contracts at a greater rate than the steel. Brazing alloys are stronger in compression, so a steel-into-copper assembly would help alleviate the problem.
- Brazing steel (or other ferrous metal) with an alloy containing phosphorous can lead to formation of a brittle phosphide prone to cracking. Braze ferrous metals with nonphosphorous-bearing alloys.
- Excessive joint clearance can lead to filler metal cracking under stress or vibration. Make sure clearances are held to 0.002 to 0.006 inch at brazing temperature (depending on the alloy).

- Too rapid quenching can sometimes cause cracking. Allow more time for the joint to cool before washing off flux residue.

Leaking joints. Sometimes the joint leaks when in service. Ninety percent of all leaking joints are the result of improper brazing techniques. The following are the most common faults:

- Improper (uneven) heating of joint. Review the proper techniques found on pages 153–154. Uneven heating causes inadequate or incomplete penetration by the filler metal.
- Overheating, causing volatization of elements (phosphorous, zinc, etc.).
- Incorrect torch flame adjustment leading to deposition of carbon or causing excessive oxidation.

Repairing Tubing Leaks

Pinhole leaks in copper-to-copper joints brazed with phosphorous-copper or copper-phosphorous-silver filler metals can often be repaired using PH. If carefully applied, you can rebraze the joint with PH without remelting the original braze. We do not recommend brazing over joints previously soldered with tin/lead solders. The low melting elements in the solder might prevent proper filler metal/base metal alloying.

Pinhole leaks in joints brazed with either the phosphorous or high-silver alloys can usually be repaired with SB solder. Be sure to clean the joint thoroughly before soldering.

Working with Tubing and Piping

On any installation, particularly those involving remote equipment or long piping runs, it is extremely important to follow good piping practices. Lines must be both properly sized and installed. Some possible consequences of poor piping are: increased oil requirements, decreased operating efficiency and loss of capacity, increased chances of fouling vital components, and failed compressors.

Insulating piping runs

Insulation of liquid and suction refrigerant lines is important. Insulation on suction and liquid lines makes the whole system more efficient. When liquid subcooling is used on a system, all liquid lines to and from the parallel rack (all the way from the building entrance through to the fixtures) must be insulated! Allowing subcooled liquid to warm in the lines cancels the energy saving advantage of subcooling the liquid and might even cause liquid to *flash*. Flashing occurs when liquid converts to gas before reaching the expansion valve and will cause erratic valve feed and subsequent loss of refrigeration.

All low-temperature suction lines must be insulated in order to ensure cool suction gas to the compressor. Cool gas is necessary to aid in cooling the motor windings. Head cooling fans also help and sometimes are required by the compressor manufacturer. Compressor motor failure can result if suction gas from fixtures warms too much on its way to the compressor.

With gas defrost, insulation on the suction line helps maintain the temperature of the hot gas flowing to the cases during defrost. Insulation on suction and liquid lines help make the whole system more efficient.

Long-radius elbows

Avoid the use of 45° elbow connectors. Use long radius elbows (els) rather than short els. Short radius els can create excessive internal stress that can lead to failure. Less pressure drop and greater strength make the long radius els better for the system. This usage is especially important on discharge hot gas lines for strength and on suction lines for reduced pressure drop (Fig. 6-14).

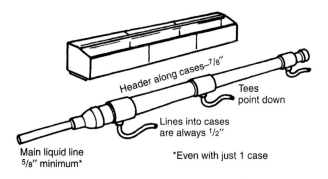

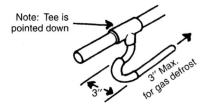

Figure 6-14 Use of long-radius elbow in gas defrost line. *(Tyler Refrigeration Corporation)*

Service valves

Install service valves at several locations in the system to provide easy maintenance and reduction of future service costs. Service valves must be UL approved for the minimum working pressure specified by the equipment's manufacturer.

Piping support

Piping must be properly supported to minimize line vibration. Vibration is transmitted to the piping by movement of the compressor and by pressure pulsations as refrigerant is pushed through the piping. Insufficient or improper support of tubing can cause excessive line vibration that can result in excessive noise, noise transmission to other parts of building, vibration transmission to floors, walls, etc., vibration transmission back to compressor and other attached components, decreased life of all attached components, and/or line breakage. Use the following guidelines when installing piping for proper support:

- A straight run of piping must be supported at each end. Longer runs will require additional supports along the length, usually at eight-foot intervals, depending on tubing size and situation. Be sure clamps are properly anchored and rubber grommets installed between the piping. Clamp to prevent line chafing (Fig. 6-15).
- Corners must be supported and cannot be left free to pivot (Fig. 6-16).
- Do not over-support piping when it is attached to the compressor rack. It must be free to float without stress (Fig. 6-17).
- Check all piping after the system has been placed in operation. Excessive vibration must be corrected as soon as possible. Extra supports are inexpensive when compared to the potential refrigerant loss caused from failed piping.

The installing contractor is responsible for proper line sizing. Horizontal suction lines should slope ½-inch per 10 feet toward the compressor to aid in good oil return (see Fig. 6-14).

Liquid lines to the cases should be branched off the bottom of the header to ensure a full column of liquid to the expansion valve. A branch line from the header to an individual case should not be over three-feet long and must have a three-inch expansion loop incorporated.

Do not run suction or liquid lines through cases that are part of a separate system, especially if either has gas defrost. If this cannot be avoided, insulate the piping for the portion that runs through the other cases.

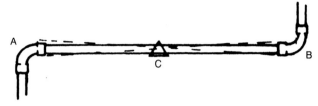

Not good:
Tubing will vibrate or rotate about point 'C'.

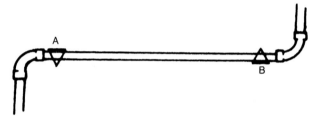

Good:
Long runs will require additional supports between 'A' and 'B'.

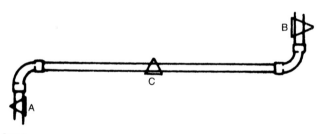

Good:
Support at 'C' may not be necessary for very short runs.

Figure 6-15 Proper positioning of clamps for straight runs of tubing. *(Tyler Refrigeration Corporation)*

Properly supporting the lines whenever they are suspended from a wall or ceiling is very important (Fig. 6-18). Line supports should isolate the line from contact with metal. When gas defrost is used, consider rolling or sliding supports, which allow free expansion and contraction. These supports would be used in conjunction with expansion loops described later in this chapter. Table 6-2 lists the recommended support spacings for various line diameters.

Figure 6-16 Proper support of tubing run corners. *(Tyler Refrigeration Corporation)*

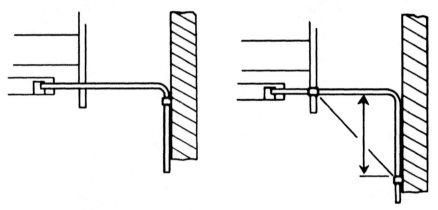

Figure 6-17 Support at the compressor rack must allow the tubing to "float" without stress. *(Tyler Refrigeration Corporation)*

Expansion of piping

Be sure to allow for expansion. Liquid lines and suction lines must be free to expand and contract independently of each other. Never clamp or solder these lines together. Pipe hangers must not restrict the expansion and contraction of piping. The temperature variations of refrigeration and defrost cycles cause piping to expand and contract. The expansion of piping must be taken into consideration to prevent a piping failure. The following are typical expansion rates for copper tubing:

- Ultra low temperatures: −40 to −100°F = 2.5 inches per 100 feet of run.
- Low temperatures: 0 to −40°F = 2 inches per 100 feet of run.

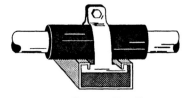

Support Detail

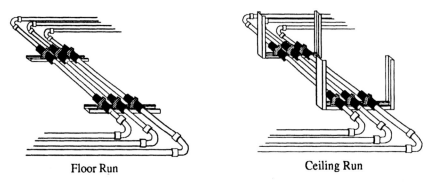

Floor Run　　　　　　　　　Ceiling Run

Figure 6-18 Supporting refrigerant lines on ceiling or floor runs. *(Hussmann Corporation)*

- Medium temperatures: 0 to 30°F = 1.5 inches per 100 feet of run.
- High temperature: 30 to 50°F = 1 inch per 100 feet of run.

Expansion loops are designed to provide a definite amount of travel. Placing the loop in the middle of a piping run will allow for maximum pipe expansion with the minimum amount of stress on the loop. Do not use 45° els for loop construction because they will not allow the lines to flex (Fig. 6-19). Tables 6-3 and 6-4 list information of expansion loop sized as keyed to Fig. 6-19.

TABLE 6-2

Line size O.D.	Maximum span
⅝"	5'
1⅛"	7'
1⅝"	9'
2⅛"	10'
3⅛"	12'
3⅝"	13'
4⅛"	14'

Working with Tubing 163

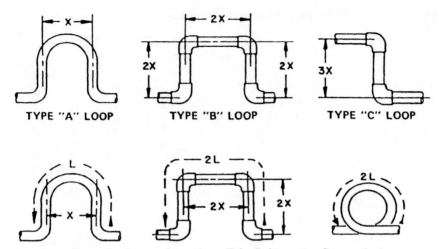

Figure 6-19 Expansion loop configurations. *(Tyler Refrigeration Corporation)*

Suction line riser recommendations

Suction line risers might not require a trap, or they might require one or two traps.

Risers without a trap. Suction line sizing is based on a design pressure drop, which relates to the velocity of the gases moving through the line. Acceptable velocities for horizontal suction lines (with proper ½" per 10-ft slope) range from 200 to more than 1500 feet per minute (fpm). A properly sized line at the low range of its capacity will have a low velocity, and one at full capacity will have velocities exceeding 1500 fpm.

TABLE 6-3 Linear Expansion of Copper Tubing (in Inches)

Tube O.D. in inches	Length in inches									
	½"	1"	1½"	2"	2½"	3"	4"	5"	6"	7"
⅞	8.0	11.0	13.5	15.5	17.5	19.0	22.0	24.5	27.0	29.0
1⅛	9.0	12.5	15.5	17.5	20.0	21.5	25.0	28.0	30.5	33.0
1⅜	10.0	14.0	17.0	19.5	22.0	24.0	27.5	31.0	34.0	36.5
1⅝	10.5	15.0	18.5	21.0	24.0	26.0	30.0	33.5	37.0	39.5
2⅛	12.0	17.0	21.0	24.0	27.0	30.0	34.5	38.5	42.0	45.5
2⅝	13.5	19.0	23.5	27.0	30.0	33.0	38.0	42.5	46.5	50.5
3⅛	15.0	21.0	25.5	29.5	33.0	36.0	41.5	46.5	51.0	55.0
4⅛	17.0	24.0	29.0	34.0	38.0	41.5	48.0	53.5	58.5	63.0
5⅛	19.0	26.5	32.5	37.5	42.0	46.5	53.0	59.5	65.0	71.5
6⅛	20.5	29.0	35.5	41.0	46.0	50.5	58.0	65.0	71.5	77.0
8⅛	24.0	33.5	41.0	47.5	53.0	58.0	67.0	75.0	82.0	89.0

TABLE 6-4 Developed Length (in Inches) of Expansion Offsets Based on 180° Loop

Tube O.D. in inches	Length in inches									
	½"	1"	1½"	2"	2½"	3"	4"	5"	6"	7"
⅞	24.5	34.5	42.5	49.0	54.5	60.0	69.0	77.5	84.5	91.5
1⅛	28.0	39.0	48.0	55.5	62.0	68.0	78.5	87.5	96.0	104.0
1⅜	30.5	43.5	53.0	61.5	68.5	75.0	86.5	97.0	106.0	114.5
1⅝	33.5	47.0	58.0	66.5	74.5	81.5	94.0	105.5	115.5	124.5
2⅛	38.0	54.0	66.0	76.0	85.0	93.5	108.0	120.5	132.0	142.5
2⅝	42.5	60.0	73.5	85.0	95.0	104.0	120.0	134.0	147.0	158.5
3⅛	46.0	65.5	80.0	92.5	103.5	113.0	131.0	146.0	160.0	173.0
4⅛	53.0	75.0	92.0	106.0	119.0	130.0	150.0	168.0	184.0	198.5
5⅛	59.0	84.0	102.5	118.5	132.5	147.0	167.5	187.0	205.0	224.0
6⅛	65.0	91.5	112.0	129.5	145.0	158.5	183.0	204.5	224.0	242.0
8⅛	74.5	105.5	129.0	149.0	166.5	182.5	211.0	235.5	258.0	279.0

The chart in Fig. 6-20 shows the size selection that will ensure oil return up a riser. This size might be the same size as the horizontal suction line selection or it might be one size smaller. If the selection point on the chart is close to the dividing line between sizes, use the smaller size. The reducer fitting must be placed after the elbow (Fig. 6-21). Use long els to make the trap or use a P-trap. Do not use short els.

Note: Medium-temperature systems can have higher risers without a trap than low-temperature systems.

Risers that require a P-trap. The design of low-temperature systems must take into consideration that oil is more difficult to move at lower temperatures and that the refrigerant gas also has a lower capacity to mix with the oil. A trap will allow oil to accumulate, reducing the cross section of the pipe, and thereby increasing the velocity of gas. This increased velocity picks up the oil. Use the velocity chart to determine if the horizontal line size has sufficient velocity in the vertical position to carry the oil along. Generally, the riser will have to be reduced one size (Fig. 6-22).

Risers that require two P-traps. When the vertical lift exceeds 16 feet, split the riser into two equal lifts, reducing one size when the chart indicates.

When a reduced riser is needed, locate the reduction coupling downstream of the P-trap (Fig. 6-23).

Vertical suction line sizing

Proper line sizing for vertical risers is very important. Equipment manufacturers provide charts to assist in the proper sizing of suction

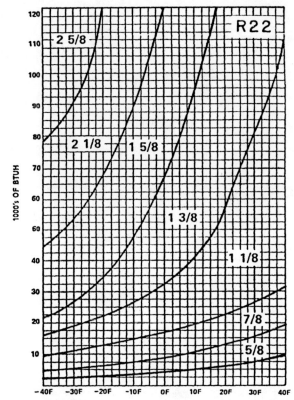

Figure 6-20 Vertical riser suctin line sizing chart for R-22 (HCFC-22) refrigerant. *(Tyler Refrigeration Corporation)*

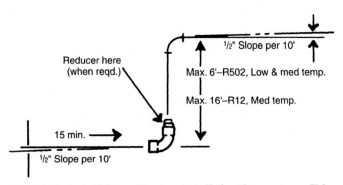

Figure 6-21 Installation with riser installed without a trap. *(Tyler Refrigeration Corporation)*

166 Chapter Six

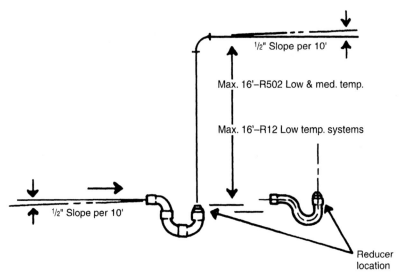

Figure 6-22 Installation with riser requiring a P-trap. *(Tyler Refrigeration Corporation)*

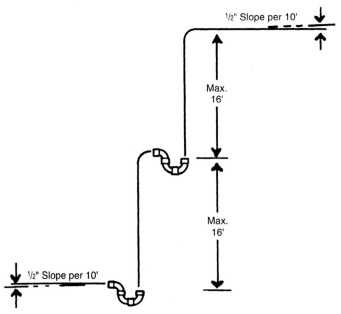

Figure 6-23 Installation of riser requiring two P-traps. *(Tyler Refrigeration Corporation)*

line vertical risers. Some sizing charts are for horizontal runs only and other charts can be used for both. So be sure you have the proper information for the job at hand.

Following the chart recommendations ensures the system will maintain minimum refrigerant velocities in the risers. Good refrigerant velocity ensures that the oil mixed with the refrigerant will return to the compressor. Improper line sizing could cause poor system performance or even compressor damage due to oil failure.

Caution: When in doubt about oil return (due to a point being near a line), use the smaller size line.

Any sizing of riser or any other suction line, or device, must be considered in view of the total system. The addition of any suction line pressure drop must not be ignored. If suction P-traps are used, they should be sized according to the horizontal line sizing chart.

Note: Do not arbitrarily reduce vertical risers without consulting these charts. Unnecessary vertical suction line reduction can cause excessive pressure drop and loss of system capacity.

Using sizing charts. Figure 6-24 illustrates a typical suction line sizing chart for R-12 (CFC-12) for lengths to 300 equivalent feet. To deter-

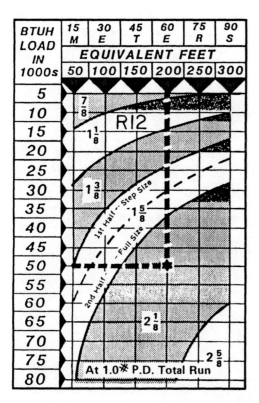

Figure 6-24 Suction line sizing charts are available for all types of refrigerants. *(Tyler Refrigeration Corporation)*

mine the *equivalent feet*, or meters, add the length of the pipe and the equivalent footage assigned for each particular fitting as listed in Table 6-5.

The graph presentation of information allows you to know just how close to full capacity a particular selection is. The suction line graphs are arranged according to temperature, and the relationship of temperature and line size becomes readily apparent. The lower the temperature, the larger the line for the same heat load.

To use the chart, simply match BTU/hour (BTU/h) load on the horizontal lines with equivalent feet on the vertical line. The point formed by the intersection will indicate the proper size unless the intersection is in a dark area.

Selections falling in the dark areas of the charts show that the gas velocity is below 500 fpm, which is too slow to ensure proper oil return, even with properly sloped lines. Reducing the line one size will increase velocity and pressure drop. Added pressure drop will require greater refrigeration capacity, so be sure the system can handle the added load. See the vertical riser charts for proper sizing of vertical suction lines.

Step sizing is suggested for selections falling in the first half of a size range in order to ensure better oil return out of the evaporators. Pipe one size smaller (than the indicated run) can be used for 50 feet of the run closest to the cases when the entire run is 100 equivalent feet or more. To show this principle, one size range on each suction chart has been bisected by a dotted line to indicate the "first half—step size" and the "second half—full size."

Example. Given a 50,000 BTU/h load with R12 at 10°F suction temperature and 200 equivalent feet of line, a 2⅛" line is required. Because the selection point is in the first half of the range, the first 50 feet of the line closest to the cases can be run in 1⅝" tubing.

The most critical of all piping in a refrigerated system is suction line sizing. Horizontal runs should be sloped ½" per 10 feet back to the compressor to assist in good oil return.

TABLE 6-5 Equivalent Length (in Feet) of Copper Valves and Fittings

Line size O.D. (inch)	Globe valve	Angle valve	90 Degree elbow	45 Degree elbow	Tee, sight glass line	T-Branch
½	9	5	0.9	0.4	0.6	2.0
⅝	12	6	1.0	0.5	0.8	2.5
⅞	15	8	1.5	0.7	1.0	3.5
1⅛	22	12	1.8	0.9	1.5	4.5
1⅜	35	17	2.8	1.4	2.0	7.0
2⅛	45	22	3.9	1.8	3.0	10.0
2⅝	51	26	4.6	2.2	3.5	12.0
3⅛	65	34	5.5	2.7	4.5	15.0
3⅝	80	40	6.5	3.0	5.0	17.0

Chapter 7

Modern Refrigerants

The chemistry of refrigeration has changed dramatically in the past decade. Chlorofluorocarbon refrigerants (CFCs) were developed over 60 years ago and have served the industry well. CFCs have the thermodynamic and physical properties that make them ideal for use as refrigerants. In addition to having the desirable boiling points and pressure characteristics needed for a good refrigerant, CFCs are low in toxicity, they are nonflammable and noncorrosive, they are compatible with other materials, and they are relatively inexpensive.

Because they were inexpensive and posed no significant health problems, the simplest way to "dispose" of old CFC refrigerants was to allow them to vent into the atmosphere. They simply boiled away and disappeared, so most people thought. The fact is CFCs do not simply disappear. They collect in the earth's outer atmosphere where they cause serious damage to the earth's protective ozone layer.

The heat in the earth's stratosphere forces CFC compounds to give up chlorine atoms. These free chlorine atoms combine with ozone to form two new compounds: chlorine monoxide and diatomic oxygen. Neither of these compounds provide the protection ozone offers against harmful ultraviolet radiation from the sun.

Table 7-1 lists the ozone depletion and global warming potentials plus the lifetime of several popular fully halogenated CFC refrigerants. Halogenated compounds contain chlorine, fluorine, iodine, or bromine.

There are several different compounds in the CFC family, each with a different proportion of carbon, chlorine, and fluorine. The hydrochlorofluorocarbons (HCFCs) and hydrofluorocarbons (HFCs) are also used as refrigerants.

HCFC refrigerants. Like CFCs, hydrochlorofluorocarbon refrigerants (HCFCs) contain chlorine atoms. However, HCFC compounds are

TABLE 7-1

Refrigerant	Ozone depletion potential	Global warming potential	Lifetime in atmosphere
CFC-11	1.00	1.00	64
CFC-12	1.00	3.05	108
CFC-113	0.80	1.30	88
CFC-114	1.00	4.15	185
CFC-115	0.60	9.60	380
Halon 1211	3.00	Unknown	25
Halon 1301	10.00	0.80	110
R-502*	0.30	5.10	NA
HCFC-22*	0.05	0.37	26

much less destructive to the ozone because the hydrogen in these refrigerants causes the chlorine to be released at lower altitudes where they cause little harm.

HFC refrigerants. Hydrofluorcarbon refrigerants (HFCs), however, have no chlorine so they do not have the potential to deplete the ozone layer.

Table 7-2 lists a number of common CFC, HCFC, and HFC refrigerants.

Other refrigerants. Ammonia (R-717) and methyl chloride (R-40) are two other refrigerants whose use has declined in modern refrigeration. Both of these refrigerants will irritate your lungs if accidentally inhaled. Both are also slightly flammable, and methyl chloride is very toxic.

Ammonia is used as a refrigerant in absorption refrigeration systems. Absorption systems do not use vapor compression. Instead, cooling is produced as the refrigerant is absorbed by another chemical substance.

Alternative Refrigerants

Because CFC compounds have been linked to the depletion of the protective ozone layer, the refrigeration industry is phasing out production of CFCs by January 1996. HCFC refrigerants are scheduled to be completely phased out by the year 2030.

TABLE 7-2

CFC refrigerants	HCFC refrigerants	HFC refrigerants
R-11	R-22	R-32, R-143a
R-12	R-123	R-125, R-152a
R-115 (51% in 502 with 49% 22)	R-124	R-134a, R-C318
	R-142b	

Modern Refrigerants

A wide variety of environmentally acceptable alternative refrigerants have been introduced in the past several years by numerous manufacturers. Others are sure to follow.

Refrigerant identification

In the past, refrigerants were simply given a number designation with the R (refrigerant) prefix, such as R-11 or R-12. Refrigerants are now identified in a new way. The prefix R is being replaced by a more descriptive term that indicates the refrigerant's compound family. For example, if the refrigerant is a chlorofluorocarbon, it must now have the prefix CFC, as in CFC-12 (formerly R-12). If the refrigerant is a hydrochlorofluorocarbon, it must now have the prefix HCFC as in HCFC-22 (formerly R-22). If the refrigerant is a hydrofluorcarbon, it must now have the prefix HFC as in HFC-134a.

ASHRAE numbering system

Figure 7-1 illustrates the ASHRAE numbering system now used for pure refrigerant compounds. The first digit indicates the number of carbon atoms; the second digit, the number of hydrogen atoms, and the third digit, the number of fluorine atoms. A lower-case letter might be placed after the third digit to indicate the atomic structure of the compound.

Alternative refrigerants might be a single pure compound or a blend of two or more compounds. Blended refrigerants can be either azeotropes or zeotropes.

Azeotropes. Azeotropes are refrigerant blends that evaporate and condense exactly like a pure refrigerant compound at one particular temperature and pressure combination. Azeotropes might not behave like a pure refrigerant compound at all pressures and temperatures, but the values will be close. Azeotropes are assigned series 500 numbers in the ASHRAE numbering system.

ASHRAE NUMBERING SYSTEM

Pure Refrigerants
(based on chemical makeup)

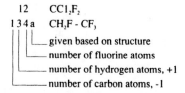

12   CCl_2F_2
134a $CH_2F - CF_3$
 — given based on structure
 — number of fluorine atoms
 — number of hydrogen atoms, +1
 — number of carbon atoms, -1

Figure 7-1 ASHRAE numbering system for naming alternative refrigerants.

With azeotropic refrigerant blends, the vapor composition in the refrigerant cylinder is different from the liquid composition. To ensure that the proper blend composition is charged in the system, the system must be liquid charged. Azeotropic refrigerant cylinders are typically equipped with dip tubes, allowing the liquid to be removed from the cylinder in the upright position. The proper position is indicated by arrows on the cylinder. Once removed from the cylinder, the azeotropic blend can be charged to the system as vapor as long as all of the liquid removed from the cylinder is transferred to the system.

Zeotropes. Zeotropes will show some amount of temperature glide when evaporating or condensing. Zeotropes with glides greater than 3°F will have one end of the evaporator warmer than the other. This might impact system performance. Zeotropes are assigned series 400 numbers in the ASHRAE numbering system.

Tables 7-3 and 7-4 list a number of the most common refrigerant blends using their ASHRAE number. Individual refrigerant manufacturers will assign their own number or letter designation to their blends, so the process of identifying the exact refrigerant in question can become confusing if care is not taken. Always rely on the manufacturer's specification sheet and recommendations for the use and retrofit of their refrigerant blends.

TABLE 7-3 Examples of Alternative Refrigerants

Refrigerant	Manufacturer	ASHRAE#	Components	Application
FX 56	Elf Atochem	R-409A+	R-22/124/142b	R-12 retrofit
FX 10	Elf Atochem	R-408A+	R-125/143a/22	R-502 retrofit
FX 70	Elf Atochem	R-404A	R-125/143a/134a	R-502 new/retrofit
FX 220	Elf Atochem		R-23/32/134a	R-22 alternative
AZ 50	Allied Signal	R-507*	R-125/143a	R-502 new/retrofit
AZ 20	Allied Signal	R-410A+	R-32/125	R-22 alternative
MP 39, 66	DuPont	R-401A, B	R-22/152a/124	R-12 retrofit
HP 80, 81	DuPont	R-402A, B	R-125/290/22	R-504 retrofit
HP 62	DuPont	R-404A	R-125/143a/134a	R-502 new/retrofit
AC 9000	DuPont	R-407C+	R-32/125/134a	R-22 alternative
(Klea) 60, 61	ICI	R-407A, B*	R-32/125/134a	R-502 new/retrofit
(Klea) 66	ICI	R-407C+	R-32/125/134a	R-22 alternative

TABLE 7-3 Examples of Alternative Refrigerants (Continued)

Refrigerant	Manufacturer	ASHRAE#	Components	Application
(Isceon) 69L	Rhone Poulenc	R-403B	R-290/22/218	R-502 retrofit
GHG 12	Indianapolis Refrigeration Products	R-406A*	R-22/142b/600a	R-12 retrofit
Greencool 12	GU/Greencool	R-405A*	R-22/152a/142b/C318	R-12 retrofit
Greencool 2018A, B	GU/Greencool	R-411A, B+	R-1270/22/152a	R-22 service
OZ 12	OZ Technologies		R-290/600	R-12 retrofit
HC 12A	OZ Technologies		Hydrocarbons	R-12 service

* These numbers have been approved for publication in an addendum to ASHRAE Standard 34-1992, Number Designation and Safety Classification of Refrigerants.
+ These numbers have been proposed for addition to ASHRAE Standard 34-1992. They will be accepted after completion of a public review process.

TABLE 7-4 Alternative Refrigerants Grouped by Application

Refrigerant	ASHRAE#	Applications	Lubricant
R-11 ALTERNATIVES			
R-123	R-123	retrofit or new centrifugals	mineral oil
R-12 ALTERNATIVES			
R-134a	R-134a	R-12 new/retrofit	polyol ester (POE)
FX 56	R-409A	R-12 retrofit	mineral oil/alkylbenzene
MP 39, 66	R-401A, B	R-12 retrofit	alkylbenzene
GHG 12	R-406A	R-12 retrofit	mineral oil/alkylbenzene
Greencool 12	R-405A	R-12 retrofit	mineral oil/alkylbenzene
OZ 12		R-12 retrofit	mineral oil
HC 12A		R-12 applications	mineral oil
R-502 ALTERNATIVES			
FX 10	R-408A	R-502 retrofit	mineral/alkylbenzene/POE
FX 70	R-404A	R-502 new/retrofit	polyol ester (POE)
AZ 50	R-507	R-502 new/retrofit	polyol ester (POE)
HP 80, 81	R-402A, B	R-502 retrofit	mineral oil + alkylbenzene
HP 62	R-404A	R-502 new/retrofit	polyol ester (POE)
Klea 60, 61	R-407A, B	R-502 new/retrofit	polyol ester (POE)
Isceon 69L	R-403B	R-502 retrofit	mineral oil + alkylbenzene
R-22 ALTERNATIVES			
FX 220		R-22 alternative	polyol ester (POE)
AZ 20	R-410A	R-22 alternative	polyol ester (POE)
AC 9000	R-407C	R-22 alternative	polyol ester (POE)
Klea 66	R-407C	R-22 alternative	polyol ester (POE)
Greencool 2018 A, B	R-411A, B	R-22 service	mineral oil/alkylbenzene

CFC/HCFC Phase-Out Regulations

The following is an overview of the regulatory actions in the United States that are impacting the use of chlorofluorocarbons (CFCs), halons, methyl chloroform, carbon tetrachloride, and their alternatives.

The Montreal Protocol

The Montreal Protocol is a 1989 international agreement regulating the production and use of CFCs, halons, methyl chloroform, and carbon tetrachloride. The Protocol is a joint effort of governments, scientists, industry, and environmental groups. This landmark agreement initially required a production and usage freeze but has since been revised to call for a complete phase-out as follows:

1. Phase-out of production and consumption of Halons as of January 1, 1994.
2. Interim reduction schedules of production and consumption as follows:
   - CFCs: 75 percent cut in 1994 and 1995
   - Methyl chloroform: 50 percent cut in 1994 and 1995
   - Carbon tetrachloride: 85 percent cut in 1995
3. Phase-out of production and consumption of CFCs, methyl chloroform, and carbon tetrachloride by January 1, 1996.
4. Methyl bromide was added to the list of controlled substances, with a production and consumption freeze in 1995 to 1991 levels, and a phase-out schedule to be determined.
5. Phase-out of hydrobrominated fluorocarbons (HBFCs) as of January 1, 1996.
6. Freeze of hydrochlorofluorocarbon (HCFC) consumption in 1996 to a calculated cap.
7. Reduction of HCFC consumption from the cap to the following schedule:
   - 35 percent cut in 2004
   - 65 percent cut in 2010
   - 90 percent cut in 2015
   - 99.5 percent cut in 2020
   - 100 percent phase-out by 2030

TABLE 7-5   Class I and II Substances

	Class I Substances
Group I	CFC-11, -12, -113, -114, -115
Group II	Halon 1211, 1301, 2402
Group III	CFC-13, -111, -112, -211, -212, -213, -214, -215, -216, -217
Group IV	Carbon tetrachloride
Group V	Methyl chloroform or 1,1,1-trichloroethane (except the 1,1,2 isomer)
Group VI	Methyl bromide
Group VII	Hydroborominated fluorocarbons (HBFCs)

Class II Substances
HCFCs and their isomers having one, two, or three carbon atoms

Clean Air Act

An amended Clean Air Act (CAA), signed into law in November 1990, includes a section titled Stratospheric Ozone Protection (Title VI). This section contains extremely comprehensive regulations for the production and use of ODS and substitutes. These regulations written by the U.S. Environmental Protection Agency (EPA) affect every industry that currently uses chlorinated and brominated substances that impact stratospheric ozone. The major provisions of the CAA include:

- Phase-out schedules for production of Class I and Class II substances, as listed in Table 7-5.
- Mandates for recycling, recovery of refrigerants in auto air-conditioning, stationary refrigeration, and air-conditioning equipment.
- Bans on "nonessential" products using CFCs and HCFCs.
- Requirements for labeling containers of products containing, or products made with, Class I or Class II substances.
- Dictates concerning safe alternatives.

The Clean Air Act now calls for a phase-out of Class I substances as of January 1, 1996, with limited exceptions for essential uses that have not yet been defined.

Recycling for stationary refrigeration and air-conditioning

The venting of CFC and HCFC refrigerants during service, maintenance, repair, or disposal of appliances and industrial process refrigeration has been illegal since July 1992. A final rule published in May 1993 added more requirements including:

- Certification for technicians servicing refrigeration systems.
- Certification for recycling and recovery equipment.
- Restrictions on refrigerant sales to certified technicians as of November 1994.
- Mandatory repair for systems greater than 50-lb. charges that leak more than 35 percent per year for commercial or industrial, or 15 percent per year for comfort cooling.

Alternative refrigerants face similar regulations as of November 1995. As of November 15, 1995, the term *refrigerant* includes all refrigerant alternatives subject to the same regulations.

Warning labels. Warning labels must be provided on containers of Class I and Class II substances, as well as products containing or made with Class I substances. The EPA now publishes lists of "acceptable" and "unacceptable" alternatives for Class I and Class II substances.

Guidelines for Recycled and Recovered Refrigerants

In the refrigeration industry, the following definitions apply to recycling and recovering refrigerants.

Recovery. Refrigerant recovery means to remove refrigerant in any condition from a system and store it in an external container without necessarily testing or processing it in any way. Recovery does not involve processing or analytical testing, as defined by the Air-Conditioning and Refrigeration Institute (ARI) of a U.S. organization.

Refrigerants can be recovered from refrigeration equipment using permanent on-site equipment or one of the portable recovery devices now on the market. The portable devices contain a small compressor and air-cooled condenser and can be used for liquid or vapor recovery. At the end of the recovery cycle, evacuate the system to remove vapors. See Chapter 8 for complete details.

In the United States, the Environmental Protection Agency (EPA) sets standards for recovery equipment. Before purchasing a specific recovery unit, check with the manufacturer to be sure it contains elastomeric seals and a compressor oil compatible with refrigerants it will be used on.

Recovery and reuse. When the recovered refrigerant's condition is good, the refrigerant can be reused. For example, when clean refrigerant is being recovered from a system because a leak must be repaired or a valve needs to be replaced, you can usually reuse the recovered refrigerant once the repair has been made. This practice is the most economical option because no new refrigerant has to be purchased.

Recovered refrigerant contaminant levels must meet an acceptable standard if it is to be put back into the system it was removed from or back into a system of the same owner. If these contaminant levels are exceeded, the refrigerant should be recycled or reclaimed, or new refrigerant should be used.

Recycle. Recycling refers to the reduction of used refrigerant contaminants using devices that reduce oil, water, acidity, and particles. Single or multiple pass through equipment using replaceable core filter-driers are often used in recycling equipment. Recycled refrigerant must be cleaned to acceptable contaminant levels. Your recycling equipment should be certified to ARI Standard 740, "Performance of Refrigerant Recovery/Recycling Equipment," and capable of consistently cleaning refrigerant to acceptable contaminant levels. It is also a good idea to periodically check the equipment's cleaning ability to ensure that its cleaning performance has not diminished.

Filter systems in recycling equipment need to be changed or cleaned regularly to properly maintain recycling equipment. When recycling equipment has been used to recycle refrigerant from a burnout, the equipment should be cleaned before being used again. Do not use recycled refrigerant if the performance of the recycling equipment is uncertain. Consult with the manufacturer before specifying a recycling device for any refrigerant.

On-site recycling. Recycling is usually a field or shop procedure with no analytical testing of refrigerant. When recycling, it is important that the entire refrigerant charge is removed from the refrigeration equipment and recycled. In most recycling situations, the recycled refrigerant is returned to the same system it was removed from or charged into compatible equipment at the same site owned by the same owner.

In some applications the refrigerant is recycled, tested to verify conformance to ARI Standard 700, and reused in a different owner's equipment providing that the refrigerant remains in the contractor's custody and control at all times from recovery through recycling to reuse. Recovering refrigerant at one site, recycling it, and installing it into a system at a second site is not advisable unless the refrigerant has been laboratory tested for purity and performance characteristics. Using recycled refrigerants of unknown purity might affect equipment warranties.

Refrigerants that have been recycled several times should be laboratory tested to verify that their chemical composition has not changed in a way that will adversely affect performance.

Caution: Used refrigerants should not be sold, or used in a different owner's equipment, unless the refrigerant has been analyzed and found to meet requirements of ARI Standard 700 "Specifications for Fluorocarbon and Other Refrigerants." These requirements exclude refrigerants contained in equipment that is sold if that refrigerant is used in the same equipment.

Reclaim. Reclaim means to process refrigerant to new product specifications by means that might include distillation. The refrigerant is then chemically analyzed to ensure that appropriate product specifications are met. The identification of contaminants, required chemical analysis, and acceptable contaminant levels will be established in the latest edition of ARI Standard 700, "Specifications for Fluorocarbon and Other Refrigerants." Reclaiming usually implies the use of processes or procedures available only at a reprocessing or manufacturing facility.

Reclamation offers advantages over on-site refrigerant recycling procedures because on-site recycling systems cannot guarantee complete removal of contaminants. Returning refrigerants that do not meet new product specifications back into equipment might cause damage.

When the refrigerant is recovered from a burnt-out or water-contaminated system, or in cases where guaranteed refrigerant quality is a must, send the recovered refrigerant for outside reclamation to ARI 700 standards.

Disposal. Disposal involves the destruction of used refrigerants. When a refrigerant becomes badly contaminated with other products and no longer meets the acceptance specifications of a reclaimer, disposal might be necessary. Licensed waste disposal firms are available to handle disposal needs. Be sure to check the qualifications of any firm before sending them used refrigerants. By law, you are responsible to see that all refrigerants are disposed of properly.

Working with recovered refrigerant

Always follow the instructions and recommendations of the compressor and equipment manufacturers. The equipment warranty might be affected if a refrigerant of poor quality is introduced into a system and causes subsequent failure of the equipment. Consider the following factors when deciding what to do with recovered refrigerant:

- The reason the system is being serviced.
- The condition of the refrigerant and system.
- The equipment manufacturer's policies.
- The feasibility and owner's preference.

Reason for system service. The reason the system is being serviced might offer insight into the condition of the refrigerant that has been recovered. Compressor failures, especially motor burnouts, might cause more contaminants to be in the refrigerant than are acceptable, indicating the need for recycling. Other system component failures not affecting the purity of the refrigerant might allow you to recover the refrigerant and return it to the system without recycling.

Condition of refrigerant and system. You should also consider the system's service history and age, the installation and service procedures used over the life of the system, and whether or not the systems were cleaned or evacuated properly from previous service problems. If the system history is not known, the recovered refrigerant should, at least, be recycled before it is put back into the system.

If the compressor has been previously replaced, determine if the system had suffered a compressor burnout. The cleanliness of the used refrigerant from a previous burnout will depend on how well the system was cleaned up when the compressor was replaced. Because systems with compressor burnouts are going to have some contaminants in the residual oil left in the system, the refrigerant used in these systems should be as clean as possible so as not to add to the contamination. In these situations, use recycled, reclaimed, or new refrigerant. When contaminant levels are uncertain, preliminary checks can be made with acid test kits and moisture indicators.

If you are confident that the refrigerant is not contaminated beyond acceptable levels, then the refrigerant can be recovered and put back into the same system without recycling, as long as you follow good service procedures.

Equipment manufacturer's policies. The manufacturer's policies should be the primary criterion in determining whether to use recycled refrigerant in that manufacturer's equipment. Take the time to understand the original equipment manufacturer's policies or recommendations concerning the use of recycled refrigerant.

Feasibility and owner's preference. Even when all other factors are favorable, recycling should be done only when it is feasible for you and agreeable to the system owner. Always consider the time and effort required to recycle refrigerant versus recovering and reclaiming it. The quantity of refrigerant might affect your final decision.

The refrigeration system must be cleaned and evacuated before refrigerant is put back into the system, regardless of whether recycled refrigerant or new/reclaimed refrigerant is used. Follow manufacturers' recommended service procedures to ensure that the system is free of contamination before adding refrigerant. At a minimum, replace all driers in the system and add suction line filters to systems with compressor burnouts to assist in removing acids that will be in oil that remains in the system.

If the plan is to remove, recycle, and return refrigerant to a system, the recovery tanks must be clean so that recycled refrigerant does not become contaminated again when it enters the tank. Clean recovery tanks by following the recycling equipment manufacturers' instructions. If recycling equipment is used for multiple refrigerants, clean it as per manufacturer's instructions because mixed refrigerants cannot be recycled. Any noncondensibles in the recovery tank must also be purged according to EPA rules and recycling equipment manufacturers' guidelines to prevent them from getting into the system when the recycled refrigerant is added.

Mixed Refrigerants

The term *mixed refrigerants* refers to refrigerants that are accidentally mixed as opposed to commercially available zeotropic or azeotropic blends. Mixed refrigerants can have an adverse impact on operating systems. Concerns include:

- The effect on performance, operation, capacity, and efficiency of the equipment.
- The effect on materials compatibility, lubrication, equipment life, and warranty costs.
- The increased service and repair requirements and higher operating costs.
- The high cost of separating or inability to separate refrigerants.
- The high cost of disposal and loss of refrigerant for future service.

Determining the presence of mixed refrigerants

It is very difficult to determine the presence of mixed refrigerants without a laboratory testing. For this reason, take every precaution available to avoid mixing refrigerants.

One method to determine if refrigerants are mixed is to check the saturation pressure and temperature of the refrigerant in the system and compare them with the published values for this refrigerant in a pressure-temperature chart. Reviewing the system's service history and determining the system's current problem might provide evidence as to the probability of having mixed refrigerants. Finally, check the nameplate for stickers that indicate if the equipment has been retrofitted with an alternate refrigerant.

Reduce the probability of mixing refrigerants

To reduce the possibility of mixing refrigerants:

1. Properly clear recovery units, or dedicate recovery units to a specific refrigerant.
2. Dedicate cylinders to a specific refrigerant.
3. Test suspected refrigerant before consolidating into larger batches and before attempting to recycle or reuse.
4. Make sure that containers are free of oil and other contaminants. Liquid recovery might increase the likelihood of contaminated cylinders because of oil entrainment.
5. Clearly mark cylinders used for recovered and/or recycled refrigerants.

Applications for Alternative Refrigerants

Currently, new systems with alternate refrigerants have been introduced and existing system are being retrofitted. These applications are of special concern because separate systems at the same location might contain a variety of refrigerants, increasing the possibility that the wrong refrigerant will be introduced into the system or storage containers at that location.

Other applications, such as domestic refrigerators, have used a single refrigerant such as R-12 (CFC-12) and will probably use a single refrigerant (HFC-134a) in the future. Domestic refrigerators are not likely be retrofitted.

Some applications, such as residential air-conditioning, currently use only one refrigerant such as R-22 (CFC-22), although this will change as the HCFC phase-out progresses into the next century.

While mixing refrigerants is a greater possibility in some systems than in others, precaution should be exercised in all situations, especially if you install and service equipment in more than one application.

Working with zeotropic blends

Zeotropic blends are subject to leakage of one or more of the constituent refrigerants. The composition of refrigerant might be different at the beginning of recovery than at the end depending on where the service ports are located. Follow these recommended practices for zeotropic blends:

1. Follow the recommendations of the system manufacturer and the refrigerant and lubricant suppliers.

2. Remove and replace with new specification refrigerant, and return the used refrigerant to the refrigerant supplier or reclaimer.
3. Attempt recycling only if the refrigerant has been analyzed for composition and it meets the new product specifications, especially when working with larger systems.

Compatibility with motor materials

Any alternative refrigerant (and its lubricant) used in retrofitting an existing refrigeration system must be compatible with the polyester insulation material used in hermetic motors. In general, most modern (10–15 years old) compressor motors used in CFC-12 compressors are compatible with the newer alternative refrigerant blends. However, check for any compatibility problems in the refrigerant manufacturer's specifications and retrofit guidelines.

Desiccants

Keeping the refrigerant and lubricant free of moisture is very important in refrigeration systems. Driers filled with moisture-absorbing desiccant are typically used to prevent moisture accumulation. For older R-12 applications, the 4A-XH-5 desiccant produced by UOP (a U.S. company, formerly Union Carbide Molecular Sieve) can be used for both loose-filled and solid-core driers. For some alternative refrigerant blends, a different desiccant might be needed for loose-filled and or solid-core driers because or the material's incompatibility with one or more of the refrigerants in the blend. Consult a drier manufacturer or equipment manufacturer to determine a suitable drier.

Hose permeation and size

Hose permeation rates of older CFC refrigerants and newer HFC-134a refrigerant and alternative refrigerant blends might not be the same for nylon-lined and nitrile (all rubber) hoses. The permeation rate of the alternative refrigerant used in the retrofit process might necessitate the switch to a different hose material and/or hose diameter. Nylon-lined hose is recommended for use with many alternative refrigerants.

Safety

It is important to handle all refrigerants safely.

Inhalation toxicity

Refrigerant blends are not hazardous when handled according to manufacturer's recommendations and when exposures are maintained at or below recommended exposure limits.

All refrigerant manufacturers provide similar safety information on their products. All pertinent data can be found on the Material Safety Data Sheet (MSDS) for the refrigerant. Always keep a copy of the MSDS sheets for all refrigerants, lubricants, solders, and other chemical compounds used in your work.

For example, the DuPont company has established Acceptable Exposure Limits (AEL) for all of its alternative refrigerants. The AEL is 800 ppm for its SUVA MP39 blend, 720 ppm for SUVA MP52, and 840 ppm for SUVA MP66.

An AEL is an airborne exposure limit established by DuPont that specifies time-weighted average (TWA) airborne concentrations, usually eight hours, to which nearly all workers can be repeatedly exposed without adverse effects. In practice, short-term exposures should not exceed three times the established exposure limit for more than a total of 30 minutes during a work day, as long as the TWA is not exceeded.

Inhaling high concentrations of refrigerant vapor might cause temporary central nervous system depression with narcosis, lethargy, and anesthetic effects. Other possible effects include dizziness, a feeling of intoxication, and a loss of coordination. Continued breathing of high concentrations of refrigerant vapors can produce cardiac irregularities (cardiac sensitization), unconsciousness, and with gross overexposure, death. Anyone who experiences any of the initial symptoms should move to fresh air and seek medical attention.

Cardiac sensitization

Cardiac sensitization is an effect that occurs with most hydrocarbons and halocarbons at high concentrations. The human heart can become sensitized to adrenalin, leading to cardiac irregularities and even cardiac arrest. The likelihood of these cardiac problems increases in someone under physical or emotional stress. High concentrations of refrigerant blends can cause these responses, but the effect level varies from person to person and has not been fully determined.

Anyone exposed to very high concentrations of refrigerant blends should not attempt to remain in the area to fix a leak or perform other duties. The effects of overexposure can be very sudden.

Medical attention is imperative if someone is having symptoms of overexposure to refrigerant blends. If the person is having difficulty breathing, administer oxygen. If breathing has stopped, give artificial respiration immediately. Follow the first aid guidelines set forth on the MSDS sheet.

Spills or Leaks

If a large amount of vapors are released, the vapors might concentrate near the floor or low spots and displace the oxygen available for breathing, causing suffocation. When a large spill or leak occurs, always wear appropriate respiratory and other protective equipment. Evacuate everyone until the area has been ventilated, and use blowers or fans to circulate the air at floor level. Anyone who enters the affected area should be equipped with a self-contained breathing apparatus.

Always use an air mask when entering tanks or other areas where vapor concentration might exist. Use the buddy system and a lifeline.

Some refrigerants have a slightly sweet odor that can be difficult to detect. Therefore, frequent leak checks and the installation of permanent leak detectors might be necessary for enclosed areas used by personnel.

Skin and eye contact

Always wear protective clothing when there is a risk of exposure to liquid refrigerant blends. Include eye protection and a face shield where splashing is possible. If eyes are splashed, flush them with plenty of water.

In liquid form, low- and medium-temperature refrigerant blends can freeze skin on contact, causing frostbite. Any skin that comes in contact with the refrigerant should be soaked in lukewarm water, not cold or hot water. If treatment cannot begin immediately, apply a light coat of a nonmedicated ointment, such as petroleum jelly. If the exposed area is in an awkward location for ointment, such as on the eye, apply a light bandage. In all cases of frostbite, seek medical attention as soon as possible.

Nonflammability

Nonflammability is essential for refrigeration and air-conditioning applications. Although some refrigerant blends might contain a flammable compound, the blends are formulated so that they are nonflammable and will not reach a flammable composition during leakage from equipment.

Refrigerant storage and shipping cylinders

Refrigerants are liquified compressed gases. According to the U.S. Department of Transportation (DOT), a nonflammable compressed gas is defined as a nonflammable material having an absolute pressure greater than 40 psi at 21°C (70°F) and/or an absolute pressure greater than 104 psi at 54°C (130°F).

Refrigerants are stored and shipped in a number of different types of containers (Fig. 7-2). All pressure-relief devices used on the containers must be in compliance with the corresponding Compressed Gas Associ-

Figure 7-2 Various capacity refrigerant storage cylinders. *(National Refrigeration Products.)*

ation (CGA) Standards for compressed gas cylinders, cargo, and portable tanks. Refrigerant cylinders are classified into three types based on how they are used: storage cylinders, service cylinders, and disposable cylinders.

Refrigerant cylinders are constructed of steel or aluminum. A valve on the top of the cylinder provides the connection required to charge or discharge the cylinder. Most cylinders also contain a threaded safety plug on the bottom of the cylinder designed to open at dangerously high internal pressures. Other cylinders have a spring-loaded relief valve.

The Department of Transportation (DOT) regulates the construction and use of refrigerant cylinders. As per DOT regulations, all cylinders used to store noncorrosive refrigerants must be inspected every 10 years. Those used to store corrosive refrigerants must be inspected every 5 years.

The appropriate DOT designations must appear on all storage and shipping cylinders. The following is a typical example for a refrigerant blend:

Proper Shipping Name: Refrigerant Gas, N.O.S. (contains chlorotetrafluoroethane and chlorodifluoromethane)

Hazard Class: Nonflammable Gas

UN/NA No.: UN 1078

Storage cylinders. Storage cylinders are used to store larger amounts of refrigerant that can then be transferred to smaller service cylinders. Storage cylinders generally have 100- to 150-lb. capacities. They are often positioned upside down with a valve at the bottom to make refrigerant transfer easier.

The 125-lb. cylinders are commonly equipped with a nonrefillable liquid-vapor valve so that the refrigerant blend can be removed from the cylinder as either a vapor or as a liquid, without inverting the cylinder. The vapor handwheel is located on the top. The liquid wheel is on the side of the valve and attached to a dip tube extending to the bottom of the cylinder. Each is clearly identified as vapor or liquid.

All storage cylinders must have a dated DOT stamp. No cylinders should be used beyond six years of the date on the stamp. Storage cylinders are also returned to the refrigerant manufacturer on a regular schedule (six months is typical) for inspection and valve testing.

Treat the cylinder valve with care. If so equipped, keep the protective screw-on cap in place over the valve when the cylinder is not in use. Keep the packing nut tight when the cylinder is not in use, and be sure to seal the valve opening with the plug or cap provided.

Service cylinders. Service cylinders are smaller (4- to 300-lb.) cylinders that are used to carry refrigerant to and from the equipment site. They are used for refrigerant charging and recovery operations.

Warning: Do not fill a service cylinder completely with liquid refrigerant. You must allow space for expansion in the event the cylinder temperature increases. The safe fill limit for service cylinders is 80 percent full by weight.

When filling a service cylinder from a larger storage cylinder, always use a scale to ensure the proper amount of refrigerant is transferred. Many refrigerant suppliers provide refrigerants in returnable service cylinders. When the service cylinder is empty, the technician simply exchanges it for a full one.

Disposable service cylinders. Refrigerants are also available in disposable service cylinders designed for one-time use only. These cylinders are only used to charge a system. They must never be used to store used refrigerant, and they must never be refilled.

The 30-lb. disposable single-use cylinder is now quite popular. The valve is a standard valve with or without a dip tube depending on whether the refrigerant requires liquid or vapor charging into the system.

Cylinder color codes. Refrigerant cylinders are color coded to aid in identifying the refrigerant stored inside. Table 7-6 lists the color codes for some of the commonly used refrigerants.

Handling precautions for refrigerant cylinders and tanks

The following rules for handling refrigerant containers are strongly recommended:

TABLE 7-6 Refrigerant Storage Cylinder Color Code

Refrigerant	Cylinder Color Code
R-11	Orange
R-12	White
R-13	Light blue (sky)
R-13B1	Pinkish red (coral)
R-14	Yellow brown (mustard)
R-22	Light green
R-23	Light blue gray
R-113	Dark purple (violet)
R-114	Dark blue (navy)
R-116	Dark gray (battleship)
R-123	Light blue gray
R-124	Deep green
R-125	Tan
R-134a	Light blue (sky)
R-401A	Pinkish red (coral)
R-401B	Yellow brown (mustard)
R-402A	Light brown
R-402B	Olive
R-404A	Orange
R-407C	Brown
R-500	Yellow
R-502	Light purple
R-503	Blue green
All others	Medium gray

- Use personal protective equipment, such as side-shield glasses, gloves, and safety shoes when handling containers.
- Avoid skin contact with refrigerants because they can cause frostbite.
- Never heat a container to a temperature higher than 52°C (125°F).
- Never apply direct flame or live steam to a container or valve.
- Never refill disposable cylinders with anything. The shipment of refilled disposable cylinders is prohibited by DOT regulations.
- Never refill returnable cylinders without the refrigerant manufacturer's consent. DOT regulations forbid transportation of returnable cylinders without this authorization.
- Never use a lifting magnet or sling (rope or chain) when handling containers. A crane can be used when a safe cradle or platform is used to hold the container.
- Never use containers for rollers, supports, or any purpose other than to carry refrigerant.
- Protect containers from any object that will result in a cut or other abrasion in the surface of the metal.
- Never tamper with the safety devices in the valves or containers.

- Never attempt to repair or alter containers or valves.
- Never force connections that do not fit. Make sure the threads on the regulators or other auxiliary equipment are the same as those on the container valve outlet.
- Keep valves tightly closed and valve caps and hoods in place when the containers are not in use.
- Store containers under a roof to protect them from weather extremes.
- Use a vapor-recovery system to collect refrigerant vapors from lines after unloading a container.

Refrigerants and moisture

Water in a refrigeration system can react with a refrigerant to form a mild acid. Over time this acid will decrease the lubricating effect of oil within the system causing excessive wear and/or premature failure. The acid will also destroy the insulation on the motor windings of hermetic compressors. As the insulation breaks down, short circuits or even a compressor burnout might occur. The acid can also attack the metal parts of the system, causing mechanical failures.

Another problem caused by too much water in the system is that this undissolved or free water can freeze and block the refrigerant flow control opening.

Refrigeration lubricants (oils)

Refrigerant lubricant or oil circulates through the system with the refrigerant. It is needed to protect and cool the internal moving parts of the compressor. The lubricant will come in direct contact with the hot motor windings of a hermetic compressor. It will also be cooled to the temperature of the evaporator.

The compressor manufacturer usually recommends the type of lubricant(s) and proper viscosity that should be used to ensure acceptable operation and equipment durability. Recommendations are based on several criteria, including lubricity, lubricant/refrigerant solubility, compatibility with materials of construction, and thermal stability and compatibility with other lubricants. It is important to follow the manufacturer's recommendations for lubricants to be used with their equipment.

Mineral oils. Traditional refrigerant lubricants have been based on mineral oil, a product refined directly from crude oil. Naphthenic and paraffinic lubricants (mineral oils) are miscible (soluble) with CFC compounds over the range of expected operating conditions.

The reduction or elimination of chlorine in HCFC and HFC compounds reduces the solubility of mineral oil lubricants. Many new alternative refrigerant blends are not miscible with mineral oils. Other lubricants that are more soluble with them, preferably miscible at evaporator temperatures, must be used. Alkylbenzene and polyol ester lubricants offer excellent miscibility with many of the new alternative lubricants.

Alkylbenzenes. Alkylbenzene oils are derived from petroleum stocks but are superior to mineral oil lubricants due to their unvarying composition, freedom from wax, and increased stability.

Polyol esters (POEs). Polyol ester lubricants are synthetic lubricants specifically developed for use with HFC compounds. These lubricants exhibit a range of critical features that include excellent solubility, viscosity, and lubrication performance over a wide range of temperatures.

Poly alkylene glycols (PAGs). These lubricants are commonly used in automotive air-conditioning systems.

Properties of a good refrigerant oil include the following.

- *Low viscosity.* The lubricant must maintain good flow characteristics through a full range of operating temperatures.
- *Low wax content.* Wax in the oil can separate from the oil and clog capillary tubes and other small passages.
- *Chemical and thermal stability.* The lubricant should not chemically react with the refrigerant or other materials found in the system. It must not form carbon deposits or hot spots on internal compressor surfaces such as valves and cylinder walls.
- *Miscibility.* Miscibility refers to the capacity for the lubricant and the refrigerant to mix. Miscibility is a very important factor in the returning of the lubricant to the compressor in a refrigeration system over its range of operating temperatures, particularly evaporator temperatures. Ideally, the lubricant/refrigerant pair is completely miscible or soluble in each other. This allows the lubricant to flow with the liquid refrigerant and return to the compressor. Even if the lubricant/refrigerant pair is not miscible in the evaporator, it might still have some degree of solubility. Solubility of refrigerant in lubricant lowers the lubricant viscosity, which helps it flow through the evaporator and return to the compressor.

For example, R134a and mineral oils are not miscible. However, polyol ester oils and R134a are miscible. The miscibility of polyol ester oils and R134a is similar to that of current refrigerant oils and R22. Some types of POEs are fully miscible with R134a (as are alkylbenzene and R22), while some POEs are partially miscible with R134a (as are mineral oils and R22).

Other factors, such as refrigerant vapor velocity, play a key role in lubricant return to the compressor. In general, lubricant/refrigerant miscibility is helpful, but not necessarily essential, for proper system operation.

- *Moisture.* Polyol ester oils are not as *hygroscopic* (able to absorb moisture) as poly alkylene glycols PAG lubricants. They are, however, 100 times more hygroscopic than mineral oils.

    Moisture can be very difficult to remove, even when heat and vacuum are used. For this reason, extreme care should be taken to prevent any moisture from entering the refrigeration system. Do not leave the compressor open to the atmosphere any longer than 15 minutes. The best method of assembly would be to remove the system components and plugs just before brazing begins. The maximum system moisture content after completing system processing should be 80 PPM. Running the system with the appropriate drier installed should reduce the system moisture down to 10 PPM or less. Always use the appropriate drier in the system when R134a is used.
- *Compatibility.* Compressor manufacturers have conducted extensive research and testing to ensure that recommended lubricants are compatible with materials used in their compressors. All publish lists of approved oils for their equipment. All approved lubricants are compatible with each other.

Retrofit concerns

When converting a system from CFCs, such as R-12, or an HCFC, such as R-22, to an HFC refrigerant or blend, the old mineral lubricant must be removed from the system and replaced with the recommended polyol ester, alkylbenzene, or poly alkylene glycol (PAG) lubricant.

Residual mineral oil left in a refrigeration system after a retrofit is performed decreases lubricant/refrigerant solubility. Even though polyol ester oils are compatible with mineral oils, they should not be indiscriminately mixed with mineral oils in retrofitted refrigerant systems. Mixing could result in the inability of the oil to return to the compressor and/or reduce heat-transfer performance in the evaporator. Small amounts (up to 1 percent) of mineral oil are acceptable in some field retrofit situations, such as when converting from R-12 to R-134a. Lubricant manufacturers often provide specific retrofit guidelines to reduce the residual mineral oil content to an acceptable level. Figure

7-3 illustrates the steps recommended by one lubricant manufacturer for successfully retrofitting from a mineral-based lubricant to a polyol ester lubricant.

Retrofit test kits. Test kits are available to provide a simple but accurate indication of whether or not the last traces of mineral oil have been cleared out to an acceptable level.

The testing usually involves adding a sample of the oil to a bottle containing a special test fluid (Fig. 7-4). When mixed together, the sample oil will change color if the mineral oil content is too high. Test kits are available for all types of retrofit lubricants.

Oil analysis service. Many lubricant manufacturers offer in-depth oil analysis to determine the condition of the compressor without a complete compressor teardown. Spectrographic analysis determines the presence of up to two dozen different types of contaminants in quantities as small as parts per million. Based on this analysis, the manufacturer or laboratory can suggest maintenance and repairs that might be in order.

Any oil removed from a refrigeration system should be clear. If the refrigerant oil is discolored, it contains impurities. It might also have a strong odor. If the oil is found to be contaminated, it must be replaced along with all system filters and driers.

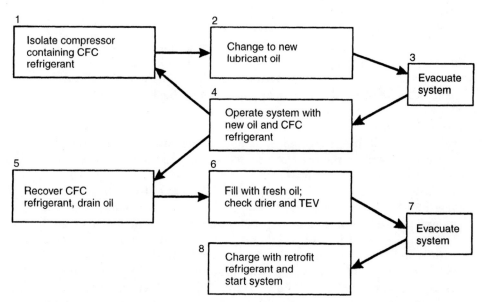

Figure 7-3 Typical step taken when retrofitting from a mineral-based lubricant to a polyol ester lubricant.

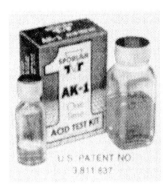

Figure 7-4 Test kit for checking for the presence of mineral oil in a retrofitted system. *(Sporlan Valve Company)*

Pressure-Enthalpy (Mollier) Diagrams

A pressure-enthalpy diagram shows graphic illustration of the thermodynamic properties of a given refrigerant. Each different refrigerant has its own unique pressure-enthalpy diagram. Figure 7-5 shows the diagram for HFC-134a. These diagrams are often called Mollier diagrams after the scientist who pioneered their use. Mollier diagrams are highly useful tools for refrigeration system analysis. Understanding how to read a Mollier diagram will give you a better understanding of what happens inside a refrigeration system.

Key diagram features

The most noticeable section of the diagram is the dome (Fig. 7-6). The dome consists of two curved lines, the saturated liquid line on the left and the saturated vapor line on the right. These two lines arch toward each other to meet at the critical point, the highest point on the dome. This point is defined by the critical temperature and the critical pressure. The critical temperature is the temperature above which the refrigerant will never condense, no matter how high the pressure. The critical pressure is the pressure corresponding to the critical temperature on the saturation curve.

The curved line of the dome defines the state of the refrigerant. To the left of the saturated liquid line, the refrigerant is subcooled. This means the refrigerant can exist as liquid only, with no vapor present. Inside the curved line of the dome, the refrigerant exists in a saturated condition. It is a mixture of vapor and liquid. To the right of the saturated vapor line, the refrigerant is superheated. This means it can exist as a vapor only, with no liquid present. In a typical vapor compression cycle, subcooled liquid, vapor/liquid mixtures, and superheated vapor are all present—and the dome defines these three areas on the diagram.

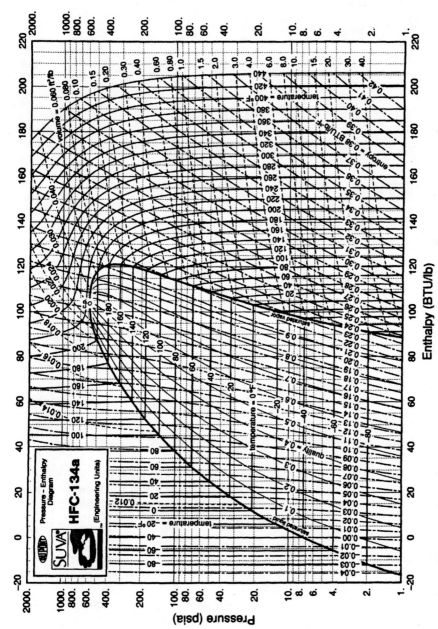

Figure 7-5 Pressure-enthalpy (Mollier) diagram for refrigerant HFC-134a. *(DuPont Fluoroproducts Corporation)*

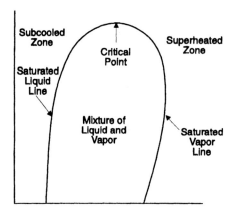

Figure 7-6 Domed section of Mollier diagram.

A Moiller diagram graphically illustrates five types of thermodynamic information:

- *Constant pressure.* Horizontal lines are superimposed on the diagram (Fig. 7-7). The pressure along any one of the lines is constant, and can be determined by following that line over to the left-hand side of the diagram and reading the pressure as listed in pounds per square inch absolute, or psia. Remember that the system pressure read on your manifold gauge is in pounds per square inch gauge, or psig. To change from psig to psia, add atmospheric pressure of 14.7 to the psig reading. Always remember that a Mollier diagram is expressed in psia, which is about 15 pounds more than the actual gauge reading.

- *Constant enthalpy.* Vertical lines representing constant enthalpy are also imposed on the diagram (Fig. 7-7). *Enthalpy* is another word for heat content, and is expressed in BTU/lb (British thermal units per pound of refrigerant). The heat content of the refrigerant at any point on the diagram can be found by following a vertical line downward to the bottom of the graph and reading the enthalpy (heat content) in BTU/lb.

- *Constant temperature.* Lines of constant temperature are also included on the diagram (Fig. 7-8). The shape of these lines varies depending on the zone. The lines are vertical in the subcooled zone, change to horizontal in the vapor/liquid zone, and finally curve downward in the superheat zone. Lines of constant temperature do not appear inside the dome on a Mollier diagram. They are indicated as horizontal dashes at the liquid and vapor saturation lines.

- *Constant specific volume.* Lines of constant specific volume are drawn in a flat arc from the saturated vapor line toward the right

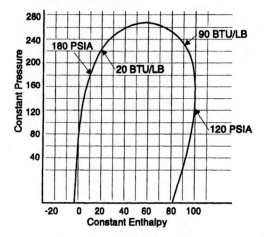

Figure 7-7 Horizontal lines of constant pressure and vertical lines of constant enthalpy on Mollier diagram.

(Fig. 7-9). At any point along one of these lines you can determine the number of cubic feet that one pound of refrigerant will occupy.

- *Constant entropy.* Lines of constant entropy move up and to the right (Fig. 7-9). *Entropy* is defined as the ratio of heat content in a pound of refrigerant vapor to its absolute temperature. The compression of the refrigerant is assumed to be an isentropic (constant entropy) process. Understanding how entropy works in the refrigeration system is important; it is explained later in this chapter.

Working with pressure-enthalpy diagrams

The component schematic for a reach-in cooler using a capillary tube flow control is illustrated in Fig. 7-10. The pressure-enthalpy diagram for HFC-134a, showing the properties of the refrigerant as it circulates through the system, is shown in Fig. 7-11.

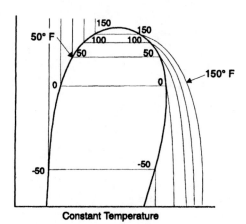

Figure 7-8 Lines of constant temperature.

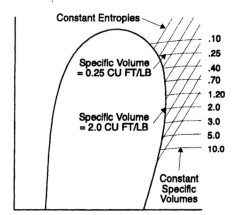

Figure 7-9 Lines of constant specific volume and lines of constant entropy.

Point A (see Fig. 7-10) is the starting point for the cycle. A mixture of vapor/liquid absorbs heat in the evaporator. The enthalpy (heat content) at A is 35.00 BTU/lb. As the refrigerant flows through the evaporator, the liquid continues to vaporize until at point B all the liquid has changed to vapor. The heat content of the refrigerant at point B is 106.00 BTU/lb. The vapor continues to flow through the evaporator, picking up additional heat, until it reaches point C at 32°F. The heat content is now 109 BTU/lb. The heat that the refrigerant picked up in the evaporator is found by subtracting the heat content of the refrigerant when it entered the evaporator at A from the heat content upon leaving the evaporator (109 − 35 = 74 BTU/lb). In this example, every pound of refrigerant passing through the evaporator has picked up 74 BTUs of heat from the refrigerated space. This value is known as the *net refrigerating effect*.

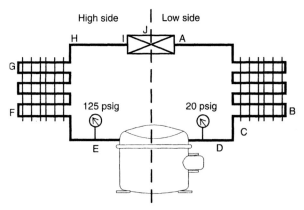

Figure 7-10 Component schematic for a reach-in cooler showing major components and pressure and temperature readings as keyed to the Mollier diagram in Fig. 7-13.

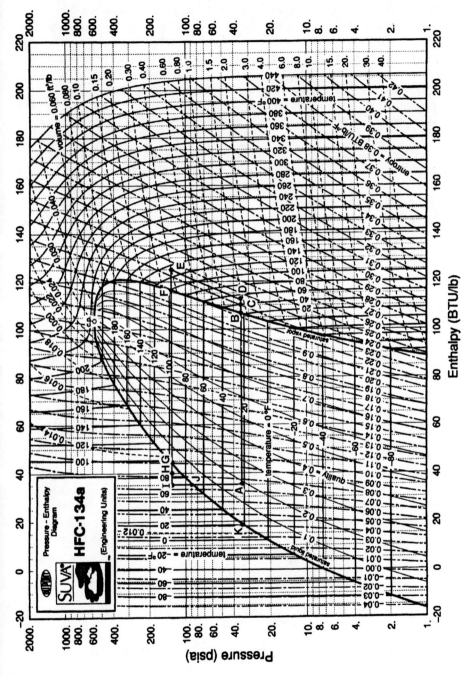

Figure 7-11 Pressure-enthalpy diagram for HFC-134a with system performance superimposed.

As the refrigerant flows through the suction line, it picks up more superheat. Refrigerant temperature increases to 34°F at the compressor, point D. Because this extra heat was not absorbed in the evaporator, it does not perform useful cooling and does not increase the net refrigerating effect.

The system pressure from A to C is constant and may increase slightly from C to D. The temperature from A to B is constant, but it increases from B to C. This increase indicates superheating is taking place.

The refrigerant enters the compressor at D and exits the compressor at E. The line from D to E is where the vapor is compressed. As mentioned in the last section, compression is a constant entropy, or isentropic process. This means the compression line must be drawn parallel to a line of constant entropy. In the example this is from point D to point E, at 135 psia, on the high-side pressure of the system.

The heat content at point D is 111 BTU/lb and, at point E, is 122 BTU/lb. This means that each pound of refrigerant has absorbed the difference between these values, or 11.0 BTU/lb, during the compression process. This final value is called the heat of compression. It represents the amount of work the compressor must do to compress the refrigerant. Remember, both heat and work are forms of energy.

Once out of the compressor, the refrigerant moves into the condenser. The compressor discharge temperature is 128°F. As it moves from E to F in the condenser, the refrigerant is desuperheated. As it moves from F to G, more and more vapor condenses to liquid until, at G, all the vapor has changed (condensed) to liquid.

As the refrigerant moves from G to H in the condenser and from H to I through the liquid line, subcooling takes place. The heat content at I is now 35.0 BTU/lb. The amount of heat rejected in the condenser and liquid line is the heat content at E minus the heat content at I or 122 − 35 = 87 BTU/lb. Notice that the heat rejected in the condenser (87 BTU/lb) is greater than the 74 BTU/lb of heat absorbed in the evaporator. This explains why condensers are larger than evaporators and normally have strong air or water cooling. Remember the following formula:

Heat rejected in the condenser
= Heat absorbed in the evaporator + Heat of compression

From the liquid line the refrigerant enters the capillary tube as a subcooled liquid. The resistance to flow in the capillary tube causes subcooling to decrease and pressure to drop until, at point J, the *bubble point,* liquid starts flashing into vapor. As the mix of vapor/liquid refrigerant flows through the capillary tube, the pressure continues to drop until the refrigerant enters the evaporator at a pressure of 35 psia (point A). The temperature dropped from 70°F to 21°F, yet the enthalpy or heat content remained unchanged.

If the liquid refrigerant could be introduced into the evaporator as 100 percent liquid with all flash gas eliminated, the resulting value would be located at point K in Figure 7-11. The heat content of the refrigerant at this point would be 20 BTU/lb. The amount of heat that this liquid could absorb, if it changed completely to vapor, is represented by a line between point K and point B. This is 106 − 20 or 86 BTU/lb. The *actual* heat absorbed in the cycle is represented by the line from A to B. This is 106 − 35 or 71 BTU/lb. Therefore, 86 − 71 or 15 BTU/lb is lost in flash gas. The percentage of refrigerant lost to flash gas can be calculated as follows:

$$15/86 \times 100 = 17.3 \text{ percent}$$

The percentage of flash gas can also be found by observing the position of point A as referenced to the quality lines located under the dome (see Figure 7-11). The quality lines give the percentage of vapor in the vapor/liquid refrigerant mixture. As you follow line A to B from left to right, the amount of vapor in the refrigerant mixture increases to 100 percent.

Using the diagrams for system troubleshooting

Once you learn how to read Mollier diagrams, they can help you to understand how the system functions under a variety of operating conditions. Pressure and temperature readings at key points in the system can be gathered and matched against the optimum performance points on the refrigerant's Moiller diagram. A few of the more common problems that can be diagnosed using Mollier diagrams are explained in the following sections.

High discharge pressures. Figure 7-12 illustrates a Mollier diagram showing elevated discharge pressure superimposed on the standard diagram used to illustrate a properly operating walk-in refrigerator running on HFC-134a. The original high-side pressure of 135 psia has increased to 210-215 psia, or about 195 psig. The quality of the refrigerant entering the evaporator as a vapor has increased from 17.3 percent to 42 percent vapor. This means that flash gas has significantly increased, decreasing useful refrigeration. The heat of compression has increased from 11.0 BTU/lb to (132 − 111), or 21.0 BTU/lb. This means

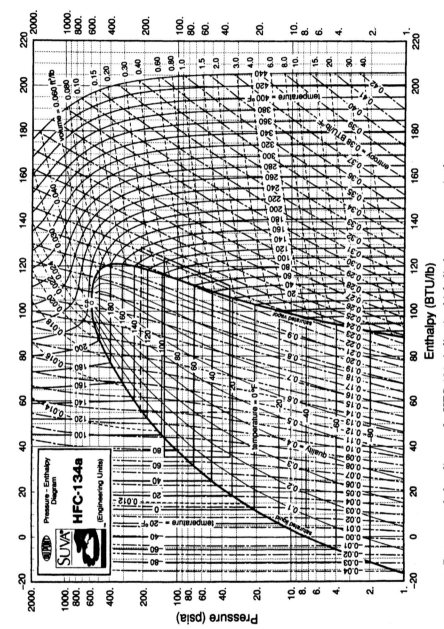

Figure 7-12 Pressure-enthalpy diagram for HFC-134a indicating high discharge pressure in the system.

the compressor has to run harder and consume more power to produce less refrigerating effect. The compressor runs hotter, which results in a shorter service life and increased likelihood of occasional overloads. The discharge temperature has also increased from about 130°F to about 160°F. This reduces the lubricating effect of the refrigerant oil, resulting in more wear on the motor.

Low refrigerant charge. When the system is undercharged with refrigerant, subcooling will decrease (Fig. 7-13). This increases flash gas, thus lowering the refrigerating capacity. The smaller amount of liquid refrigerant in the evaporator vaporizes sooner, resulting in a higher superheat at the compressor. Higher superheat reduces the motor cooling done by the refrigerant. The higher compressor operating temperatures shorten motor life. Vapor passing through a TEV wears its internal parts.

Higher system temperatures also break down the refrigerant oil, decreasing its lubricating ability.

Excess refrigerant charge. Slight overcharging of refrigerant in the system, will cause liquid refrigerant to back up in the condenser, resulting in more subcooling. However, if the system is significantly overcharged, liquid will continue to build up in the condenser until condenser discharge pressure increases. An increased condenser discharge pressure will force more refrigerant through a capillary tube flow control. This will flood the evaporator and all liquid refrigerant might not vaporize before it reaches the compressor. If the liquid refrigerant enters the compressor cylinders it might cause a mechanical breakdown.

Low discharge pressure. As discharge pressure decreases, the net refrigerating effect increases, increasing the capacity of the unit (Fig. 7-14). The heat of compression will decrease, reducing operating current and improving overall system efficiency. Therefore, lower discharge pressure is generally considered beneficial within the refrigeration system. However, should the pressure drop too low, the force pushing the refrigerant through a capillary tube would decrease to the point where an insufficient amount of refrigerant is being metered into the evaporator.

For example, if the system has 135 psia on the high side and 35 psia on the low side, a pressure differential of 100 psia exists that would still allow the refrigerant to be pushed through the tube. However, if the high-side pressure should drop to 85 psia, the differential decreases to only 50 psia, causing a significant decrease in refrigerant flow.

Low suction pressure. The suction pressure in a refrigeration system is directly related to the load on the evaporator. As the cooling load

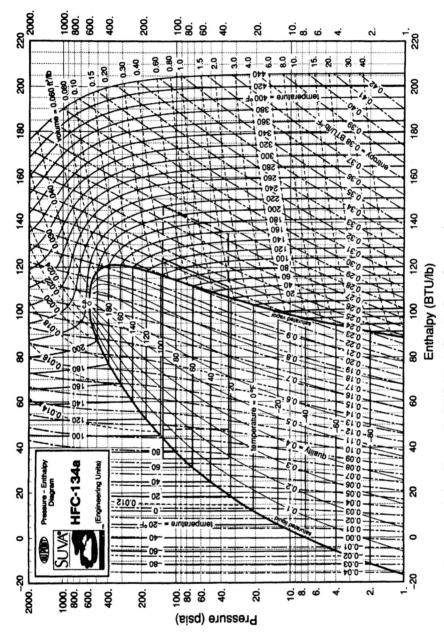

Figure 7-13 Pressure-enthalpy diagram for HFC-134a indicating low refrigerant charge.

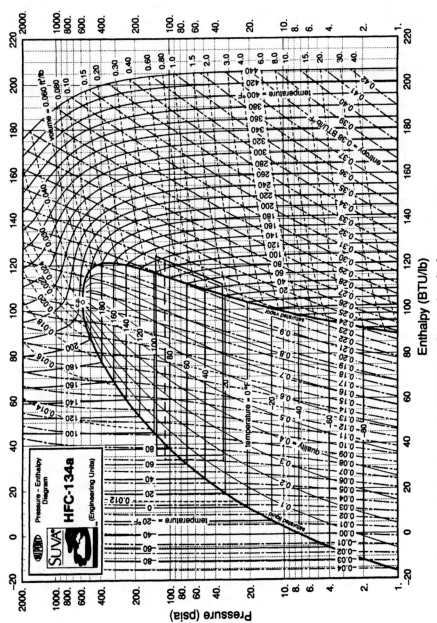

Figure 7-14 Pressure-enthalpy diagram for HFC-134a indicating low discharge pressure.

decreases, less vapor is generated in the evaporator and the pressure drops. Suction pressure is also related to discharge pressure. The increase in pumping capacity along with the decrease in head pressure increases compressor efficiency, resulting in a pressure drop.

Figure 7-15 shows that at lower pressure the amount of flash gas increases and the net refrigerating effect decreases. The heat of compression also increases indicating that compressor efficiency has dropped off. The specific volume of the refrigerant (in ft^3/lb) also increases to indicate that while the compressor is pumping the same amount of refrigerant, the refrigerant weighs less. With fewer pounds of refrigerant in circulation, the overall cooling capacity is lowered.

To avoid low suction pressure, prevent the high-side pressure from getting too low, and keep the evaporator clean to maximize heat transfer. Also make sure that evaporator fans are well lubricated and in good working order.

Low suction pressures normally are not a problem with reach-in refrigerators unless they are operating in low ambient conditions. Low suction pressure is much more common with walk-in units.

High suction pressure. A properly operating refrigeration system should have the highest suction pressure possible while keeping the evaporator coil cold enough to perform properly (Fig. 7-16). High suction pressure can benefit the system in the following ways:

- Low heat of compression.
- Low flash gas.
- High efficiency.
- Low specific volume with more pounds of refrigerant in circulation.

When a refrigeration system starts up for the first time, the temperature inside the unit can be 80°F or higher. This warm air puts a heavy load on the system, resulting in high suction pressures. As the cooled space temperature drops toward normal operating levels, the suction pressure and running current will gradually decrease to normal.

It is not uncommon for problems to occur when a unit has been left off for a period of time. This is because on start up, the system must work very hard to bring the temperature down to operational levels. The stress on the system created by this extra workload can aggravate mechanical or electrical weak points in the compressor, resulting in compressor failure.

TEV system considerations

When plotted out on a pressure-enthalpy diagram, the performance characteristics of a refrigeration system having a capillary tube flow

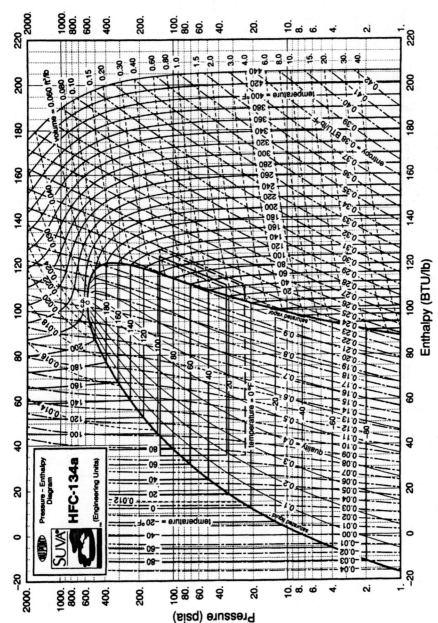

Figure 7-15 Pressure-enthalpy diagram for HFC-134a indicating low suction pressure.

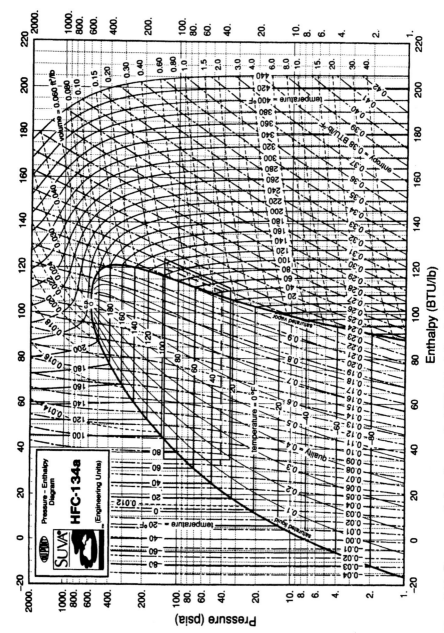

Figure 7-16 Pressure-enthalpy diagram for HFC-134a indicating high suction pressure.

control are very predictable. However, the addition of a refrigerant receiver or accumulator after the condenser in systems controlled by a thermostatic expansion valve creates some slight changes that must be understood. Any subcooling that has occurred in the condenser will be lost as the refrigerant enters the receiver. The liquid leaving the receiver will be subcooled as it flows through the liquid line to the TEV. The end result is that the refrigerant entering the TEV will have several degrees less subcooling than would occur in a similar system equipped with a capillary tube flow control. In all other respects, analysis of the performance plots is the same for both systems. In a TEV controlled system, the high discharge pressures caused by overcharging the system will not occur until the receiver has been completely filled with liquid refrigerant.

Chapter 8

Working with Refrigerants

Working with refrigerants involves a number of different procedures. These include:

- Recovering refrigerant from the system.
- Purging the system lines of contaminants, particularly gases generated by a compressor burnout.
- Testing the system for leaks.
- Evacuating the refrigeration system to remove all air, oil, and contaminants.
- Charging the system with the correct amount of refrigerant.

All of these procedures are used during refrigerant system retrofits. A retrofit involves converting a refrigeration system from an older CFC refrigerant to a new ozone-safe refrigerant, such as an HFC refrigerant compound. A retrofit also involves additional steps and precautions that are outlined later in this chapter.

Recovering Refrigerant

As mentioned in Chapter 7, as of July 1, 1992, the Federal Clean Air Act requires that refrigerants must be recovered and not vented. Many local regulations also prohibit venting of CFCs, and many environmentally responsible companies are now recovering refrigerants during service of equipment. In addition, the recovery and reclamation of refrigerant saves money by saving refrigerant.

Refrigerant recovery equipment

Equipment manufacturers now offer a wide range of compressor-driven refrigerant recovery equipment to fit all applications. Compressors for recovery equipment can be electrically or pneumatically driven. Small portable units are ideal for recovering refrigerant from domestic refrigerators and freezers, commercial reach-in and walk-in units, ice machines, and roof-top units. Larger cart-mounted units are used for supermarkets, large field-erected systems, high-pressure chillers, and other systems requiring large vapor capacities.

All recovery equipment should be ARI 740 certified. This certification from the Air-Conditioning and Refrigeration Institute guarantees that the claims made by the recovery or recycling unit's manufacturer have been tested and verified by an independent laboratory.

Equipment features

A good recovery unit will have a compressor oil sight glass to allow proper maintenance of the oil level. A suction pressure regulator is a feature that prevents compressor overheat and shutdown due to thermal overloads. This excess temperature is eliminated, protecting the compressor windings. Another feature to look for when selecting equipment, is easy-access oil fill and drain ports to allow for quick replacement of contaminated oil after a burnout of the existing system.

Speed is a factor in refrigerant recovery equipment. Good recovery rates are 10 pounds per minute for liquid refrigerants and $3/4$ to 2 pounds per minute for refrigerant vapor.

The recovery unit should pull a disabled unit to 20-in. Hg or more of vacuum, as required by EPA regulations. A good recovery unit can also recover 99 percent of the refrigerant charge.

Units that use the "push-pull" method of refrigerant recovery direct the refrigerant into the cylinder without transfer through the recovery unit. Because no liquid or contaminated refrigerant travels through the unit, it extends the life of the recovery unit. The refrigerant inlet pressure is reduced at the compressor suction section to avoid thermal overload while providing faster recovery.

Portable vapor recovery units

The system shown in Fig. 8-1 is a portable vapor recovery unit. However it can be used to recover liquid refrigerant if used with a service cylinder with a two-port valve. The cylinder must be weighed to avoid overfilling or have a gauge that indicates fill level. Because it is much faster to recover liquid than vapor, liquid should be recovered before the vapor.

Liquid recovery. For liquid recovery (Fig. 8-2), the recovery unit pumps refrigerant vapor from the top of the recovery cylinder to the disabled refrigeration unit. The pressure differential between the refrigeration unit and the recovery cylinder transfers the liquid refrig-

Figure 8-1 A typical small- to mid-size capacity recovery unit. *(National Refrigeration Products)*

erant to the cylinder. The liquid refrigerant is recovered directly from the unit to the cylinder. To prevent compressor damage, do not allow the liquid to go through the recovery unit.

When the scale shows that the recovery cylinder is 80 percent full, the unit must be turned off to avoid over-filling the cylinder and damaging the recovery unit.

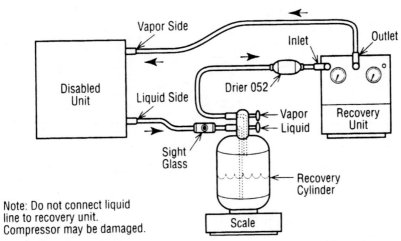

Figure 8-2 The setup for liquid refrigerant recovery using a portable recovery unit. *(National Refrigeration Products)*

212   Chapter Eight

Vapor recovery. The remaining vapor should be recovered after all the liquid has been recovered (Fig. 8-3). The vapor recovery rate depends on the suction pressure, but ¾ to 1 pound/minute is typical. The vapor recovery cycle is complete when the suction gauge on the unit reaches 4-in. Hg vacuum. The LV1 can recover systems to 15-in. Hg vacuum, should the EPA require this level of vapor recovery.

Self-pump out. The recovery unit shown in Fig. 8-1 has a self-pump out feature, which allows the refrigerant trapped in the condenser coil and in the piping of the recovery unit to be transferred to the recovery cylinder, completing the recovery of the refrigerant charge. This operation reduces the risk of mixing refrigerants and refrigerant venting.

Commercial liquid and vapor recovery equipment

The recovery system shown in Fig. 8-4 is an air-driven positive displacement pump, which includes a 1.5-HP oilless air compressor to power it. This recovery unit includes a condenser/subcooler coil, which will pump vapor and condense it for faster transfer. Any recovery cylinder with one or two valves can be used.

This recovery unit pumps either refrigerant liquid or vapor in the same way without changing piping or setting of the unit. There is no risk of slugging the pump or risk of damage by running the pump dry. This unit will recover systems to 20-in. Hg vacuum.

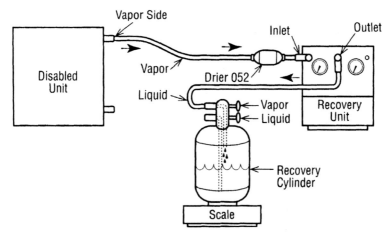

Figure 8-3 The setup for vapor refrigerant recovery using a portable recovery unit. *(National Refrigeration Products)*

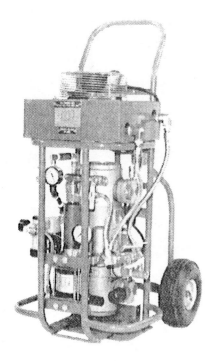

Figure 8-4 Cart-mounted large-capacity refrigerant recovery unit. *(National Refrigeration Products)*

The transfer unit might also be used very effectively to charge systems with refrigerant. The unit is supplied with a unique condenser pump out system to avoid unnecessary venting of refrigerant.

The unit shown in Fig. 8-5 is designed for commercial installations. It consists of a belt-driven open-type compressor, which will recover systems to 20-in. Hg vacuum. The unit is mobile with all components mounted on a steel frame. The open compressor eliminates all problems related to the handling of contaminated refrigerant.

Recovery of different refrigerants

To avoid mixing refrigerants, change the filter drier in the recovery unit and evacuate the recovery unit and hoses before recovery from a new system. Although it is not necessary to change the compressor oil before recovering different refrigerants, the oil must be changed after recovery from a burned-out system.

Compressor oil

During normal recovery operations, a very small amount of compressor oil will be carried out of the recovery unit. Before each use, check the compressor oil sight glass and adjust the oil level in the recovery unit.

Figure 8-5 Commercial recovery unit with open-belt-driven compressor. *(National Refrigeration Products)*

Recovery cylinders

All recovery cylinders must meet ARI guidelines and DOT specifications for the refrigerant being recovered from the system. It is very important to evacuate the recovery cylinders and purge the hoses to avoid introducing noncondensibles in cylinders that would increase the discharge pressure. Also, use a separate cylinder for each type of refrigerant to avoid cross contaminations.

Caution: For safety reasons, be sure to fill all cylinders by weight in accordance with the cylinder supplier's instructions and ARI guidelines. See Chapter 7 for more details on service cylinders and working safely.

Moisture and Contaminants

Ideally, the recirculation lines of a refrigeration system should contain two things only: a refrigerant and a refrigerant oil. But in virtually all applications, air finds its way into the system when the system is assembled or serviced. Air contains contaminants such as oxygen, nitrogen, hydrogen, and water vapor. These gases are noncondensible. This means the gases will take up valuable space in the condenser that could otherwise be used for condensing refrigerant vapor into liquid.

The fact that the condenser is now working at less than its designed capacity can lead to a rise in head pressure.

All of these gases, with the exception of nitrogen, can react with refrigerants and refrigerant oils to form acids that can damage system components. Very small volumes of noncondensible gases are found in most refrigeration systems and do not adversely affect system efficiency. But as the volume of gases increases, both service life and system efficiency are adversely affected. To prevent these problems, the refrigeration system is purged (sometimes) and evacuated (always) to remove noncondensible gases and moisture.

Purging and evacuation are procedures that remove unwanted gases, contaminants, and moisture from a refrigeration system. Whenever a sealed refrigeration system is opened to the atmosphere for any reason, or has experienced a compressor burnout, purging and/or evacuation is required to restore system integrity.

Purging Refrigerant Lines

Purging a refrigeration system consists of applying a pressurized neutral gas, such as nitrogen, to one point while allowing the gas to escape at some other opening (Fig. 8-6). This procedure will sweep out unwanted system contaminants. Any system in which a burned-out compressor has been replaced should be purged. The gases generated by burned insulation or oil should not be allowed to remain in the system.

Purging a system can be performed by applying the purging gas from a cylinder under pressure to one of the access valves and allowing it to escape from the other. Using the low- and high-side access valves, this will purge only the portion of the system between the source of the purging gas and the exhaust port, and not the compressor. An alternative method would be to apply pressure to one of the access valves after temporarily disconnecting an appropriate connection or pipe. This method allows the purging gas to flow through any desired portion of the system. After sufficient purge gas has passed through, the disassembled connection can then be restored.

Using the manifold gauge set for the purging operation allows the cleansing gas or liquid to be easily controlled. It permits the evacuation procedure to be performed immediately afterwards by disconnecting the gas cylinder from the gauge set and connecting the vacuum pump. After the purging operation, the refrigeration system must be completely evacuated to remove the purging agent and all traces of contamination that might remain in the system.

If a refrigeration system has failed "clean" and does not contain contaminants or excessive moisture, purging might not be needed. Simply evacuate the system after repairs have been made.

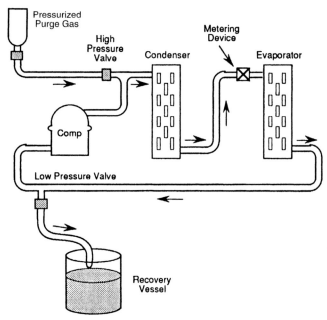

Figure 8-6 Typical setup for purging refrigerant lines. *(Reprinted from* Practical Air Conditioning Equipment Repair *by Anthony J. Caristi, McGraw-Hill, copyright 1991.)*

Evacuation of Refrigeration Units

Evacuating a refrigeration system of gases and moisture is done by creating or "pulling" a vacuum inside the refrigerant lines.

Measuring vacuum

Vacuum levels are expressed in several different ways. As explained in Chapter 3, atmospheric pressure at sea level at a temperature of 68°F is 14.696 psia. Atmospheric pressure will support a column of mercury 29.92 in. (760 mm) tall. As atmospheric pressure is removed, this column of mercury would decrease in height until it ceases to exist at perfect vacuum. So, in order to pull a perfect vacuum in a refrigeration system under atmospheric conditions, the pressure in the refrigerant lines must be reduced to 29.92-in. Hg vacuum. The vacuum scale on a compound manifold gauge begins at 0-in. Hg vacuum and reads down to 29.92-in. Hg vacuum.

Very low vacuum levels are normally expressed in microns of mercury. One millimeter equals 1000 microns. So if the 760 mm of mercury supported at atmospheric pressure is reduced below 1 mm, it is expressed in microns. Vacuum levels from about 200 microns ranging down to as low as 50 microns are recommended when evacuating most modern refrigeration systems. A perfect vacuum is not needed to evacuate gases and moisture to acceptable levels. In fact, at the very low pressures approaching perfect vacuum, the refrigerant oil in the system can boil slightly, leaving a vapor.

Water and vacuum

At extremely low pressures, any moisture in the system will boil to a vapor. If there is a small amount of moisture in the system, the vapor can then be removed by a vacuum pump fairly quickly. However, a vacuum pump is not recommended to remove large amounts of water from a system. Even moderate amounts of water will evaporate into very large volumes of vapor. For example, just one pint of water in the system will produce over 1000 ft^3 of vapor when boiled at 65°F. If possible, manually drain large amounts of water from the system before connecting the vacuum pump for final evacuation.

Vacuum pumps

A special vacuum pump is used to remove the air, gases, and moisture from the refrigeration system down to very low levels. Vacuum pumps used to evacuate refrigeration systems are typically driven by compact rotary compressors. A two-stage rotary vacuum pump can pull the lowest vacuum. It is capable of lowering the pressure in a leak-free unit down to 0.1 micron.

Setup for evacuation

A refrigeration system can be evacuated by connecting a manifold gauge set between the vacuum pump and the system, as shown in Fig. 8-7. Small refrigeration systems that do not contain large amounts of moisture can be successfully evacuated in one step if the vacuum pump is capable of reaching a high vacuum, such as 29.5 inches of mercury at sea level. The maximum level of vacuum obtainable by any pump is about one inch less for each 1000 feet (305 meters) of altitude. When the vacuum pump is not capable of such a high level, or if there is considerable moisture to be removed, a two- or three-step evacuation is needed. If the refrigeration system has both a low-pressure and high-pressure access valve, the manifold gauge set can be connected to both ports to facilitate the pumping operation.

Multi-step evacuation

Multi-step evacuation is used to achieve an extremely low level of contamination in a system. Multi-step evacuation, as the name suggests,

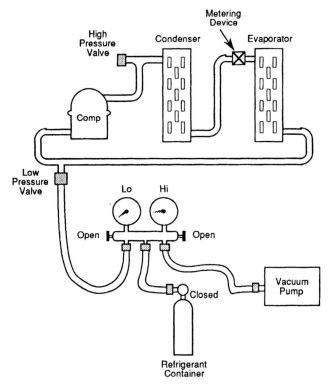

Figure 8-7 Typical setup for evacuating a refrigeration system using a vacuum pump and manifold gauge set. *(Reprinted from* Practical Air Conditioning Equipment Repair *by Anthony J. Caristi, McGraw-Hill, copyright 1991.)*

is a repeated evacuation of the system. The system is first evacuated to a low vacuum of about 1 or 2 mm. A small amount of refrigerant is bled into the system, and the system is again evacuated down to 1 mm Hg.

To perform a multi-step evacuation procedure, run the vacuum pump until the system vacuum is equal to the limit obtainable by the vacuum pump. If moisture is present in the system, additional running time will be required for the pump to vaporize the water and remove it. This process will be indicated by the low-pressure manifold gauge reading, which reaches a plateau lower than vacuum pump capability, as the moisture is being removed.

When the vacuum pump limit has been reached, the valve to the vacuum pump is closed. Refrigerant (or nitrogen) from the cylinder is allowed to enter the system until the vacuum reaches 20-in. Hg on the manifold gauge. The small amount of refrigerant vapor will penetrate the system and mix with other vapors in the lines. The cylinder valve is closed and the vacuum pump valve is opened. The vacuum pump will then remove the new charge of gas along with the residual contaminants left over from the previous evacuation.

This procedure can be repeated one more time. The vacuum pump must be allowed to operate until its vacuum limit is attained. This condition is known as *flat out*.

Once the vacuum has been pulled a third time, close the high-pressure manifold valve and vacuum pump valve and fully charge the system with refrigerant.

Detecting leaks within a vacuum

If there is a leak in the system while it is under vacuum, the vacuum gage will rise quickly. However, using a vacuum to determine if the system is leaking is not recommended for three reasons. First, the vacuum will pull air into the system, which is exactly what you do not want. Second, the location of the leak cannot be found. Third, if the system does not leak (no pressure rise) when the vacuum is pulled, this only proves that the system will not leak under a pressure differential of 14.696 psi. This is the pressure difference between the vacuum inside the system and atmospheric pressure outside of the system.

However when a system is fully charged the pressure differential between the highly pressurized refrigerant vapor inside the system and the 14.6-psi atmospheric pressure outside the system is typically 300 psi or more.

Standing pressure leak test

The recommended method of leak testing a refrigeration system is by using a standing pressure test. In this test a small amount of refrigerant is added to the system to raise system pressure to around 10 psig. Pressure inside the system is then raised to a higher pressure (around 150 psig) using a second inert gas, such as nitrogen.

Caution: Never use ordinary compressed air as a pressure source. Never pressurize the system above its working pressure as specified on the label attached to the equipment.

When the system reaches about 150 psig, tap the gauge lightly to ensure the needle is free and then note the exact pressure reading. With the system holding at this pressure, inspect for leaks using a detector designed for the refrigerant used.

After the system has been fully checked for leaks, note the reading on the gauge once more. If it has fallen, then there is a leak somewhere in the system that has not been detected. Always remember that the gauge manifold and/or the connections might be the source of a slow leak.

Even when there is no drop in the pressure reading, it is good practice to allow the system to stand for a longer period of time. The time should vary based on the system's size. A small reach-in cooler might only need to stand for 45 minutes or so. A 10-ton system would most likely need 12 hours or so to ensure that no leaks are present in the system.

Charging the System

Adding the correct amount of refrigerant to the system is known as *charging the system*. Correct charging is particularly important in systems that do not have a receiver to store excess liquid refrigerant. For example, capillary tube controlled systems require a critical charge to operate effectively.

Any refrigeration system that has been opened to the atmosphere during repairs, or that is suspected of containing air, moisture, or other contaminants must be properly evacuated prior to charging with refrigerant.

A refrigeration system can be recharged without evacuation if you are certain the system does not contain moisture, air, or other contaminants. For example, a small domestic unit that has never been recharged (evident by the absence of any access valves), can often be brought up to full charge by the addition of refrigerant. One exception to this rule would be a unit with a leak on its low-pressure side that is relatively low on refrigerant. In this case, the compressor suction pressure might have reached vacuum level, drawing in air and moisture through the leak. Such a unit should be repaired and evacuated prior to recharging.

Refrigerant can be added to a system as either a vapor or liquid; each case requires its own procedure. The correct charge can be determined in a number of ways:

- Weighing the refrigerant on a scale.
- Metering out a specific volume of refrigerant using a graduated cylinder.
- Determining correct charge by system operating pressure.

Vapor refrigerant charging

Vapor charging involves moving refrigerant in a vapor state from a charging cylinder or drum into the refrigerant system. When the system is not operating (compressor off) refrigerant vapor can be added to either the low- or high- pressure side of the system.

When the system is operating (compressor running), refrigerant vapor can only be added to the low-pressure side of the system. This is because the high-side pressure is usually greater than the refrigerant pressure in the charging cylinder and vapor cannot flow into the system. For example, an HFC-134a system, might have condensing temperature of about 130°F on a hot summer day (95°F ambient temperature plus 35°F heat of compression). The pressure of HFC-134a at 130°F is 200 psig. The charging cylinder has the same ambient temperature of 95°F, so the refrigerant inside the cylinder is at a lower pressure of 115 psig. The higher high-side system pressure makes it impossible for the refrigerant to flow from the cylinder into the high side of the system.

But the low side of the system is operating at a much lower pressure. On a 95°F day, the evaporator pressure might only be around 20 to 25 psig. The refrigerant will easily move from the charging cylinder to the lower pressure evaporator areas (Fig. 8-8).

During colder months, the charging cylinder might be exposed to much colder temperatures (when stored in a truck outside). In this case, its pressure might be lower than the low side of the system (Fig. 8-9). It is then necessary to allow the cylinder to warm up to increase its internal refrigerant pressure. The cylinder can also be placed in a container of hot water to speed up the process.

Warning: Never use a torch to heat a refrigerant cylinder. A warm tub of water is recommended, with a maximum water temperature of 90°F. Move the cylinder around in the water to allow the liquid in the center to contact the warm surface area of the cylinder.

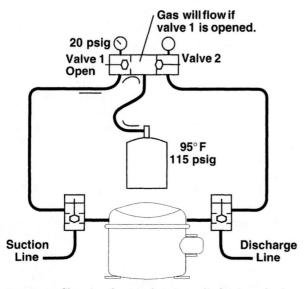

Figure 8-8 Charging from a charging cylinder into the low side of the system.

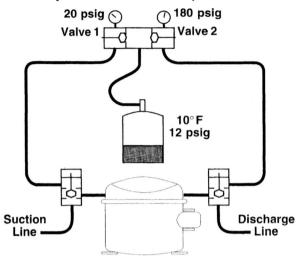

Figure 8-9 If the temperature of the charging cylinder is very cold, pressure in the cylinder will be so low that refrigerant will flow out of the system and into the cylinder.

As refrigerant vapor is pulled from the charging cylinder, the remaining liquid boils to fill the space left by the pulled vapor. As more vapor is released, the liquid in the bottom of the cylinder continues to boil and its temperature decreases. If sufficient vapor is pulled, the temperature and pressure inside the cylinder could fall to the same levels found on the low-side of the refrigeration system. The vapor movement from the charging cylinder to the system would then stop. More heat is needed to raise the temperature of the liquid refrigerant to help keep the pressure up.

A relatively full cylinder will hold its pressure longer than one that is almost empty. For this reason, when large amounts of refrigerant must be added to a system it is best to use the largest charging cylinder or drum available. Don't use a 30-lb cylinder to charge 20 lb of refrigerant if a 125-lb cylinder is available.

During vapor charging it is possible to monitor both low and high system pressures on units equipped with two access valves (Fig. 8-10). All hoses should be purged of air by temporarily loosening the hose connections and allowing a very small amount of refrigerant gas to escape as the air in the hoses is forced out by either the residual refrigerant pressure in the refrigeration system or cylinder pressure.

A high-pressure connection is not necessary during the recharging procedure. However, if a high-pressure access valve is available, it can be used to check for excessive compressor head pressure during opera-

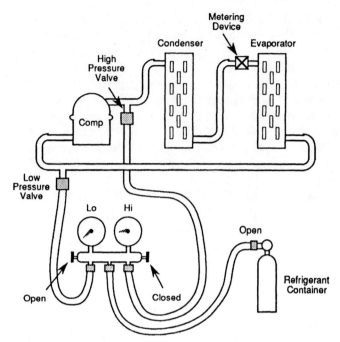

Figure 8-10 Monitoring both low- and high-side pressures during vapor charging. *(Reprinted from* Practical Air Conditioning Equipment Repair *by Anthony J. Caristi, McGraw-Hill, copyright 1991.)*

tion. Excessive head pressure can be caused by insufficient cooling of the condenser coils.

When the high-pressure gauge connection of the manifold gauge is used during vapor charging, the hose line will accumulate liquid refrigerant. To prevent liquid from spewing out when the hose is disconnected, use a check-valve adapter placed between the high-pressure hose and access valve. An alternate method is to allow the accumulated liquid in the hose to vaporize after the system has been shut down and before the hoses are disconnected. To vaporize the accumulated liquid, open both low- and high-pressure gauge valves and allow time for the liquid refrigerant to vaporize.

The charging operation begins with the manifold gauge set and refrigerant charging cylinder connected to the system. After all hoses have been properly purged of air, bring the system pressure up to cylinder pressure by opening the low-pressure manifold valve and refrigerant charging cylinder valve.

With the refrigeration system thermostat set to its minimum setting, turn on the compressor and monitor the suction line pressure as refrigerant vapor enters the system.

Only vapor refrigerant should be allowed to enter the system. Keep the refrigerant container upright for gaseous refrigerant. During the vapor charging operation, leave the high-pressure manifold valve in the closed position at all times.

Check the compressor suction pressure during the charging process by temporarily closing the low-pressure manifold valve. Monitoring this pressure is important when charging units with capillary tubes because the magnitude of pressure will be a satisfactory indication of the percentage of charge. Systems that use HFC-134a as a refrigerant will have a compressor suction pressure of approximately 50 to 70 PSI when fully charged.

In refrigeration systems equipped with a sight glass in the liquid line feeding the evaporator, full charge is indicated by the absence of gas bubbles flowing with the refrigerant, indicating pure liquid refrigerant in the line.

When a refrigeration system is fully charged and has been in operation long enough to be stabilized, the temperature of the evaporator return line to the compressor can also be used as an indicator of full charge. The difference in temperature between the return line and the boiling temperature of the refrigerant (indicated by the appropriate scale on the low-pressure gauge), is called *superheat* and can be specified by the manufacturer of the system. Fully charged systems will usually have evaporator return lines that feel cool to the touch and can collect condensation on the outside.

Liquid refrigerant charging

If large amounts of refrigerant are needed, such as when there is not refrigerant in a large capacity system, liquid charging will save time. Liquid refrigerant is typically added via the "king" valve on the liquid line between the condenser and the flow control device. Draw a vacuum on the system, connect the charging cylinder, and allow the liquid to flow until it has almost stopped. As the liquid enters the system, it will flow toward the evaporator and condenser and be equally divided between these two components. There is no possibility of liquid refrigerant entering the suction line of the compressor when the system starts up (Fig. 8-11). Unlike vapor charging, charging with liquid refrigerant does not lower the charging cylinder pressure.

On some systems the king valve can be front seated while the compressor is running. This causes the low-side system pressure to decrease. If this is the case, liquid refrigerant can be added to the system through an auxiliary charging port located between the king valve and the refrigerant flow control. This liquid directly feeds the flow control device so it is important not to overcharge the system using this method (Fig. 8-12). It might be necessary to temporarily bypass the

Working with Refrigerants 225

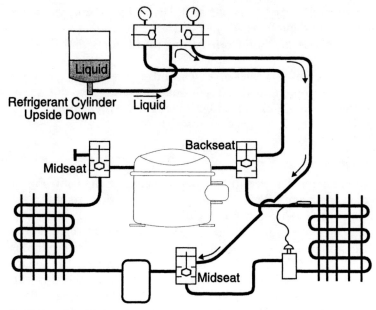

Figure 8-11 Liquid charging of an evacuated system through the king valve in the liquid line.

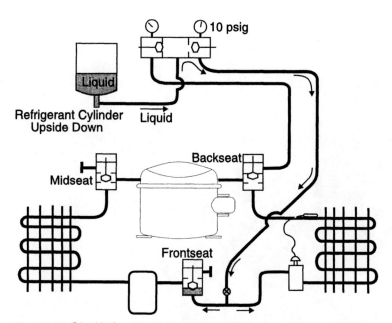

Figure 8-12 Liquid charging through a liquid-line charging valve with the king valve front seated and liquid refrigerant feeding the flow-control device.

low-pressure control during charging so it does not shut the system off. Remember to remove the bypass when charging is complete.

Caution: Never charge liquid refrigerant into the suction line of a compressor.

Smaller units, such as reach-in cases and domestic refrigerators, should have the recommended charge amount printed on the nameplate. The refrigerant should be added during a deep vacuum to help ensure that it is the correct amount.

Weighing refrigerant

Special high-precision scales can be used to weigh refrigerant. Figure 8-13 shows an electronic digital display scale used for weighing the refrigerant charge dispensed from a cylinder.

Electronic scales are highly accurate. Many can be set to zero after the full cylinder is placed on the scale. As refrigerant exits the cylinder, the scale reads a positive value. So if 48 ounces of refrigerant is needed to charge the system, the cylinder is first set on the scale and the scale adjusted to zero. The scale will count upward as the refrigerant flows from the cylinder into the cylinder. Once the proper weight is dispensed from the cylinder, refrigerant flow is shut off. With this type of scale, it is not necessary to account for the weight of the cylinder in any calculations.

Figure 8-13 Electronic scale used to weigh refrigerant. *(Fluoro Tech, Inc.)*

Measuring with graduated cylinders

A graduated cylinder can also be used to measure out the refrigerant charge. Graduated cylinders are calibrated containers that allow you to view the liquid level of refrigerant inside the cylinder (see Figure 2-17). A pressure gauge on the top of the cylinder reads refrigerant pressure inside the cylinder. The refrigerant pressure is used to calculate the temperature of the liquid. Determining the temperature is important because liquid refrigerant has different volumes at different temperatures. The pressure/temperature is dialed on the graduated cylinder to set the proper volume scale.

For example, say a graduated charging cylinder contains HCFC-22 refrigerant at 100 psig. When the dial is turned to 100 psig, the volume or level of HCFC-22 refrigerant reads 5 lb 2 oz on the proper scale. If 32 oz of refrigerant is needed to charge the system, this amount is subtracted from the 5 lb 2 oz to obtain the final cylinder reading after charging is complete.

$$(5 \text{ lb} \times 16 \text{ oz/lb}) + 2 \text{ oz} = 82 \text{ oz in cylinder at start of charge}$$

So,

$$82 \text{ oz} - 32 \text{ oz} = 50 \text{ oz or 3 lb 2 oz in cylinder at completion of charge}$$

An advantage to the graduated cylinder is that the liquid can be seen as the level drops. Some graduated cylinders have a heater in the bottom to keep the temperature from dropping when vapor is pulled from the container. It is important to choose a graduated cylinder that is large enough to accommodate the system being charged. Cylinders are typically used only once for an accurate charge. In systems with more than one refrigerant, one cylinder is used for each type of liquid. Using a cylinder once increases accuracy and helps eliminate mistakes.

Some system manuals include charts that detail typical operating pressures. These charging charts can be compared with gauge readings to determine the correct charge for a particular system.

Retrofits with Alternative Refrigerants

A system retrofit involves switching from a CFC refrigerant to a more ozone-safe alternative refrigerant. Before performing a system retrofit, consult with the manufacturer of the system or the system's compressor concerning any specific lubricant recommendations for their compressors.

The filter-drier must also be changed during the retrofit, as is routine practice following system maintenance. Use the exact type of drier recommended by the alternative refrigerant manufacturer.

Caution: Many of the newer alternative refrigerants were not designed to be mixed with other refrigerants or additives. Doing so could have an adverse effect on performance. Therefore, unless the alternative refrigerant manufacturer specifically approves of the practice, do not top off the existing refrigerant with new alternative refrigerants.

Overview of the retrofit process

Retrofit of an existing system to alternative refrigerants and lubricants is done using standard refrigeration service practices and equipment. The following steps are involved in retrofitting:

1. Evacuate the old refrigerant charge from the system.
2. Remove the mineral oil from the compressor and replace it with a lubricant recommended for the retrofit refrigerant.
3. Replace the filter-drier with a new drier compatible with the retrofit refrigerant blend.
4. Charge the system with the new refrigerant.
5. Start the system and adjust the charge and/or controls to achieve the desired operation.

For most systems, changing the compressor lubricant, changing the filter-drier, and (in systems with expansion valves) adjusting the superheat setting are the only system modifications required in a retrofit to a new refrigerant. If the system is still under warranty, it is best to contact the equipment or compressor manufacturers concerning warranty terms before performing the job. Some equipment or compressor warranties might be voided if the refrigerant originally specified for the system or compressor is changed.

Equipment and supplies

You should have the following equipment and supplies on hand to perform the retrofit:

- Safety equipment (gloves, glasses)
- Manifold gauges
- Thermocouples to read line temperature
- Vacuum pump
- Leak detection equipment
- Scale
- Recovery unit

- Recovery cylinder
- Container for recovered lubricant
- Replacement lubricant
- Refrigerant cylinder with retrofit refrigerant
- Replacement filter-drier
- Labels indicating the refrigerant and lubricant charged to the system

Typical retrofit procedure

The following procedure is typical for many refrigerant changeover retrofits. Prior to the retrofit, be sure to review the refrigerant manufacturer's retrofit guidelines, plus the Material Safety Data Sheets for safety information on the use of these products.

1. When performing the first retrofit with a given alternative refrigerant, collect system performance data while the old refrigerant is in the system. This baseline data of temperatures and pressures at various points in the system (evaporator, condenser, compressor suction and discharge, expansion device, etc.) at normal operating conditions can be useful when optimizing operation of the system with the retrofit refrigerant. Figure 8-14 shows a system data sheet for recording this baseline data.

2. Recover the original refrigerant such as CFC-12 or R-500 from the system as outlined earlier in this chapter. If the recommended refrigerant charge size for the system is not known, weigh the amount of refrigerant removed if possible. The initial quantity of replacement refrigerant charged in the system should equal this amount.

3. In systems where mineral oil is the existing lubricant, drain it from the compressor. This might require removing the compressor from the system, particularly with small hermetic compressors that have no oil drain. In this case, the lubricant can be removed from the compressor by using an oil pump or by draining through the suction line of the compressor. In most small systems, 90 to 95 percent of the lubricant can be removed from the compressor in this manner. Larger systems might require drainage from additional points in the system, particularly low spots around the evaporator, to remove the majority of lubricant. In systems with an oil separator, any lubricant present in the separator must also be drained.

   Always measure and record the amount of lubricant removed from the compressor and compare the volume to the compressor/system

System Data Sheet

Type of System/Location: _____

Equipment Mfg.: _____   Compressor Mfg.: _____
Model No.: _____   Model No.: _____
Serial No.: _____   Serial No.: _____
R502 charge size: _____   Original Lubricant:
    Type/mfg: _____
    Charge size: _____
New Lubricant:
    Type/mfg: _____
    1st Charge size: _____
    2nd Charge size: _____
    Additional Charge size: _____
Drier Mfg.: _____   Drier type (check one):
Model No.: _____   Loose fill: _____
    Solid core: _____

Condenser cooling medium (air/water): _____
Expansion Device (check one): Capillary tube: _____
    Expansion valve: _____
If Expansion valve:
    Manufacturer: _____
    Model No.: _____
    Control/set point: _____
    Location of sensor: _____
Other System Controls (ex.: head press control), Describe: _____

(circle units used where applicable)

Date/Time				
Refrigerant				
Charge Size (lb, oz/grams)				
Ambient Temp. (°F/°C)				
Relative Humidity				
Compressor:				
Suction T (°F/°C)				
Suction P (psig, psia/kPa, bar)				
Discharge T (°F/°C)				
Discharge P (psig, psia/kPa, bar)				
Box/Case T (°F/°C)				
Evaporator:				
Refrigerant Inlet T (°F/°C)				
Refrigerant Outlet T (°F/°C)				
Coil Air/H_2O In T (°F/°C)				
Coil Air/H_2O Out T (°F/°C)				
Refrigerant T @ Superht. Ctl. Pt. (°F/°C)				
Condenser:				
Refrigerant Inlet T (°F/°C)				
Refrigerant Outlet T (°F/°C)				
Coil Air/H_2O In T (°F/°C)				
Coil Air/H_2O Out T (°F/°C)				
Exp. Device Inlet T (°F/°C)				
Motor Amps				
Run/Cycle Time				

Comments: _____

Figure 8-14 A data sheet used to record baseline data during a system retrofit. *(DuPont Fluoroproducts Corporation)*

specifications. Retrofit check sheets are often used to record lubricant amounts and other important information during the retrofit. A sample check sheet is shown in Fig. 8-15).

At least 80 percent of the lubricant must be removed from the system to ensure proper oil return and lubrication. If poor system performance is noted on start-up, an additional lubricant change might be required (this should occur in less than 1 percent of retrofits). Record the amount of lubricant removed on the attached retrofit checklist, as this will be needed in the next step.

4. Charge the compressor with the same volume of retrofit lubricant as the volume of mineral oil lubricant removed in step three. Where possible, check with the compressor manufacturer for specific lubricant recommendations. Reinstall the compressor if removed in step three. Use normal service practices.
5. Replace the filter drier with one recommended for use with the retrofit refrigerant.
6. Reconnect the system and evacuate using the procedure outlined earlier in this chapter. To remove air or other noncondensibles in the system, the system should be evacuated to a deep vacuum (500 microns). Leak check the system using a standing pressure test. Reevacuate the system following the leak check.
7. Charge the system with retrofit refrigerant. With azeotropic refrigerant blends, the vapor composition in the refrigerant cylinder is different from the liquid composition. To ensure that the proper blend composition is charged in the system, the system must be liquid charged. Azeotropic refrigerant cylinders are typically equipped with dip tubes, allowing the liquid to be removed from the cylinder in the upright position. The proper position is indicated by arrows on the cylinder and cylinder box. Once removed from the cylinder, the azeotropic blend can be charged to the system as vapor as long as all of the liquid removed from the cylinder is transferred to the system.

The refrigeration system might require a smaller charge size with the new blend, as opposed to CFC-12 or other CFC refrigerants. The optimum charge will vary depending on the operating conditions, size of the evaporator and condenser, the size of the receiver (if present), and the length of the pipe or tubing that runs in the system. For most systems, the optimum charge will be 75 to 90 percent by weight of the original CFC-12 charge.

When retrofitting R-500 equipment, the refrigeration system might require a slightly larger charge with the new blend. For most systems, the optimum charge will be about 105 percent by weight of the original R-500 charge. It is recommended that the system be ini-

Checklist for Suva HP62 Retrofit

_____ Establish baseline performance with R502. (See data sheet for recommended data.)

_____ Consult the original equipment manufacturer of the system components for their recommendation on the following:
- Plastics compatibility
- Elastomers compatibility
- Lubricant (viscosity, manufacturer, additives)
- Retrofit procedure to sustain warranty

_____ Drain lubricant charge from compressor (unless polyol ester lubricant is already in the system).*
- Remove 90–95% of lubricant from the system
- Measure amount of lubricant removed and record_____

_____ Charge polyol ester lubricant. Run system for 48–72 hours *minimum*.
- Recharge with amount equivalent to amount of mineral oil removed.

_____ Repeat lubricant drain and ester charging until mineral oil content is less than 5%.

_____ Remove R502 charge from system.
(Need 10–20 in. Hg vacuum [34–67 kPa, 0.34–0.67 bar] to remove charge.)

_____ Reinstall compressor (if removed).

_____ Replace filter drier with new drier approved for use with Suva HP62.
- Loose fill driers: use XH-7 or XH-9 desiccant or equivalent
- Solid core driers: check with drier manufacturer for recommendation

_____ Reconnect system and evacuate with vacuum pump. (Evacuate to full vacuum [29.9 in. Hg vacuum/0.14 kPa/0.0014 bar]).

_____ Leak check system. (Reevacuate system following leak check.)

_____ Charge system with Suva HP62.
- Initially charge 80% by weight of original equipment manufacturer specified R502 charge
- Amount of refrigerant charged: _____

_____ Start up equipment and adjust charge until desired operating conditions are achieved.
- If low in charge, add in increments of 2–3% of original R502 charge
- Amount of refrigerant charged: _____

Total Refrigerant Charged (add 9 and 10)   _____

_____ Label components and system for type of refrigerant (Suva HP62) and lubricant (polyol ester).

_____ Conversion is complete!!

*R502 charge should only be removed if compressor must be taken out of system to drain oil, such as for small hermetics.

Figure 8-15 A sample check sheet used during system retrofit. *(DuPont Fluoroproducts Corporation)*

tially charged with 95 to 100 percent by weight of the original R-500 charge.

Add the initial charge to the high-pressure side of the system until the system and cylinder pressure equilibrate. Position the refrigerant filling connections to the low-pressure side of the system, start up the compressor, and load the remainder of the refrigerant to the suction line of the system. Because liquid must be removed from the refrigerant cylinder, it is important to charge the refrigerant slowly into the suction line to allow it to flash before it enters the system in order to avoid damage to the compressor from liquid refrigerant entering the suction side of the compressor. A throttling device might also be used to cause the refrigerant to flash before entering the system.

8. Let the conditions stabilize. If the system is undercharged, add additional refrigerant in small amounts (while continuing to remove liquid from the charging cylinder) until the desired operating conditions are achieved.

When the system is lined out, compressor suction pressures for the blends should be within about 1 psi of normal system operation with CFC-12 for most medium-temperature applications. Compressor discharge pressures will typically be about 10 to 20 psi higher than normal system operation with CFC-12. For retrofits of R-500 equipment, compressor discharge pressures will be about 5 psi higher, and compressor suction pressures about 3 psi lower than with R-500.

The new blends are often more sensitive to charge size than CFC-12, therefore system performance will change more quickly if the system is overcharged (or undercharged) with the alternative blends. Review working with pressure-enthalpy diagrams in Chapter 7 to see how suction pressure, superheat, and subcooling all affect system performance.

Note: Sight glasses in the liquid line can be used for charging in most systems; however, it is recommended that the initial system charge be determined by measuring operating conditions (discharge and suction pressures, suction line temperature, compressor amps, superheat, etc.). If the sight glass is close to the exit of the condenser or there is very little subcooling prior to the sight glass, bubbles might still be observed in the sight glass when the system is properly charged. Attempting to charge until the sight glass is clear might result in overcharging of refrigerant.

Note: After retrofitting the system with an alternative refrigerant blend, always label the system components to identify the type of refrigerant and lubricant in the system, so that the proper refrigerant and lubricant will be used to service the equipment in the future.

Tips for retrofitting from CFC-12 (R-12) to HFC-134a

HFC134a has emerged as the industry's choice for an alternative refrigerant for CFC-12, commonly called R12. Unfortunately, HFC-134a is not a "drop-in" replacement for R12. There are significant differences between R12 and HFC-134a that must be considered when handling, processing, applying or retrofitting with HFC-134a.

Refrigerant properties. The pressure/temperature plot of HFC-134a vs. R12 is shown in Fig. 8-16. The cross-over point is approximately at 64°F. Above this temperature, the saturation pressure of HFC-134a is higher than that of R12; below, it is lower. The capacity curve, shown in Fig. 8-17, indicates relative capacity of HFC-134a vs. R12 for evaporating temperatures from 0°F to 50°F. This curve is based on 120°F condensing temperature. The cross-over point will move depending on the condensing temperature. The higher the condensing temperature, the higher the cross-over point.

Table 8-1 compares some properties of R12, HFC-134a, and HCFC-22 (R22). This data is taken at the standard refrigerant conditions of 5°F evaporating or 86°F condensing. For the same amount of subcooling, HFC-134a produces greater refrigerating effect.

Water solubility. Both liquid HFC-134a and R22 can absorb much more water than R12. This makes it less likely that a low-temperature system would exhibit capillary tube blockage due to ice buildup. It does not, however, eliminate the need for a filter-drier.

Extensive testing and research has shown that HFC-134a is compatible with all materials used in most hermetic compressors and condensing units. Consult with specific manufacturer's to ensure compatibility.

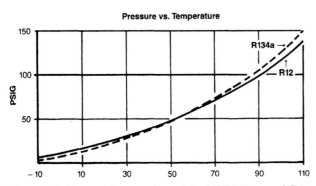

Figure 8-16 Pressure/temperature plot of HFC-134a and R12. *(Tecumseh Products Company)*

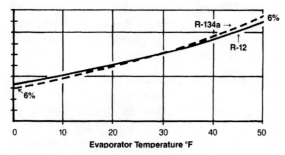

Figure 8-17 Relative capacity versus evaporator temperature for HFC-134a and R12. *(Tecumseh Products Company)*

Lubricants for HFC-134a. HFC-134a and mineral oils are not miscible. However, polyol ester (POE) oils and HFC-134a are miscible. The miscibility of polyol ester oils and HFC-134a is similar to that of current refrigerant oils and R22. Some types of POEs are fully miscible with HFC-134a (as are alkylbenzene and R22), while some POEs are partially miscible with HFC-134a (as are mineral oils and R22).

Moisture. Polyol ester oils are not as hygroscopic (able to absorb moisture) as the previously considered poly alkylene glycols (PAGs). They are, however, 100 times more hygroscopic than mineral oils.

Moisture can be very difficult to remove, even when heat and vacuum are used. For this reason, extreme care should be taken to prevent any moisture from entering the refrigeration system. Do not leave the compressor open to the atmosphere any longer than 15 minutes. The best method of assembly would be to remove the system components and plugs just before brazing begins. The maximum system moisture content after completing system processing should be 80 PPM. Running the system with the appropriate drier installed should reduce the system moisture down to 10 PPM or less. Always use the appropriate drier in the system when HFC-134a is used.

Compatibility. Even though polyol ester oils are compatible with mineral oils, they should not be indiscriminately mixed with mineral oils in

TABLE 8-1

	R12	HFC-134a	R22
Chemical formula	CCL_2F_2	CF_3CH_2F	$CHCLF_2$
Evaporating pressure (psig)	11.8	9.1	28.2
Condensing pressure (psig)	93.3	97	158.2
Sat. vapor density @ 5°F (lbs/cu. ft)	0.6859	0.5128	0.8042
Sat. liquid density @ 86°F (lbs/cu. ft)	80.7	74.3	73.3
Latent heat of vaporization @ 5°F (BTU/lb)	68.2	89.3	93.2

HFC-134a refrigerant systems. Mixing could result in the inability of the oil to return to the compressor and/or reduce heat transfer performance in the evaporator. However, small amounts (up to 1 percent) of mineral oil are acceptable in field retrofit situations.

System design. Compressors designed specifically for use with HFC-134a are designed with compatibility of oil, refrigerant, and materials in mind. These compressors are also designed to closely match the capacity of their corresponding R12 compressors at their individual rating point. In many cases, the same displacement is used. In some cases it is necessary to use the next larger displacement. Testing each compressor selection in its application is recommended to ensure compatibility, because system operating conditions can vary greatly.

HFC-134a typically has a greater refrigerating effect than R12, which reduces the required mass flow for a given capacity. As a result, the nonheat exchange capillary tubing might have to be changed. New tubing could either be more restrictive or less restrictive, depending on the application. A heat exchange capillary tube might not require any change at all. Be sure to test the system when selecting tubing.

Expansion valve manufacturers have designed products specifically for use with HFC-134a. Consult them for their recommendations. Be sure to use an appropriate drier with any HFC-134a system.

Recommended return gas/discharge temperatures. The theoretical discharge temperature for HFC-134a is slightly lower than that of R12 in similar conditions. Therefore, existing compressor guidelines for return gas and discharge temperatures for R12 can be used with HFC-134a compressors as well. It is best to keep the return gas cool without flooding to help limit compressor discharge and keep motor temperatures at acceptable levels.

Refrigerant charge. This depends on the system components. As a general guideline, based on limited application data, 5 percent to 30 percent less HFC-134a will be needed compared to R12.

Filter-driers. The polyol ester oils used with HFC-134a are prone to hydrolyze with moisture, resulting in the formation of acids. Therefore, an appropriate drier must be used in every application. Molecular sieve driers, presently compatible with R22, are recommended for use with HFC-134a as well. The XH-6 (bonded core), XH-7, and XH-9 types are recommended. The XH-6 (loose fill) type is not recommended due to high attrition rates.

If a solid-core drier made with bauxite is used, it would have the tendency to absorb both polyol ester oil and moisture. The ester could hydrolyze and form acidic materials. If the drier was then overloaded due to excessive moisture in the system, it could release the acidic materials back into the system. This would not be healthy for the compressor. For this reason, the use of solid-core driers made with bauxite for systems containing polyol ester oils is not recommended by many compressor manufacturers. Contact your drier supplier for specific drier recommendations.

System processing compatibility. R12 has more tolerance for system processing materials, such as drawing compounds, rust inhibitors, and cleaning compounds, than HFC-134a. These materials are not soluble in HFC-134a. If they were washed from system surfaces by polyol ester oils, they could freely accumulate at the capillary tube or expansion valve and plug it. For this reason, care should be taken to remove such processing materials from all system components.

Earlier investigated PAG oils were found to be totally incompatible with chlorinated materials. The current polyol ester oils, however, do not behave in this way. As with R12 systems, residual chlorinated materials must be considered as system contamination and eliminated from all internal surfaces of the refrigeration system.

Evacuation. Evacuation levels for HFC-134a systems are the same as the R12 requirements—a minimum of 200 microns at the system is pulled from both the low- and high-pressure sides of the system. If care is not taken to prevent moisture from entering the system before assembly, it could take longer to evacuate to within acceptable limits of system moisture and noncondensibles. A maximum of 2 percent noncondensibles and 80 PPM moisture is recommended. The completed system should have a moisture level of 10 PPM or less after running the appropriate drier.

Polyol ester oils vaporize much less than mineral oils, for the same level of heat and vacuum. Therefore, if oil vaporization was not a problem with the R12 system processing, it should not be a problem with the HFC-134a system processing. Consult your vacuum pump manufacturer to see if your existing equipment needs to be converted for use with HFC-134a polyol ester systems.

Leak testing. Many leak-detector manufacturers have HFC-134a detectors on the market, and more are currently being developed. Use only equipment approved for HFC-134a use by its manufacturer.

Caution: Do not attempt to use HFC-134a as a mixture with air to pressure test for leaks.

Refrigerant charging. Most R22 refrigerant charging equipment—charging boards, valves, and hoses—should also be compatible with HFC-134a. (This equipment is considered more aggressive with gaskets and plastics than R12.) The equipment would need to be recalibrated, however, before use. Once it is designated for HFC-134a use, it should only be used in HFC-134a applications. Converted R12 equipment should be clean of all residual R12. Pulling a deep vacuum (25 to 50 microns) and repeated flushing with HFC-134a should be sufficient. Consult the equipment manufacturer for specific recommendations for converting R12 equipment for use with HFC-134a.

HFC-134a can be charged in either the liquid or vapor state. If refrigerant charging is done in the liquid state, it should be done into the liquid line. Vapor charging can be done into the suction line while the compressor is running.

Caution: Always break the vacuum with refrigerant vapor before applying power to the compressor.

Retrofitting. Ideally, HFC-134a would be limited to use with new equipment only. This way all system components could be selected and tested by a system designer keeping all the necessary considerations regarding HFC-134a and polyol ester oils in mind. However, there are still in existence today millions of R12 systems, even though the R12 supply is rapidly declining due to increasing CFC restrictions.

Compressor replacement. When replacing an R12 compressor with an HFC-134a compressor, take special care not to leave the system or the HFC-134a compressor open to the atmosphere for more than 15 minutes. To replace the compressor, first recover the R12 refrigerant and any residual mineral oil left in the system using the proper recovery equipment. Next, refit the system with the correct capillary tube or expansion valve. Install a drier suitable for HFC-134a use that is the correct size for the system being retrofitted.

Install the HFC-134a-compatible compressor containing polyol ester oil. The compressor should have a white label stating that it is charged with polyol ester oil, and a light-blue label stating it uses only HFC-134a refrigerant. Be sure to check that the electrical components are correct; they might differ from those used with the original R12 compressor.

Next, thoroughly evacuate the system. Break the system vacuum with HFC-134a vapor. Charge the system with the correct amount of HFC-134a using the charging methods outlined earlier in this chapter. Generally the amount will be less than the R12 system requirements. Finally, check the system for proper operation.

Chapter

9

Understanding Refrigeration Electrical Systems

Almost half of all refrigeration system service calls involve electrical troubleshooting and repair. You will not succeed in your career without a clear understanding of how electricity works and the ability to perform the basic electrical tests and repairs. In addition to the electrical circuits used to power compressors, fan motors, lights, and control devices, many refrigeration systems are equipped with electronic control systems. These control systems use electronic temperature sensors and solenoid valves along with a microcomputer to provide precise, programmable system control. Troubleshooting these systems requires the technician to take accurate voltage and resistance readings using electronic diagnostic equipment.

Electricity

The movement-free electrons is called *current*. Electric current is the controlled flow of electrons from atom to atom within a conductor. To produce this flow of current, there must be an outside source of energy, or electromotive force, acting on the electrons in the conductor.

Measuring electricity

The flow of electricity through a wire or conductor is similar to the flow of water through a hose. A certain volume of water exits the hose in a given time period. We can measure this volume in gallons per minute. If the hose becomes kinked, the flow of water meets resistance and less water exits the hose nozzle. Water pressure at the faucet will also affect the volume of water exiting the hose. High water pressure results in high water volume exiting the hose with great force. A lower water pressure will result in a smaller volume of water exiting the hose during the same time period. The same principles of volume, resistance, and pressure can be applied to electricity.

Current. The unit for measuring the rate of current flow is the ampere or amp. An ammeter is used to measure current. There are two types of current.

Direct current. When electrons flow in only one direction, it is called *direct current*. Direct current can be stored in a battery or converted from alternating current using a power supply. It is the type of current often used in computer control circuits.

Alternating current. Alternating current is produced by huge generators at the power utility. Due to the operating characteristics of the generator, alternating current continuously changes its direction of flow. With every revolution of the generator armature, the ac current flows in one direction, stops, and flows in the opposite direction. This is known as a *cycle*. Alternating current used for residential and commercial applications is generated at a cycle frequency of 60 cycles per second, commonly stated as 60 hertz. Alternating current is used to power virtually all compressors and motors used in commercial and residential refrigeration systems.

Voltage. *Voltage* is the pressure or electromotive force (emf) that causes current flow. It is measured in volts. One volt is the amount of pressure needed to move one amp of current through a resistance of one ohm.

Resistance. All conductors offer some resistance to current flow. The basic unit of resistance is the *ohm*. An ohm is defined as the resistance that will allow one ampere to flow when the potential of emf equals one volt. Resistance is measured using a ohmmeter and is indicated by using the symbol Ω. Many factors contribute to the total resistance in a given circuit. Wire and terminal connections at relays, motors, and the like produce the biggest source of unwanted resistance in a circuit. The wire's length and gauge (thickness) also affect resistance.

Ohm's Law

In a given circuit, voltage (E), current (I), and resistance (R) have a very specific relationship that is stated in Ohm's Law:

$$E = I \times R$$

Ohm's Law states that the voltage in a circuit is equal to the current (in amps) multiplied by the resistance in ohms. So when any two factors are known, the third unknown factor can be calculated using simple mathematics. To solve for current when voltage and resistance are known, Ohm's Law is expressed as:

$$I = E/R$$ or current equals voltage divided by resistance.

When voltage and current are known, resistance is calculated by Ohm's Law in the following format:

$$R = E/I$$

or the resistance of a circuit in ohms equals the voltage divided by the current in amps.

Ohm's Law explains how an increase or decrease in voltage, current, or resistance will affect a circuit. Ohm's Law can be applied to an entire circuit or to any part of a circuit.

Power. The total power consumed in an electrical circuit is called wattage. Power is measured in watts. It is calculated by multiplying the voltage by the amperage.

$$P = E \times I$$

As with Ohm's Law the basic power formula can be adapted to calculate for the other two factors as follows:

$$E = P/I \text{ and } I = P/E$$

Single-phase and three-phase (poly) installations

Alternating current electrical power can be supplied from the utility company as either single-phase or three-phase power.

Single-phase ac. Single-phase power is used in virtually all residential and many light commercial installations. In a single-phase installation, ac power is delivered in a single continuous pattern or waveform that varies from +120 V to −120 V peak to peak during each cycle (Fig. 9-1a). This means that for 60 times each second, voltage delivered is zero.

A three-wire system is used in modern installation with two 120-V lines and a neutral line. Connecting a load, such as a single-phase fan or compressor motor between the neutral and one of the 120-V lines provides 120 Vac. Connecting the load between the two 120-V lines provides 240 Vac (Fig. 9-2).

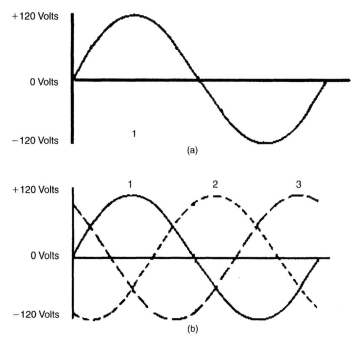

Figure 9-1 Graphic representation of (a) single-phase and (b) three-phase alternating power.

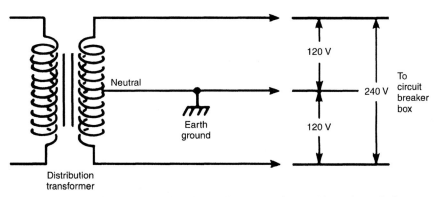

Figure 9-2 Single phase power of 120/240 volts supplied to residential and light commercial installations. *(Reprinted from* Practical Air Conditioning Equipment Repair *by Anthony J. Caristi, McGraw-Hill, copyright 1991.)*

Three-Phase ac. In heavy commercial and industrial installations three-phase alternating current is supplied (Fig. 9-3). This means the generator produces three separate ac waveforms and delivers them at equally spaced intervals. This means the line voltage never drops to zero at any point in the cycle (Fig. 9-1b). The end result is that

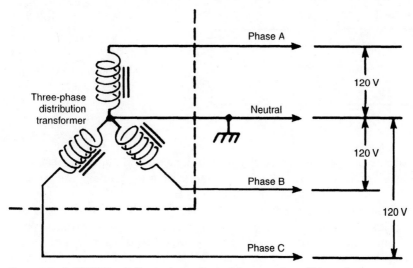

Figure 9-3 A 120/208-volt three-phase, four-wire power drop. Voltage between any two-phase wires is 208 volts. *(Reprinted from* Practical Air Conditioning Equipment Repair *by Anthony J. Caristi, McGraw-Hill, copyright 1991.)*

polyphase motors operate much more smoothly than single phase motors and load down the electrical line at a more uniform rate. They also have greatly increased starting torque.

All equipment powered by three-phase ac must be of the three terminal type specifically designed for polyphase ac. Additional information on polyphase electric motors can be found in Chapter 10.

Electrical circuits

An electrical circuit is simply a continuous path along which current flows. Current flows from the source of electromotive force (breaker panel or distribution box) through the conducting wires, through the components connected to the circuit such as the compressor, fan motors, or lights and back to the source (ground). There are three basic types of circuits found in refrigeration electrical systems:

- Series circuits
- Parallel circuits
- Series-parallel circuits

Series circuits. Figure 9-4 shows a circuit diagram for a simple series circuit that includes a power source, a switch, conducting wires, and a resistor (load). The circuit components of the circuit are represented by symbols, many of which are standard throughout the industry. These

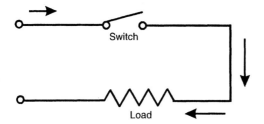

Figure 9-4 Simple-series circuit that provides a single path for current to flow.

circuit diagrams are commonly called *schematics*. More information will be given concerning electrical schematics and symbols later in this chapter.

In all series circuits, there is only one path for the current to flow. The series circuit shown in Fig. 9-5 has two resistors placed in the path of the current flow. All of the current supplied to the circuit must flow through each of these resistors before it can return to ground. In a series circuit, total circuit resistance is found by adding together the resistance values of the individual resistors in the circuit. Our sample circuit has one 3-Ω resistor and one 2-Ω resistor. So total resistance in this series circuit is:

$$3\,\Omega + 2\,\Omega = 5\,\Omega$$

Voltage is lost (or dropped) each time it pushes current through a resistor or load. To determine the voltage drop across any resistor or load, simply use Ohm's Law. Use only the amperage and resistance at that particular resistor or load.

$$\text{Voltage drop} = \text{Resistance (load)} \times \text{Amperage}$$
(at any one resistor or load)

In testing refrigeration electrical systems, the amount of voltage required to overcome the resistance in different parts of the circuit is measured using a voltmeter. The two voltmeter leads are attached on either side of (or "across") each of the resistors or loads. This is known as connecting the voltmeter in parallel with the load. The amount of volts used by each resistor or load is read on the voltmeter. As explained by Ohm's Law, the high-resistance loads require more voltage to push a given amount of current than do low-resistance loads.

Voltage drop is the actual voltage measurement across a resistor or load. When resistance is high, the voltage measurement and hence the voltage drop will be high. When resistance is low, the voltage required to push this current through (and the voltage drop) will be low.

The following points summarize the characteristics of a series circuit:

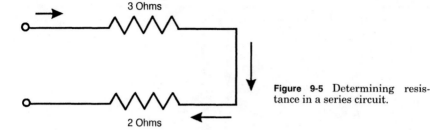

Figure 9-5 Determining resistance in a series circuit.

- Total resistance is calculated by adding all individual resistances or load in the circuit:

$$R = R1 + R2 + R3 \text{ etc.}$$

- Current (amps) passing through each resistor is the same.
- The voltage drop across each resistor (load) is different if the resistance values are different.
- The sum of the voltage drops equals the source voltage.

Parallel circuits. A parallel circuit provides two or more separate paths or branches for current flow. Each branch of a parallel circuit operates independently of the others and can have its own resistors and loads connected to it (Fig. 9-6). If an open condition occurs in one branch of the total circuit, the other branches will continue to operate.

In a parallel circuit, current can flow through more than one resistor at a time. Unlike a series circuit. total resistance in a parallel circuit is not additive. Instead it is found using the following formula:

$$\text{Total resistance} = \frac{1}{1/R1 + 1/R2 + 1/R3, \text{ etc.}}$$

So in the parallel circuit shown in Fig. 9-6:

$$\text{Total resistance} = \frac{1}{½ + ¼ + ¼}$$

$$= \frac{1}{2/4 + ¼ + ¼}$$

$$= \frac{1}{4/4 \text{ or } 1}$$

$$\text{Total resistance} = 1 \, \Omega$$

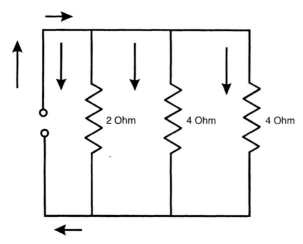

Figure 9-6 A parallel circuit providing several different paths for current flow.

In a parallel circuit, total resistance is always less than the smallest resistor in the circuit. When more resistors are added to the parallel circuit, total resistance decreases. The following points summarize the characteristics of a parallel circuit:

- Total resistance is always less than the lowest resistor.
- Current (amps) flowing through each resistor is different if the resistance values are different.
- Voltage drop across each resistor is the same. This also equals the source voltage.
- Total amperage in the parallel circuit is equal to the sum of amperages on the branch circuits.

Series-parallel circuits. A series-parallel circuit has both series and parallel paths or branches. Figure 9-7 illustrates four resistors connected in a series-parallel circuit. Total resistance in a series-parallel circuit is calculated by first adding together the resistance in the series portion of the overall circuit. The resistance in the parallel portion of the circuit is then calculated and added to the series resistance to determine total resistance in the entire circuit. So for the series-parallel circuit shown in Fig. 9-7:

$$\text{Series resistance} = R1 + R4$$

$$= 5 + 3$$

$$= 8 \text{ ohms}$$

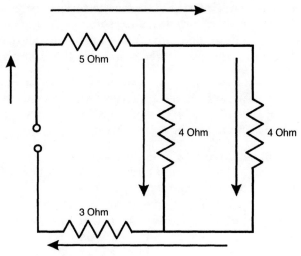

Figure 9-7 Series-parallel circuits are the most complex circuits used in electrical work.

$$\text{Parallel resistance} = \frac{1}{\frac{1}{4} + \frac{1}{4}}$$

$$= \frac{1}{\frac{1}{2}}$$

$$= 2 \text{ ohms}$$

$$\text{Series-parallel resistance} = 8\ \Omega + 2\ \Omega$$

$$= 10\ \Omega$$

Circuit conditions

There are four basic circuit conditions: closed, open, shorted, and grounded.

Closed circuits. A properly wired and operating circuit is closed. That is, it has a complete path from its power source through its intended loads and back to its source through the intended ground circuit. Problems exist in a circuit when it becomes open, shorted, or improperly grounded.

Open circuits. An open circuit occurs when there is no longer a complete path for the current to follow. The break might be caused by a corroded connector, a broken wire, a wire that burned open from too much current, or a faulty component. An open circuit acts like a circuit having a switch in the open position. Unfortunately for the service technician, locating this unwanted "switch" can be troublesome.

Voltage drop across an open circuit is always equal to the source voltage. An open circuit condition is also called a break in the circuit's continuity.

Short circuit. A short circuit occurs when the current bypasses part of the normal circuit. This bypassing can be caused by wires touching or solder melting and bridging conductors in a component.

Grounded circuit. A grounded circuit acts like a short circuit except that the current flows directly into a ground circuit that is not part of the original circuit. An unintentional ground circuit can be caused by a bare or broken wire rubbing against a pipe or other conductor. A grounded circuit can also be caused by deposits of oil, dirt, or moisture around connections or terminals, which provide a good path to ground.

High resistance. In addition to opens, shorts, and grounds, electrical circuits might experience unwanted resistance caused by loose, corroded, or damaged connectors. An increase in circuit resistance causes a decrease in current supplied to loads such as compressor motors, electric defrost heaters, etc. The voltage available to the load also decreases because there is now an additional voltage drop across the unwanted resistance. The net result is less available voltage for desired loads such as motors, lights, etc.

Electromagnetism

Electromagnetics is used in the operation of many actuators such as the solenoids used to control valve operation and relays used to control the flow of current to a given circuit. The principles of electromagnetism are also used in the design and operation of ac motors.

Magnets have the ability to attract substances known as magnetic materials, such as iron, steel, and nickel. Every magnet has two points of maximum attraction, or poles. These poles, named North and South, are located at the ends of the magnet. Opposite poles attract one another while similar poles repel one another.

Electricity and magnetism are related; one can be used to create the other. Passing current through a wire creates (or induces) a magnetic or flux field around that wire. This flux field is made of invisible lines along which magnetic force acts. All lines of force form a complete loop and never cross or touch one another (Fig. 9-8a).

The polarity of a current carrying wire's magnetic field reverses itself when the current flow changes direction. This important fact is used in the design and operation of ac electric motors.

Coils and flux density. Looping the wire into a coil also concentrates the lines of magnetic force inside the coil (Fig. 9-8b). The magnetic field generated equals the sum of all individual loops added together. Coiling wire is a simple, effective method of generating a strong magnetic field with small amounts of current and conducting wire (Fig. 9-8c).

Reluctance. The lines of magnetic force created by a magnet can only occupy a complete circuit or path. The resistance this circuit or path offers to the line of flux is called *reluctance*. Reluctance can be compared to electrical resistance. Every material has a certain reluctance value.

When a wire is looped into a coil, the air inside the coil offers a very high reluctance. This limits the strength of the magnetic field produced. However, when a bar of soft iron is placed inside the coil, reluctance is dramatically reduced and the strength of the magnetic field increases tremendously, sometimes by as much as a factor of 2500.

So when a coil of conducting wire is wound around a soft iron core it becomes a usable electromagnet. This is the principle used in the manufacture and operation of solenoids, relays, and many other actuators used in refrigeration and electrical systems. The magnetic attraction or repulsion of the electromagnet (solenoid) is used to perform work (Fig. 9-9).

Electrical System Components

The following sections describe the most common types of electrical devises used in modern refrigeration equipment electrical systems.

Wiring

Copper wire is an excellent conductor of electricity. It also bends easily, has good mechanical strength, has the ability to resist corrosion, and can be easily joined together. These characteristics make copper the metal for choice for wiring and other conducting components.

Warning: Your body is an excellent conductor of electricity. Always remember this when working on the electrical system of heavy-duty diesel. Always observe all electrical and storage battery handling safety rules.

Wire size. Wire size in the United States is defined by the American Wire Gauge (AWG) numbering system. The American Wire Gauge lists the largest wire, number 0000 (4/0), down to the smallest wire, number 50. In refrigeration work, wire sizes from number 20 to number 4/0 are the most common. Average sizes range from number 16 to number 4.

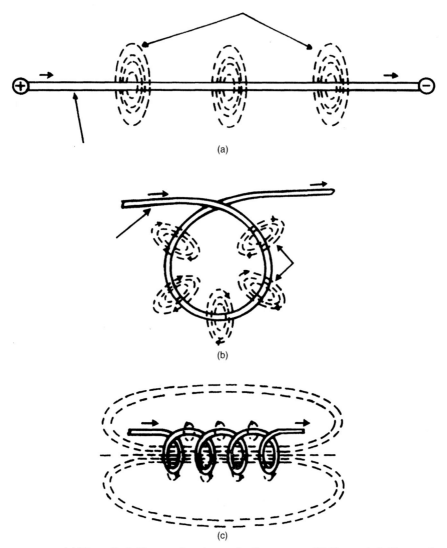

Figure 9-8 (a) Magnetic field around a wire conducting current; (b) Magnetic field around a looped wire; (c) Magnetic field around a coiled wire.

When wire larger than 4/0 is required, the circular mil sizing system is used. Most circular mil wire ranges in size from 250 MCM (about ½ inch in diameter) to 750 MCM (about 1 inch in diameter). MCM stands for 1000 circular mils.

Insulation. Insulators are materials that have very high resistance to current flow. Wire insulation is available in different grades for differ-

Understanding Refrigeration Electrical Systems 251

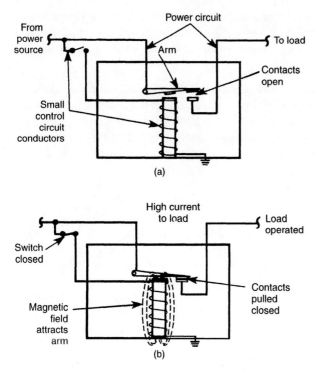

Figure 9-9 Construction of a simple solenoid-activated relay.

ent purposes. For example, depending on its location, wire insulation might need to be heat resistant, moisture resistant, heat/moisture resistant, or oil resistant.

Thermoplastics are the most commonly used insulating materials found in refrigeration electrical systems. Cracked, frayed, or otherwise damaged insulation can cause problems in the electrical system such as shorts. It can also be dangerous if inadvertently touched or handled.

Sizing circuit wiring. When sizing circuit conductors, you must consider the voltage drop plus the type of insulation, enclosures, and safety devices to be used. Voltage drop is determined largely by the length of the wire run. When performing installation work, it is your responsibility to calculate it. The insulation type, the enclosure, and safety procedures can be determined using the National Electrical Code as a guide.

Voltage drop in a conductor largely determines the size of the wire needed in a given installation. For most refrigeration installations, wiring runs should not exceed 100 feet. Any voltage drop between the supply and the equipment is lost to the equipment. If the voltage drop is large enough, it will seriously affect equipment operation. Even a slight drop, however, could cause damage. The standard acceptable voltage drop limit is 5 percent.

National Electrical Code. Always refer to the most up-to-date National Electrical Code Handbook for information on sizing electric conductors. Wire size values and voltage drops are listed in table format. The voltage drop can be ignored on short-distance circuits. There is a correction factor in NEC tables in cases where the temperature is greater than 30°C.

Safety devices

Motors and other larger loads are designed to operate at a maximum current draw. If this maximum current rating is exceeded, due to a short or overload, excess heat generated by the high current can melt insulation and surrounding materials, burning up a compressor or fan motor and creating a fire hazard. To provide protection against current flow higher than the circuit wiring or motors are designed to handle, protective devices, such as circuit breakers, overload devices, and fuses are designed into every circuit. These devices can be used alone or in combination with one another.

Fuses. Fuses are the simplest of all overload devices. They are best used to detect large overloads, not small ones. If current through the fuse exceeds the fuse's amperage rating, the metal element melts, breaking the circuit's continuity and protecting wiring and components located further along in the circuit. Fuses can be mounted in a fuse panel or block or they can be mounted in-line.

Overload devices. An overload device can also be used to protect the refrigeration system against either large or small overloads. The *thermal* overload is triggered by heat, while a *magnetic* overload uses the induced magnetic field of the current draw to mechanically open the circuit.

Thermal overloads are often used as a *pilot-duty* safety device. When the overload trips, it breaks the control circuit, which in turn shuts down the motor or load. Pilot duty thermal overloads are typically found on motors rated at two horsepower or larger. A thermal overload also might be installed directly in the main power circuit that drives the motor or load.

Circuit breakers. A circuit breaker is simply a set of electrical contacts connected by a strip or arm made of two different types of metal. If excessive current passes through this bimetal arm, it heats up and the

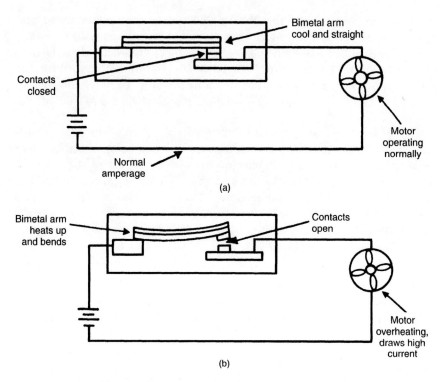

Figure 9-10 (a) A bimetal switch mechanism is used in circuit breakers and thermal overload switches. (b) The heat from excessive current opens the contacts via thermal expansion.

two different metals begin to expand. One metal expands faster than the other, causing the arm to bend, opening the contacts and tripping the breaker switch (Fig. 9-10). This breaks the circuit continuity. The breaker must then be manually reset before power is restored to the circuit.

Circuit breaker panels have replaced fuse panels for the past twenty-five years, so you will find them in virtually all commercial installations and all but the oldest residential applications. A breaker has one, two, or three poles. The poles indicate the number of hot legs being fed from the breaker to the appliance. A one-pole breaker supplies one hot leg and comprises a 110-V, single-phase circuit. A two-pole breaker supplies two hot legs, and provides a 230-V, single-phase circuit. And a three-pole breaker supplies three hot legs, and provides a three-phase circuit at the supplied voltage.

Like fuses, circuit breakers are rated in amperes. Circuit breaker size ranges between 1 and 100 amps. Breaker panels are rated based on the number of panels the main lugs can carry and according to the rating of the main breaker. Breaker panels are used with single-phase or three-phase systems, for 250 or 600 volts. Most commercial and industrial panels use breakers that attach to main lugs with screws.

Breakers connect in some cases to the main lugs with clips. Others are connected to the main lugs with screws. When installing the screw-type breaker, be sure that the connection is tight. Before beginning any installation work on a breaker panels, be sure to turn off the power.

Circuit breakers are generally very reliable. Two types of problems can occur. Either the breaker cannot be reset or it opens at a lower amperage than its rating. To check a breaker for resetting, take a voltage reading between the ground and the breaker. If voltage is present on the load side of a breaker under load, the breaker is good. If a breaker is stuck in the closed position, it must be replaced. A breaker that trips at lower-than-rated amperage should be checked by reading the actual amps of the circuit.

Distribution centers

A distribution center distributes the electrical supply to several areas within a large system. Distribution centers are typically used in commercial/industrial applications.

Disconnect switches

All refrigeration equipment should be equipped with a disconnect switch that can be used to cut the power supply to the unit. The disconnect switch should be housed in a protective enclosure near the equipment.

Some disconnect units have built-in circuit breakers or fuse blocks, so they can serve as a safety device as well. If a switch is needed only to break off the power supply, then a nonfusible disconnect switch should be used.

Loads

A *load* refers to any electrical device that uses electricity in order to perform work. A load could be a simple device that draws little current, such as a light bulb or a solenoid, or it could be a large motor that draws over 50 amps. Loads are important in the refrigeration system because they perform the work. They operate compressors, fans, and the solenoid part of a relay, which starts and stops loads or opens and closes valves. The following are several types of loads commonly used in a refrigeration system.

Motors. A motor uses electric energy to rotate a component within an electric system. In refrigeration systems, a motor is used to rotate many different components, including the compressor, the condenser, and the evaporator fans. For this reason, the motor itself constitutes the largest, most important load in a refrigeration system. Additional information on motors is found in Chapter 10.

Solenoids. As explained earlier in the chapter, a solenoid creates a magnetic field when it is energized. The force of the magnetic field is used to create mechanical movement in a refrigerant line valve, control circuit relay, or other device.

Signal light. A signal light is a light that illuminates to indicate the existence of a specific condition within a system. A signal light can be designed to show that a piece of equipment is operating. They can also be used as alarm indicators or warning lights.

Resistors

Resistors are electrical components that provide a predetermined amount of electrical resistance in a circuit. They limit current flow, and thereby voltage, in circuits where high current and full voltage are not needed. Resistors can have a fixed, stepped, or variable resistance value. They are considered loads in a circuit.

Switches

A switch provides a means of opening or closing electrical contacts within the circuit so the current flow can be safely controlled. Switches can be used to turn current on or off or direct it to a specific branch of the circuit. Switches can be either operator controlled, or they can be self-activating. Self-activating switches can be controlled by temperature, pressure, humidity, or flow.

Switches are named using the terms *pole* and *throw*. Simply remember that the pole refers to the number of input circuits connected to the switch, while the throw refers to the number of output circuits controlled by the switch. So if a switch has one conducting wire running into it and a single wire running out of it, it is called a single pole, single-throw (SPST) switch. If the switch has one wire running into it and two circuits running out of it, it is called a single pole, double throw (SPDT) switch. A double-pole, double-throw (DPDT) switch has two input circuits running to it and controls current flow to two output circuits. Figure 9-18 (on page 265) illustrates the electrical symbols of several popular switch configurations.

Bimetal switches. These are temperature-sensitive switches that are most commonly used in thermostat controls or thermal overload protection devices (Fig. 9-11).

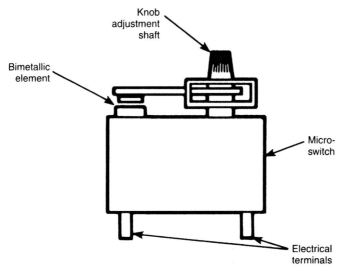

Figure 9-11 Bimetal switch used in a thermostat control. *(Adapted from* Practical Air Conditioning Equipment Repair *by Anthony J. Caristi, McGraw-Hill, copyright 1991.)*

Relays and contactors

A relay (Fig. 9-12) is an electric switch that allows a small amount of current to control the flow of a much larger amount of current. A relay relies on the principles of electromagnetism for its operation. When a small amount of current is passed through the relay's control circuit, it turns a piece of soft iron in the relay housing into a electromagnet. This magnet attracts or pulls an armature into position, closing the contacts

Figure 9-12 Typical compressor contactor used in refrigeration electrical systems. *(Watsco Components Inc.)*

inside the relay and completing the relay's power circuit. This power circuit allows a much greater amount of current to flow to loads downstream of the relay. When the small amount of current is switched off in the control circuit, the electromagnet loses its magnetic force, the armature returns to its open position, and current is cut off in the power circuit of the relay.

The main difference between a relay and a contactor is size. A contactor is simply a large relay. A contactor is typically rated to carry 20 amperes or more. Relays are typically rated at less than 20 amps.

Magnetic starters

A magnetic starter has an ampere rating that is similar to the contactor rating, but the magnetic starter has an integral overload protection device that the contactor does not have. This overload device protects the motor from high current draws.

Capacitors

A capacitor is a device that stores a voltage potential releasing it when needed (Fig. 9-13). While it does not actually create voltage, a capacitor is a source of the voltage it stores.

Capacitors are used to increase the starting torque or running efficiency of a single-phase motor. They are used to provide a phase shift of the current so that the start winding of a single-phase induction motor can be excited by an out-of-phase current. This produces an offset magnetic field, which provides the starting (and running) torque for compressors as well as blower motors.

A capacitor has two aluminum plates with an insulator between them. The insulator stores electrons, preventing electron flow from one plate to another. Two types of capacitors are used in refrigeration units: the electrolytic, or starting capacitor, and the oil-filled, or running capacitor.

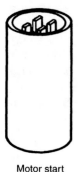
Motor start

Motor run

Figure 9-13 Typical motor-start and motor-run capacitors.

Starting capacitors. The starting capacitor has two aluminum electrodes or plates. Paper treated with a nonconductive electrolyte is sandwiched between the plates. Start capacitors are typically rated from 75 to 600 microfarads and from 110 to 330 volts. A start capacitor is housed in a compact housing with a dielectric, which is a nonconductor of electric current. It is used only briefly on each cycle of the motor.

Running capacitors. The running capacitor is built to remain in the motor circuit for its entire cycle of operation. Heat is dissipated by means of oil found in the capacitor case.

Single-phase compressors always employ running capacitors that are connected between the "run" and "start" windings of the motor. Many blower motors also are designed as permanent split capacitor (PSC) types, which are more efficient than the cheaper, split-phase designs. These capacitors are nonpolarized and are usually rated at 4 microfarads capacitance or more. In general, the larger fan motors and compressors will have larger values of run capacitors.

Troubleshooting. A malfunctioning or short-lived capacitor could be the result of a number of conditions. High voltage could cause the capacitor to overheat, damaging plates and shorting electrodes. If the starting components are faulty, the capacitor could remain in the line circuit too long, which would result in damage. High temperatures caused by poor ventilation, starting cycles that are too long, or too frequent starting cycles can also cause problems. For more on the use and troubleshooting of compressor motor start and run capacitors, see Chapter 10.

Hard-start circuits

Hard-start circuits are used in some refrigeration systems to help avoid the problem of a locked rotor condition when the compressor is called upon to restart without total pressure equalization. A locked condition can occur if the thermostat is manipulated or if there is a momentary loss of line power. Hard-start circuits are usually limited to single-phase systems, because three-phase compressors have sufficient starting torque to overcome a locked rotor.

The additional starting torque is provided by a start capacitor that is temporarily connected across the run capacitor during the start sequence. Once the compressor has reached speed, the start capacitor is switched out of the circuit, usually by means of a set of relay contacts. A current, potential, or thermally activated relay can be used for cutting out the start capacitor (Fig. 9-14).

Current-type start relays. The start current of a compressor can be 10 times greater than its run current. Current relays sense the starting current drawn by the compressor and open a set of contacts when the

Understanding Refrigeration Electrical Systems 259

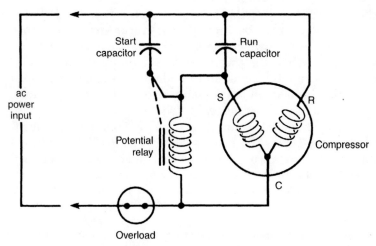

Figure 9-14 Schematic of a hard-start circuit controlled by a potential relay. *(Reprinted from* Practical Air Conditioning Equipment Repair *by Anthony J. Caristi, McGraw-Hill, copyright 1991.)*

start sequence is completed. The coil resistance of a current relay is extremely low, much less than 0.1 Ω. When the circuit is idle, the contacts of the current relay are open and the start capacitor is not connected to the circuit. When power is applied between the run and common leads of the compressor, the high current demanded by the compressor is fed through the coil of the relay, causing its contacts to close. This connects the start capacitor to the circuit.

When the compressor reaches operating speed, its current draw is reduced. This forces the relay contacts to open and disconnect the start capacitor from the circuit.

Potential-type start relays. Potential starting relays differ from current relays in that the contacts are closed prior to applying power to the compressor. Closed contacts eliminate the arcing that can take place when power is applied to the normally open contacts of a current relay.

The coil of the potential relay consists of many turns of fine wire that produces a relatively high resistance. It responds to voltage and is connected across the start winding of the compressor. When power is applied, the closed contacts of the potential relay connect the start capacitor into the circuit. As the motor builds up speed, the voltage across the start winding increases rapidly. This causes the relay circuits to open, which disconnects the start capacitor from the circuit.

Thermal-type start relays. Thermal relays do not have electromagnetic coils; they use the current drawn by the compressor to heat a bimetallic strip or resistance wire. Because compressor current must flow through the "coil" of the relay, its resistance must be extremely low as it is in a current relay. At the beginning of the start sequence, the thermal relay contacts are closed, connecting the start capacitor into the compressor circuit. When the thermal relay heats up due to a small amount of power being dissipated in the relay coil, the contacts open. The thermal relay connection to the compressor and start capacitor is very similar to that of the current relay.

Alternate methods include connecting a positive temperature coefficient (PTC) resistor across the run capacitor (Fig. 9-15). This component exhibits a low resistance value when cool and quickly increases in resistance as it heats up during compressor operation. These circuits operate without the need of relay contact or moving parts and can be more reliable.

Troubleshooting hard-start relays. Check current, potential, and thermal relays by examining the contacts for pitting and the mechanical parts for excessive wear. Contact resistance should also be checked with an ohmmeter set to its lowest scale. Normal readings should be 10 milliohms (0.01 ohms) or less. The coil resistance of the relay should also be measured with the ohmmeter. Normal readings will be virtually zero ohms for current and thermal relays, and several hundred ohms for potential relay coils.

The best method of checking hard-start relays is under true working conditions where the component is subject to the voltage and current it experiences during everyday operation. To simulate real working conditions, connect an ammeter in series with the starting capacitor so you can monitor its current during the start sequence. An analog clamp-on

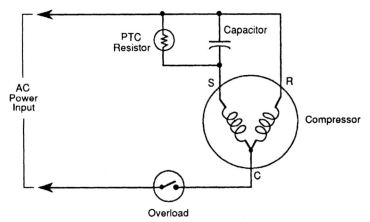

Figure 9-15 Compressor wiring schematic with PTC hard-start circuit. *(Reprinted from* Practical Air Conditioning Equipment Repair *by Anthony J. Caristi, McGraw-Hill, copyright 1991.)*

ammeter is ideal because it allows the test to be made without breaking the circuit.

Set the ammeter to a range comparable to the expected starting current of the circuit, which is usually less than 10 amperes. When power is applied to the refrigeration unit, the ammeter should immediately indicate the surge of motor starting current, then fall to zero as the compressor starts. This sequence will occur very fast, but it will provide verification that the hard-start relay or solid-state circuit is doing its job.

Solid-state controls

Solid-state controls are becoming more and more commonplace in refrigeration. Solid-state controls might employ thermistors, which are variable resistors that change in value with temperature. Other, more sophisticated controls might employ thyristors (triacs or SCRs) or other solid-state components to control the starting circuit of the compressor. More elaborate refrigeration systems use microprocessor controls, which have logic circuits that automatically take into account many operating factors.

The advantage of solid-state controls is the elimination of contacts and moving parts, thus eliminating arcing and wear-out mechanisms found in electromechanical components. However, solid-state devices are subject to catastrophic failure, especially when subjected to excessive current or voltage due to a line voltage transient or failure of some component in the system. As a service technician, you will probably replace any solid-state module that does not perform properly; field repair is generally not feasible.

Many manufacturers now employ solid-state time-delay controls in both large and small systems, which automatically prevent compressor operation until a predetermined time elapses after the compressor shuts down.

Wiring schematics and service manuals

Electrical wiring diagrams, commonly called schematics, contain essential information about a system's electrical layout and operation (Fig. 9-16). A schematic diagram details how the electrical components in a refrigeration system are wired to one another and operate (Fig. 9-17). This makes a schematic an excellent troubleshooting tool.

As a refrigeration technician, you must often rely on wiring diagrams to correctly install, troubleshoot, and repair both large and small refrigeration systems and their accessories. Schematic diagrams combine lines and symbols to trace out all circuits and show the location of all components including motors and loads, switches, relays, coils, resistors, fuses, and lamps.

262 Chapter Nine

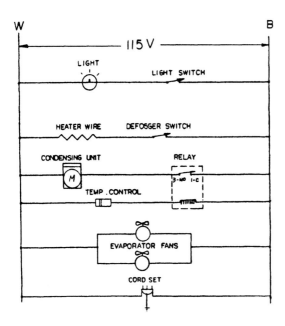

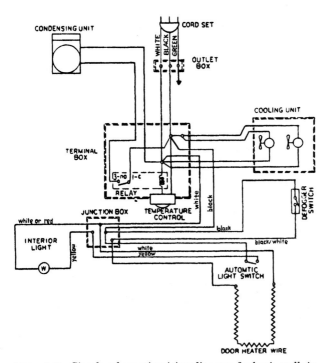

Figure 9-16 Simple schematic wiring diagram for basic walk-in cooler electrical circuit. *(Traulsen & Co., Inc.)*

Understanding Refrigeration Electrical Systems 263

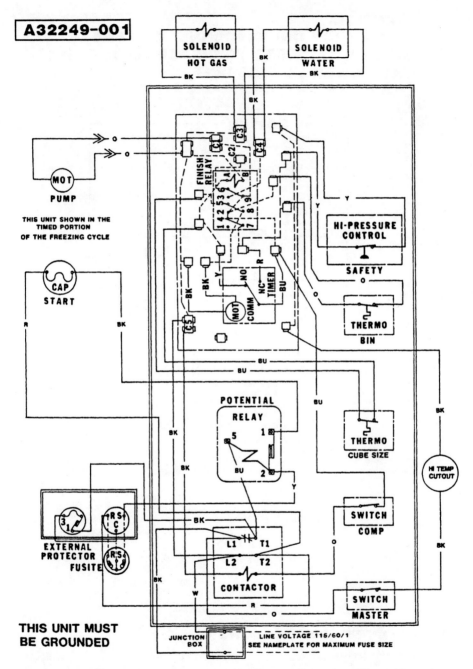

Figure 9-17 More involved schematic wiring diagram for an ice maker with a water-cooled condenser. *(Courtesy of Scotsman®, Vernon Hills, IL)*

**CAUTION:
MORE THAN ONE DISCONNECT MEANS MAY BE REQUIRED TO DISCONNECT ALL POWER TO THIS UNIT.**

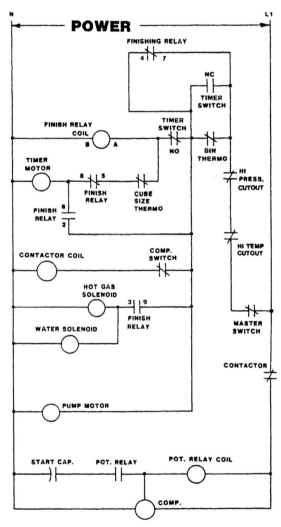

USE COPPER CONDUCTORS ONLY

Figure 9.17 (*Continued*)

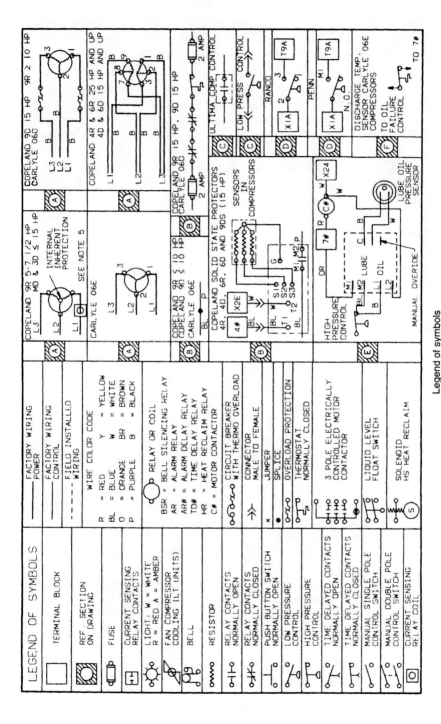

Figure 9-18 Typical electrical symbols used in refrigeration electrical work. (*Hussmann Corporation*)

The refrigeration industry uses a somewhat standard set of symbols to visually represent different system components (Fig. 9-18). Select manufacturers might slightly vary some symbols, so always double-check the symbol key provided in the equipment's service manual.

The increased use of more complex electronic/electric control systems and more sophisticated safety systems makes it difficult for the service technician to read schematic diagrams. Most reach-in coolers or walk-in units have relatively simple schematics with a few components. As more and more components are added to the control system of a commercial or industrial unit, its schematic diagram becomes increasingly complex. A complex diagram might contain circuits for defrost, timers, and alarm circuits. Regardless of a schematic's complexity, the basic layout or arrangement of a schematic diagram remains the same.

Familiarize yourself with the system's schematic before attempting to troubleshoot the unit. Break down large schematics into individual circuits to make them easier to understand. Remember the rules that govern current and resistance in series, parallel, and series-parallel circuits outlined earlier in this chapter.

Service manuals provided by refrigeration equipment manufacturers are a valuable source of troubleshooting and general information. They are needed to obtain desired specifications on refrigerant pressure and temperature, piping specifications, and other valuable installation and start-up procedures. Manufacturer's service manuals also provide drawings and photographs that show where and how to perform many major service and preventive maintenance procedures. The quality of the service literature and schematic drawings provided varies greatly between equipment manufacturers. However, you should always check out the service literature available for the units you are working on. A little time spent reading or studying a wiring layout can save hours of troubleshooting time. Learn to effectively use the information available to you.

Special tools or instruments are also listed in the manuals, and they show when they are required. Precautions are also given to prevent injury or damage to parts.

Although the manuals from different publishers vary in presentation and arrangement of topics, all service manuals are easy to use after you become familiar with their organization.

Chapter 10

Electric Motors Used in Refrigeration

Electric motors are used to create a rotating motion and drive components that need to be turned with a rotating motion. Refrigeration system motors are either open-design motors or sealed (hermetic) motors. Open motors are used to drive fans, pumps, timers, and other accessories. They can also be used to drive a compressor either via a belt, a mechanical coupling, or directly off the motor shaft. Sealed or hermetic motors are only used to directly drive compressors that are housed inside a sealed protective dome.

Motors used throughout the industry vary in running and starting characteristics. For example, compressors require a motor with a high starting torque and a good running efficiency. Small propeller fans use motors with a low starting torque and average running efficiency. Matching the correct motor with the job at hand is extremely important for system efficiency.

Small fractional horsepower motors are normally designed so that the flow of air generated by the fan blade also cools the motor. Such motors are referred to as *open* or *air over* motors. Other motor designs are completely enclosed to keep dirt and dust from entering the unit. Such motors are called totally enclosed nonventilated (TENV) motors.

Induction Motor Design

Refrigeration system motors are induction motors (Fig. 10-1). These motors are either single-phase or three-phase motors. An induction motor uses electricity to induce two separate magnetic fields into its two major components: the stator and the rotor. By using the principle that like poles of magnets repel and unlike poles attract, the magnetic fields generate the force needed to turn the motor shaft.

The stator is actually the motor frame. It is normally cylindrical in shape. The field poles with field windings on them are part of the stator.

The rotor (also called the armature) is mounted on the drive shaft. This shaft has two journal bearings, one at each end. End bells or plates that hold the bearings attach to the motor frame. The rotor shaft journals mount in the bearings and the bells support the rotor.

In most motors, the stator is the primary magnetic field. The rotor produces a movable secondary magnetic field (Fig. 10-2). There are no windings on the rotor. Instead, thin copper bars are inlaid into the surface of the rotor. These bars run parallel to the motor drive shaft. The stator has one or more field windings of copper wire. When alternating current is passed through the stator windings, it generates a magnetic field that changes polarity 60 times per second.

Current in the stator windings induces magnetism into the copper bands of the rotor. The magnetic poles of the induced field in the rotor do not change polarity. This means the magnetic fields of the stator and rotor repulse each other one-half of the time. This force causes the rotor to rotate on the drive shaft as it is continuously repelled by the ever changing polarities in the field winding of the stator.

Motor speed. The rpm of an induction motor is determined by the number of poles of the motor and its line frequency (see Table 10-1). Actual rpm of an induction motor is affected only slightly by load. The

(a) Split phase

(b) PSC

Figure 10-1 Typical electric-induction-type motor used in refrigeration systems (a) split-phase and (b) PSC motors.

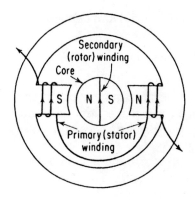

Figure 10-2 Simplified layout of an induction motor.

actual rpm is slightly less than the motor's synchronous speed due to a normal operating condition, called *slip*. Synchronous speed is the rpm of the magnetic field generated by the stator winding. It is equal to 3600, 1800, and 1200 rpm in 2, 4, and 6 pole Hertz motors respectively.

Caution: A motor designed for a 60-Hz line should not be used on a 50-Hz line unless specified by the manufacturer.

Start winding. A single-phase induction motor by itself cannot provide starting, because the single phase of current that passes through the starter coil cannot produce the torque needed to start rotation. When a second winding (or start winding) is added to the stator, however, a magnetic field is created that offsets the field created by the main winding. The two windings, each with their own magnetic field at an angle to each other, create torque within the rotor. This causes the shaft to rotate in the right direction, and it soon reaches full speed. Some motors have a centrifugal switch that disconnects the starting circuit when the motor reaches its rated rpm.

Start relay. Hermetic motors often employ a start relay located outside of the sealed dome. This relay temporarily connects the start winding of the motor to the power circuit. When the motor reaches about 75 percent of its running speed, the start relay cuts power to the start

TABLE 10-1

Cycles	Speed Type	Two Pole	Four Pole
60 Hertz	Synchronous	3600 rpm	1800 rpm
	Operating	3425 rpm	1725 rpm
50 Hertz	Synchronous	3000 rpm	1500 rpm
	Operating	2825 rpm	1425 rpm
25 Hertz	Synchronous	1500 rpm	750 rpm
	Operating	1425 rpm	690 rpm

winding. See Chapter 9 for more details on potential and current-type start relays.

Motor Strength

Motors are generally classified by starting methods and/or power rating. Five general types of motors are used in refrigeration systems: shaded-pole, split-phase, permanent split-capacitor, capacitor-start-capacitor-run, capacitor-start, and three-phase.

The starting torques of the five general types of induction motors can be expressed as a percentage of their running torques. They are:

- Shaded-pole motors—100 percent
- Split-phase motors—200 percent
- Permanent split-capacitor motors—200 percent
- Capacitor-start-capacitor-run, motors—300 percent
- Three-phase motors—600 percent

For example, a split-phase motor must generate twice its normal running torque to get itself running. A three-phase motor must generate six times its normal running torque to overcome its inertia and begin rotating.

Shaded-Pole Motors

A shaded-pole motor is a single-phase induction motor that uses a starting winding to generate the starting torque needed to begin rotation. Its starting winding is normally a closed turn of heavy copper wire that is banded around a section of the motors stator poles.

Shaded-pole motors are ideal when very small starting and running torques ($\frac{1}{5}$ to $\frac{1}{100}$ hp) are needed, such as in a small condensing unit or evaporator fan. Because of its low running torque, a shaded pole motor stalls easily. But the small locked rotor current draw of the motor normally ensures the windings will not burn out.

Operation

As shown in Fig. 10-3, each pole of a shaded-pole motor has a small groove routed into the stator and banded by a solid copper band. When the motor is starting, a current is induced into the shaded pole from the main windings. The shaded poles generate a magnetic field that is out of phase with the magnetic field of the main winding. This produces a rotating magnetic field that is strong enough to supply the needed starting torque. When the motor nears its full speed, the effect of the shaded pole becomes negligible. The shade-pole motor rotates from the unshaded pole to the shaded pole.

Service tips

A single-speed, shaded-pole motor is easy to troubleshoot due to its simple winding patterns. Multispeed shaded-pole motors use additional speed windings that can make them slightly tougher to troubleshoot.

Check the condition of the windings with an ohmmeter as described on page 280. If a shaded-pole motor stalls and its windings are not faulty, the problem is probably mechanical, such as bad bearings or poor lubrication.

Split-Phase Motors

The split-phase motor is another type of single-phase electric motor. These economical fractional-horsepower motors are used when high starting torque is not required as in belt-driven evaporator fans, small hermetic compressors, and exhaust fans.

In a split-phase motor, rotation is started by splitting the phase to produce two-phase current. The single-phase current is split between the run and the start windings. This places one of the windings 45 to 90 degrees out of phase. This out-of-phase condition generates the needed starting torque. The start windings remain energized until the motor has accelerated to approximately 75 percent of its normal operating speed. At this point a centrifugal switch or start relay opens to drop the starting windings out of the circuit. The motor continues to operate at full speed solely on the running winding.

A split-phase motor can run on 110 volts, single-phase, 60 hertz or 208/230 volts, single-phase, 60 hertz. Dual voltage split-phase motors

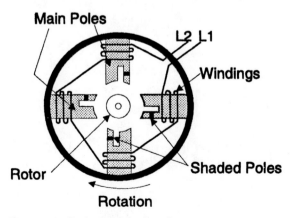

Figure 10-3 Design of a shade-pole motor.

can operate on either voltage by making simple changes in their wiring. Split-phase motor rotation is reversible by simply switching the leads of the starting winding at the terminals in the motor.

Operation

To split the phases in a split-phase motor, the starting winding is designed with smaller wire and more turns than the running winding (Fig. 10-4). When power is applied to the motor, it flows through both the start winding and run winding. Both windings are inductive and cause the currents flowing through them to lag behind the current of the power mains. But because the start winding has higher resistance than the running winding, its current will lag behind less than the current of the run winding. This lag or displacement creates the phase difference that produces the needed rotating magnetic field.

Some split-phase motor designs allow current to reach the run winding first by designing an increased resistance into the starting winding and a decreased induction into the running winding. In either case, the motor operates on the same split-phase principle.

Service tips

Split-phase motors are highly reliable and are used in many refrigeration applications having single-phase equipment. The split-phase motor is simple to troubleshoot once you understand how it operates. Problems normally occur with the bearings, the windings, and the centrifugal switch.

Normal wear and poor maintenance can lead to trouble with the bearings of any type of motor. A motor with bad bearings will run noisy, and have trouble turning. In some cases, the motor might freeze completely. Rotate the motor by hand and note rough spots or points

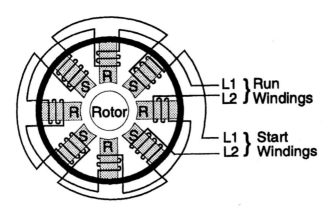

Figure 10-4 Design of a split-phase motor.

where the motor tends to lock up. Service the bearings as outlined on page 284.

Problems in the motor windings include shorts, opens, or grounds. An ohmmeter can easily diagnose any of these conditions. See page 280 for details.

Problems with the centrifugal switch are difficult to pinpoint because the switch is active in the circuit for only a short time. A worn centrifugal switch will often stick in an open or closed position. Listen for the click of the switch when it drops in after the motor is cut off.

Permanent Split-Capacitor Motors

A permanent split-capacitor (PSC) motor offers moderate starting torque and good running efficiency, facts that made them popular choices for both compressor and fan motors. A PSC motor can be used to drive the compressor in systems where high- and low-side refrigerant pressure equalize during the off cycle. It does not generate sufficient starting torque to overcome high head pressures. The aftermarket typically supplies the split capacitor motor as a replacement when needed.

Operation

The run and start winding of a PSC motor are wired in parallel. The start winding and run capacitor of the PSC motor are wired in series. The capacitor causes the start winding to shift out of phase with the run winding. This results in motor rotation.

Multi-speed PSC motors have additional run windings connected in series. For example, a two speed motor will have high- and low-speed run windings. The start winding is in series with the run capacitor and in parallel with the run windings. For high-speed operation, only the start and high-speed windings are energized. For low-speed rotation, the start, high-speed, and low-speed winding are energized (Fig. 10-5).

Service tips

The PSC motor is generally highly reliable. Problems might occur with the windings, bearings, or run capacitor. Troubleshoot winding problems with an ohmmeter as describe on page 280. Keep in mind that winding can be added to provide additional speeds.

A faulty run capacitor might keep the motor from starting or might cause it to draw high current while running. PSC motors are normally changed when they are bad. Because most motors of this type are designed to operate at multiple speeds, the motor must be connected correctly. Follow the wiring diagram and directions supplied with the replacement unit. Miswiring can result in permanent damage to the motor.

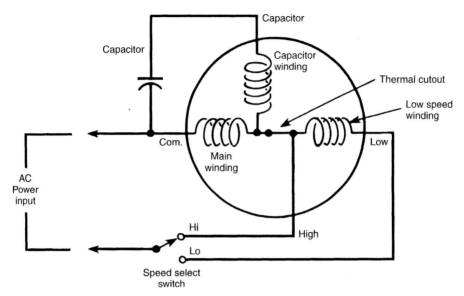

Figure 10-5 Wiring diagram for PSC induction motor. *(Reprinted from* Practical Air Conditioning Equipment Repair *by Anthony J. Caristi, McGraw-Hill, copyright 1991.)*

Capacitor-Start Motors

Capacitor-start motors are used to drive large fans and small hermetic compressors that must overcome high-side pressure during start-up. A capacitor-start motor operates like a split-phase motor except that it has a capacitor wired in series with the start relay or centrifugal switch and the start winding (Fig. 10-6). The capacitor increases the phase difference between the start and run winding increasing the start-up torque generated. The switch opens to stop current to the start capacitor and start winding when the motor reaches about 75 percent of its rated speed.

Capacitor-start motors can be open in design as when used with fans, or enclosed as when used to drive a hermetic compressor. Troubleshooting the bearings, windings, and capacitor of an open type capacitor-start motor is similar to the split-phase and other motors described earlier in this chapter.

When the capacitor-start motor is used in an enclosed hermetic compressor, an external relay replaces the centrifugal switch to cut power to the start winding and capacitor.

The hermetic compressor is equipped with terminals on its external casing that make it easy to check internal windings using an ohmmeter. Frozen or poor rotation is indicated by a high current draw. Use an ammeter to check for this condition or listen carefully for the hum of a locked motor.

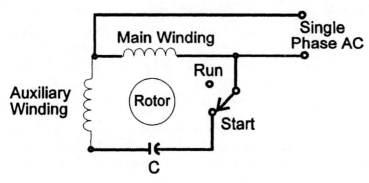

Figure 10-6 Wiring diagram for PSC induction motor.

Capacitor-Start-Capacitor-Run Motors

The capacitor-start-capacitor-run (CSR) motor combines the good running characteristics of a permanent split-capacitor motor with the capacitor-start design of a capacitor-start motor. CSR motors generate high starting torque and offer good running efficiency. A CSR motor is simply a capacitor-start motor with a run capacitor added to its start winding (Fig. 10-7). The start winding is energized whenever the motor is running.

Capacitor-start-capacitor-run motors are almost always used to drive hermetic or semi-hermetic compressors. They are seldom used as an open-type motor because of their relatively high cost.

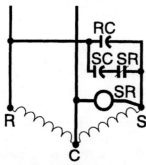

SC: Starting Capacitor
SR: Starting Relay
RC: Running Capacitor
S: Starting Winding Terminal
R: Running Winding Terminal
C: Common Terminal

Figure 10-7 Wiring diagram for a capacitor-start, capacitor-run motor.

Operation

A phase displacement between the starting and running windings of a CRS motor allows rotation to begin. The running capacitor contributes slightly to motor start up, but its main job is to increase running efficiency.

Service tips

The number of components found in a capacitor-start-capacitor-run motor makes it more difficult to troubleshoot. The windings, bearings, start relay, starting capacitor, and running capacitor can all cause problems.

Use an ohmmeter at the hermetic casing terminals to check windings for shorts, opens, or grounds as described on page 280.

Worn bearings can cause rough turning or complete freeze-up. The bearings of hermetically sealed motors are inaccessible and obviously impossible to inspect. However problem conditions are indicated by a whining sound during operation or if the motor pulls a larger-than-normal ampere draw. Keep in mind, though, that a high ampere draw can be caused by other problems besides worn bearings.

Troubleshoot the starting relay by inspecting the contacts and the coil. There should be no pitting or damage. To visually inspect, first disassemble the relay, then examine the contacts for sticking, pitting, or misalignment. Also check the contacts with an ohmmeter. The contacts should show zero resistance.

Check the coil of the relay for shorts and opens in the same manner as the windings of a motor. Keep in mind that a capacitor can be destroyed if the contacts or coil of a starting relay are bad.

The starting and running capacitors are easily checked with an ohmmeter. See page 280 for details.

Three-Phase Motors

Three-phase motors are strong, durable motors. The main advantage of a three-phase motor is its high starting torque. Three-phase motors are used to power compressors in systems where the high- and low-side pressures do not equalize on compressor shutdown. The squirrel cage induction type three-phase motor is the most popular design of three-phase motor used in refrigeration systems. These motors are available in almost any voltage range required. In addition, dual voltage three-phase motors are available that operate on two different voltages with simple modifications in the wiring.

Operation

Three-phase motors require the use of a three-phase power supply. Because three phases of electrical power are supplied to the motor instead of just one phase, three-phase motors are considerably stronger than single-phase motors. Three-phase power provides three hot legs to the motor instead of the two hot legs provided by a single-phase power supply. If you review the sine wave of a three-phase power supply (Fig. 9-1b), you will see that none of the phases peak at the same time. Each phase is located 120 electrical degrees out of phase with the others. At any given time, one of the phases is in position to supply high starting torque. This eliminates the need for start windings or capacitors to create a phase displacement.

Operation of three-phase motors is similar to single-phase motor with the exception of the three-phase displacement (see page 242). Each of the phases can have two or four poles. A 3600 rpm three-phase motor has three sets, each with two poles, and an 1800 rpm motor has three sets each with four poles. Each phase changes its direction of current flow at a different time, but the order is always the same.

The rotational direction of a three-phase motor can be changed by switching any two of the motor leads, so care must be taken when three-phase fan motors are used. Fans must be installed to rotate in the correct direction.

Service tips

Check the resistance of a three-phase motor winding using an ohmmeter. A resistance reading of 0 ohms indicates a short in the motor. If resistance is infinite, the winding is open. The correct resistance reading will range from 1 ohm to 40 ohms, based on motor size. Larger motors will read smaller resistances. The smaller the motor, the larger the resistance of the winding will be.

Hermetic Compressor Motors

Hermetic compressor motors are induction-type motors designed for single- or three-phase current. Four types of single-phase motors are used in hermetic compressors. Split-phase and capacitor-start motors of less than 1 horsepower are used on small compressors. Permanent split-capacitor motors and capacitor-start-capacitor-run motors are used on application that require moderate to good starting and running torque. Virtually all hermetic compressors used on larger systems are built with three-phase motors having a horsepower rating of three or greater.

Hermetic motors operate the same as open compressor motors with the exception of their enclosure or shell. Because hermetic compressor motors are totally enclosed inside a shell and are constantly exposed to refrigerant and oil, they require special considerations.

Centrifugal switches cannot be used to cut power to the start windings.

Nothing can be used inside the shell that can create a spark. For example, electrical arcing of relays will cause the refrigerant and oil to deteriorate. This means that no starting components can be housed inside the compressor shell. Starting relays and capacitors must be mounted and wired on the outside of the shell enclosure, remote from the motor.

Wiring terminal identification

Every single-phase hermetic motor has a common, a start, and a run terminal. (On an open-type motor these terminals can be wired directly into the motor making them difficult to locate.)

On a hermetic compressor, a terminal box on the outside of the dome houses the three terminals. One terminal is for the start winding, one is for the run winding, and one is for the line common to the start and run windings (Fig. 10-8). The motor leads are insulated from the steel compressor shell using ceramic or neoprene insulation. Neoprene insulation can deteriorate over time.

If the start, run, and common terminals are not properly labeled on a single-phase motor, they can be identified by using an ohmmeter to read the resistance of each winding with respect to the common terminal (Fig. 10-9). Locate the largest resistance reading between any two terminals. The remaining terminal is the common. The largest reading between common and the other two terminals is the start terminal. The remaining terminal is the run terminal.

For a three-phase motor, the resistance readings between all three terminals should be identical because no starting apparatus is required and the windings are identical.

Service tips

Troubleshooting hermetic compressor motors is limited to electrical troubleshooting because the motor is totally enclosed in a housing and cannot be visually inspected. But as described earlier in this chapter, most single- or three-phase hermetic compressor motors have three

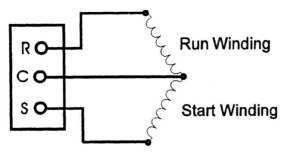

Figure 10-8 Wiring diagram and terminal identification of a single-phase compressor motor.

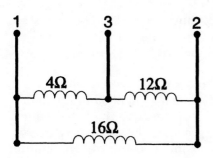

Figure 10-9 Identifying single-phase motor terminal by resistance readings.

terminals on the outside of the casing that connect the motor to the external power wiring. Some large hermetic compressors have more than three terminals, such as dual-voltage, part winding motors, or two-speed motors.

Single-phase motors. The winding layouts of all single-phase hermetic compressors are similar. The only difference is the size of the windings, which will vary the resistance readings of the motor windings. Electrical troubleshooting is performed by taking a resistance reading of the windings with an ohmmeter. The resistance readings of single-phase motor windings are not the same because the compressor has a start winding and a run winding connected by a common wire (see Fig. 10-9).

On a single-phase hermetic compressor motor, the sum of the resistance readings of the start to common terminals and the run to common terminals should equal the resistance reading obtained between the run and start terminals. This fact allows you to match the resistance readings to determine the condition of the windings. If the readings do not match, a spot burnout of the winding is likely.

Three-phase motors. Three-phase motors should show the same resistance in each winding. If resistance varies, a spot burnout in the motor is likely. But before changing the motor, be certain the problem is not simply bad or corroded terminal connections. Double-check the readings and make certain the ohmmeter is properly set.

Electrical Troubleshooting

Motors that rotate freely when turned by hand but run slowly or not at all likely have electrical problems. Motor windings can be open or shorted. A short circuit between a relatively few turns of the stator coil, or from the stator winding to the frame, can be sufficient to cause high motor current draw or prevent a motor from starting or running at proper rpm. Electrical leakage between turns of deteriorated wire will increase at high operating temperatures. The motor might run normally when first operated from a cold start, but when it heats up to its normal operating temperature, it might slow down or even stop operating. Motors that exhibit electrical faults, such as shorted or open stator windings, are usually not repaired. They are replaced.

Line voltage. Before replacing a motor that is suspected of having a defective winding, check the line voltage being supplied to the motor to be sure that it is within at least 10 percent of the rated value. Diagnosing an open, a shorted, or a grounded hermetic compressor motor is easy because the resistance readings obtained are definite and exact.

Testing for open windings

An open winding in the compressor motor means there is no continuity or no complete circuit. There is a break in the circuit. Open windings can be easily detected using an ohmmeter to measure winding resistance.

Stator coils of single-phase motors are connected in series, so on a good motor the ohmmeter will indicate some resistance from any wire to each of the others. Normal resistance readings of motor windings might be from a few ohms to 100 or more, depending on the size of the motor. Some PSC motors might have isolated start or capacitor windings that will indicate no connection to the run windings.

When an open is present in the coils, the ohmmeter will read an infinite resistance reading (Fig. 10-10). Test for open windings by reading on a low-ohm scale ($R \times 1$).

An open reading might be caused by a faulty thermal overload switch, which is usually connected in series with the common circuit of the motor. Small hermetic compressors usually have some type of external overload device. Larger hermetic compressors normally are

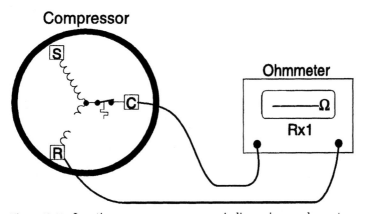

Figure 10-10 Locating an open-compressor winding using an ohmmeter.

designed with internal overloads. If the compressor shell is hot to the touch, it is a good indication that the internal overload is open.

The overload device might open due to a low refrigerant charge, a locked compressor motor, faulty starting components, or a high discharge pressure. Keep in mind that mechanical failures, such as a frozen bearing, can result in a "no-run" condition that appears to be electrically related.

When the internal overload opens the electrical circuit, the unit must cool down before the overload relay closes and operation can resume. Cool down normally takes 20 to 30 minutes.

Field-replaceable external overloads mount against the compressor shell so they can sense shell temperature. Always use the exact replacement overload device. Switching to a different overload device can compromise compressor protection.

Check an external overload relay with an ohmmeter or continuity checker. The relay contacts should be closed when the relay is at normal (ambient) operating temperatures. If the relays become pitted or corroded, the compressor might periodically stall during operation. Replace the overload relay if this occurs.

Testing for shorts

A motor can also fail if it develops shorted windings. A shorted motor winding has burned together. The resistance reading will be zero ohms (Fig. 10-11). Test for shorted windings by reading on a low-ohm scale ($R \times 1$). However, because short circuits might involve only a few turns of wire and/or be temperature dependent, it might not be possible to determine if a motor winding is shorted by use of the ohmmeter. If the motor does not have an open winding yet fails to run properly or draws excessive current, it most likely has a shorted winding and must be replaced. Always check leakage to the frame, which should be at least 20 megohms. A megger is the best instrument for this purpose.

Testing for improper grounds

A grounded winding in a compressor motor occurs when part of the winding touches the compressor body. The ohmmeter will read a resistance reading between the shell and the terminals of a compressor (Fig. 10-12). Test for grounded winding by reading on the $R \times 10,000$ scale or higher. Resistance readings as high as 500,000 ohms indicate a grounded compressor that requires replacement. Grounded compressors, if allowed to run, will likely operate at higher than normal temperatures. As the windings heat up, resistance to ground becomes less.

Good contact on the compressor shell must always be maintained if the motor is grounded. To provide for this ground, remove all paint from a small section of the compressor.

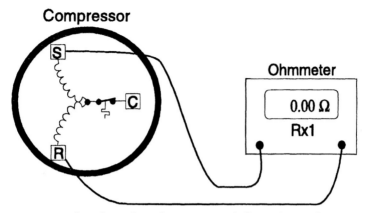

Figure 10-11 Locating a shorted compressor winding using an ohmmeter.

A grounded compressor can be dangerous. Touching the casing of a slightly grounded compressor can result in electrical shock.

Troubleshooting capacitors

When a compressor or permanently split capacitor fan motor fails to operate properly, begin troubleshooting by first eliminating the capacitor as the possible source of the problem. Simply substitute a known good capacitor for the capacitor in question. This method is quicker and more fail-safe than testing the capacitor.

Capacitors fail in a number of ways. They become open, shorted, or have excessive electrical leakage. Capacitors also have built-in fuses that blow in the event of an excessive current surge through the device.

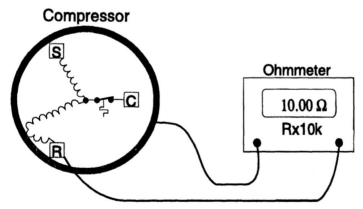

Figure 10-12 Locating an improperly grounded compressor winding using an ohmmeter.

Such a surge can occur if the capacitor shorts out. A shorted capacitor will indicate an open circuit when tested.

Capacitors can hold an electrical charge long after the refrigeration unit has been switched off. Some large-value capacitors have an external resistor connected across their terminals. This is a safety feature that automatically discharges the capacitor within a reasonable time after power to the circuit is removed. To prevent a shock hazard, all capacitors without bleed resistors should be treated as though they are charged.

Testing capacitors. A capacitance meter can be used to measure the value of capacitance, and its leakage, on any capacitor that is suspect.

An ordinary ohmmeter, set to the 100,000 ohm scale, also can be used to test if a capacitor is shorted or open. Before testing a capacitor, discharge it so that it has no residual charge. To do this, connect a low-value power resistor, such as 100 ohms 5 watts, across the terminals for a few seconds. Once the capacitor is discharged, disconnect the resistor (and bleed resistor if so equipped). Now connect the leads of the ohmmeter in place across the capacitor terminals.

If the capacitor is operating properly, the ohmmeter reading should quickly go to a low resistance reading and then slowly increase to infinity as the capacitor charges up. Capacitors that are shorted or leaky will read a finite resistance value that does not change. Capacitors that are open will not produce any reaction at all when first connected to the ohmmeter.

The ohmmeter test outlined above does not check the capacitor at its normal operating voltage of 120 Vac or more. To do so simply requires an ordinary ac ammeter capable of measuring 10 amperes. This ammeter test also provides a reasonable accurate measure of the value of capacitance (Fig. 10-13).

In this test, the capacitor is subjected to a potential of about 120 Vac from the power line, and the current draw is measured by the ammeter. To protect the ammeter in the event the capacitor is shorted, a 10-ampere fuse must be placed in series with the ammeter. A good capacitor will draw approximately 0.045 amperes from a 120-volt, 60-Hertz power line for each microfarad of capacitance. At 50 Hertz, the current will be 0.038 amperes.

The actual value of the capacitor can be checked against its indicated value by using the following formula:

UFD = (Reading in amperes)/0.045 for 120 volt 60 Hertz,
or UFD = (Reading in amperes)/0.038 for 120 volt 50 Hertz
where UFD is the value of capacitance in microfarads.

If the capacitor does not draw the expected current, within plus or minus 20 percent, replace it. If the motor uses a start and/or run capacitor, substitute this component to determine if the capacitor is the problem. Motor circuits that employ relays or starting switches should also be checked to be sure that these components are not defective.

Mechanical Problems

Today, fewer motors used in refrigeration systems are made with bearings that can be lubricated. It is more common for a manufacturer to use a lesser quality bearing that cannot be serviced in order to lower the overall cost of the motor. Steel bearings, which do not have as long a service life as bronze bearings, are frequently used as well. A motor with steel bearings inside it can cause excessive noise after just a few years service.

On lower quality units that have long service hours or service life, it is not unusual for fan motor bearings to fail.

Mechanical failures of blower and compressor motors will almost always involve dry, frozen, or worn bearings that are the result of lack of lubrication. Some motors that are equipped with steel bearings might have worn shafts. Loose bearings or worn shafts will result in excessive radial play and often cause noisy operation. Small motors with worn components are usually discarded because the cost of replacement bearings is not financially feasible.

Motors with worn shafts can sometimes be returned to service by repositioning the shaft on the rotor. To do this, remove the rotor assembly from the motor. Place the rotor section on a support and use a soft-

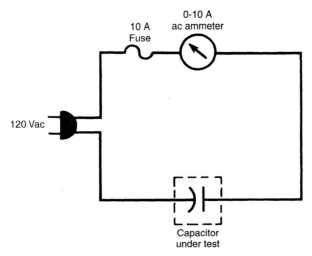

Figure 10-13 Testing a capacitor under 120-Vac excitation. *(Reprinted from* Practical Air Conditioning Equipment Repair *by Anthony J. Caristi, McGraw-Hill, copyright 1991.)*

head mallet to force the shaft to a new position so that the worn sections of the shaft will not be located at their original positions inside the bearings when the motor is reassembled.

Bearings also fail because they dry out or "freeze." This means that the motor will not come up to speed, or in some cases, rotate at all. A high-pitched squeal from a motor usually indicates that one or more bearings have lost all lubrication. Such motors require bearing lubrication and can often be repaired at far lower cost than replacement.

Lubrication

Most small-fan motors used in domestic refrigerators and small commercial systems employ sleeve bearings, often with no provision for field lubrication. Commercial refrigeration motors usually employ higher quality motors, which often include oil ports for periodic lubrication (usually once a year). If the motor is equipped with oil ports, lubricate it with a generous amount of nondetergent number 20 SAE oil. Allow time for the oil to seep into the bearing housing while the shaft of the motor is either rotated by hand or coaxed into self-operation. If the motor bearings have not been damaged by the lack of lubrication, the motor can often be returned to service after sufficient oil has penetrated and lubricated the bearings.

If the motor does not have oil ports, oil can be applied to the shaft; allow it to seep into the bearings. A better method is to completely disassemble the motor, remove the rotor from the unit, and soak the bearing housings in oil. When disassembled, inspect the rotor shaft for wear and polish it with fine steel wool prior to reassembly. A motor refurbished in this manner will likely operate for a long period of time.

A blower motor with a completely frozen shaft is not necessarily ready for replacement. Work a penetrating oil such as WD-40 into the bearing while rotating the shaft by hand. This might result in sufficient play to allow the motor to rotate under its own power. Once unfrozen by the penetrating oil, the bearings must be fully lubricated with SAE 20 oil before the motor can be returned to service.

If the motor shaft cannot be rotated by hand at all or is sluggish, a sharp blow to the end of the shaft with a rubber mallet will sometimes break the bearing loose sufficiently to begin the penetrating oil treatment.

A properly lubricated motor will rotate freely when turned by hand, will start promptly when power is applied, and will coast for some time after power is removed. Any motor that has been brought back to use by lubrication should be operated for at least an hour to ensure that it has been properly repaired.

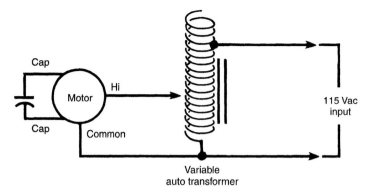

Figure 10-14 Using a variable autotransformer to increase or decrease input voltage to a motor. *(Reprinted from* Practical Air Conditioning Equipment Repair *by Anthony J. Caristi, McGraw-Hill, copyright 1991.)*

A good method of checking a motor repaired by lubrication of dried bearings is to attempt to start it using low-line voltage generated by a variable transformer or Variac (Fig. 10-14). A motor in good working order will start (in low speed) under the low-torque conditions of 20 percent low-line voltage. Using the Variac to temporarily raise the voltage supplied to a sluggish motor is also an excellent way to get a motor started under its own power.

Motor Replacement

Motor replacement in a refrigeration system can run from the simple to the very complex, depending on the size of the system and its design. When a blower (noncompressor) motor replacement is required, always determine the voltage, current, and rpm rating of the defective motor, especially if the replacement motor is ordered from any supplier other than the factory-authorized parts distributor. Check the motor nameplate for pertinent information concerning the blower motor specifications. Replacing one motor with another one of a different rpm or too small an HP rating will cause major problems. Also note the motor's physical size and mounting dimensions. It is possible to obtain a properly rated motor that will not fit the space requirements. Blower motors are fastened to the structure of the unit by various means, and the replacement motor might or might not have the exact same mounting bracket. If the new motor has been obtained from an authorized parts distributor, it usually will contain an adapter kit (if necessary) to permit proper assembly of the motor into the unit. If the new motor must be obtained from an aftermarket distributor, obtain one that has a compatible mounting bracket for the unit under repair. Many refrigeration units are designed with multiple mounting holes drilled in the motor support assembly, that allow mounting two or more different motor styles.

Fractional HP motors can be single or multiple speed and can be categorized as split-phase or permanent-split-capacitor (PSC) types. PSC aftermarket motors are often used for replacement of lower efficiency split-phase units. The replacement motor might or might not have exact equivalent color-coded leads as compared to the original unit. If possible, refer to the schematic diagram of the refrigeration system and replacement motor to determine the correct connections to be made. The schematic diagrams often illustrate the color coding of the wires, which is helpful information if the new motor is color coded differently.

It is usually possible to remove a blower motor from a unit by performing some amount of disassembly. At the very least, the fan blades must be disconnected from the shafts. These might be secured with Allen set screws, which might be accessible only with special Allen wrenches long enough to reach the hub of the blade from the outside.

When disconnecting a defective motor, always leave a small portion of the original wires connected to the system to provide reference in the event of a problem when rewiring the circuit for the new motor.

Be very careful when wiring the replacement motor to the original circuit, especially if it is color coded differently. A miswire can easily result in the burnout of the new motor. The common lead of the blower motor is usually color coded white and connected to one side of the power line. Line power must never be applied between the high- and low-speed connections.

If the replacement motor has more speeds than the original, use high speed for single-speed units or high and low for two-speed units. Wires that are unused should be taped off to eliminate the possibility of a short circuit to the chassis or any other component. All wires should be secured to prevent any possible interference with the moving parts of the motor.

Motor rotation is also very important. Motors are classified as clockwise (CW) and counterclockwise (CCW) rotation. Unfortunately, not all manufacturers use the same reference point when determining rotation. Some look at the shaft end of the motor while others look at the back end of the motor housing. This means one manufacturer's CW rotation might be another's CWW rotation. So make certain you know which way the motor should be rotating. Do not install a motor with the wrong rotation direction; it can severely decrease system performance. It is also not possible to turn the fan blade around.

When operating a unit for the first time after motor replacement, monitor power line current. If excessive current draw occurs due to a miswire, the motor can quickly be turned off to prevent damage.

Troubleshooting Hermetic Compressor Motors

When the compressor does not operate, the problem might be with the compressor. Because the compressor is the most expensive item in most refrigeration systems, test it completely before declaring it unserviceable.

If the compressor is mechanically stuck or has a shorted winding, its overload will likely cut out. The circuit breaker or fuse protecting the compressor circuit might also trip or blow. As mentioned earlier, monitoring the line current drawn by the refrigeration unit can provide valuable information. Keep in mind that shorts or mechanical problems are more common in failed compressors than open windings. When this is the case, the line current to the system will remain substantially below normal when the unit is turned on.

Measure the line voltage applied between the common and run terminals of the unit to be sure that it is within 10 percent of the rated value. Also, the run capacitor, connected to the start winding, can be replaced with a new unit to be certain that this component is not the cause of the fault. Be sure that there is no open, short, or ground leakage by taking a resistance reading (with all wires removed from the compressor terminals) between the terminals of the compressor, and between any winding and case as described earlier in this chapter.

If these tests point to a failed compressor, disconnect all wires from the compressor terminals and wire the compressor to the power line using a new capacitor which is known to be in perfect condition. A compressor that still fails to run is bad.

Warning: The protective cover should always be placed over the terminals before applying line power to prevent injury from a possible expelled terminal.

Predicting remaining compressor life

A megger is a test instrument that is capable of determining electrical leakage between the windings and shell of a hermetic compressor (Fig. 10-15). Its reading can be used to predict remaining compressor life.

The megger impresses a high voltage, usually 500 volts, between the compressor stator winding and frame to determine the amount of current flow and electrical leakage, measured in megohms (millions of ohms). An ordinary ohmmeter, which employs a very low voltage for resistance measurements, cannot properly determine leakage in a hermetic compressor.

Warning: The 500-volt output of the megger is a potential shock hazard. Follow the megger manufacturer's directions for making connections and unit operation. Also, never test hermetic compressors with a megger when the refrigeration system is under vacuum conditions.

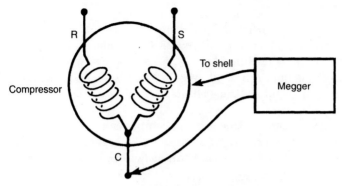

Figure 10-15 Using a megger to check the electrical condition of the compressor. *(Reprinted from* Practical Air Conditioning Equipment Repair *by Anthony J. Caristi, McGraw-Hill, copyright 1991.)*

The resistance between the windings and frame of a compressor are listed in Table 10-2. Test the compressor when fully warmed up after operation for 10 minutes or more to help drive out refrigerant from the windings for a more accurate reading. All wires must be disconnected from the compressor terminals prior to the megger test to prevent a false reading from some other part of the system. The connector block of the compressor should be thoroughly cleaned before taking a measurement. Dirt and grease on the connector block can increase resistance readings.

Resistance readings of 50 megohms or less indicate a problem in the system. Moisture might be the cause, and the system should be fully evacuated after replacing or installing a drier. If this procedure does not improve the resistance reading, the oil in the system might be contaminated. Commercial air-conditioning systems might have provisions to change compressor oil if contaminated.

Resistance readings of less than 20 megohms taken after thorough evacuation and recharging indicate severe system contamination. Remaining compressor life could be relatively short.

TABLE 10-2 Hermetic Compressor Condition Based on Winding to Frame Resistance

Resistance Reading	Compressor Condition
Greater than 100 megohms	Excellent. No signs of moisture. No PM needed.
Between 50 and 100 megohms	Low levels of moisture present. Replace drier and evacuate system.
Between 25 and 50 megohms	Contaminated oil and high levels of moisture are present. Perform multiple drier change-outs.
Between 0 and 25 megohms	Severe contamination. Change oil and drier.

Freeing a stuck rotor

If a compressor rotor is "stuck" or locked due to a minor mechanical problem, it might be possible to free it by electrical means. But before trying any of the methods outlined below, make absolutely certain the cause of the problem is not high head pressure due to trapped refrigerant in the high-pressure side of the system.

Start capacitor. One method to free a stuck compressor is to temporarily replace the start capacitor with another of much greater value. This provides additional starting torque that might get the motor turning.

Increasing line voltage. Another method of freeing up a stuck compressor motor is to apply a higher-than-normal voltage to the compressor (such as 240 volts for a 120-volt unit) for only one or two seconds. The temporary increase in power might break loose the obstruction.

Reverse direction. A third method of freeing a stuck compressor is to run the compressor in a reverse direction for a short period. This is done by interchanging the start and run connections in the circuit. This method might allow the rotor to turn a sufficient amount and push away whatever contaminant is preventing its normal forward rotation.

If these methods fail to restore the compressor back to normal operation, the unit is beyond repair and must be replaced. When compressor replacement is necessary, it is easiest if an exact replacement is used, because the inlet, discharge, and electrical connections will be located in the same position as the original.

Chapter

11

Domestic Refrigerators and Freezers

The primary function of a refrigerator or freezer for domestic or commercial service is to store food at reduced temperatures. Fresh foods last longer when kept at temperatures just above freezing. While unfrozen meats must be stored at as close to 32°F (0°C) as possible, most fresh foods can be kept from three days to a week in temperatures between 35° and 45°F (2° and 7°C). Ideally, for the sake of ease in spreading, butter should be stored at a slightly warmer temperature. The air in the fresh-food compartment of a refrigerator is quite dry because any moisture tends to condense on the surfaces of the evaporator. Yet fresh fruits and leafy vegetables are best preserved in a high-humidity environment. Most refrigerators are designed to handle the special requirements of the various foods stored in them and to provide a favorable storage condition for each of them. These design features are discussed in greater detail in the section on cabinets. To preserve foods for longer than a week, they must be frozen. Most frozen foods can be kept for several weeks if they are stored in temperatures of 0° to –10°F (–18° to –23°C). When foods are to be stored for a year or more, they must be frozen at –20°F (–29°C) or lower. Food freezers and combination refrigerator/freezers are designed to hold these temperatures. The frozen food space in a single-door refrigerator, however, is usually warmer. It is not designed to store food for any extended period of time.

Domestic Refrigerator/Freezer Cabinet Designs

Domestic refrigerators are often classified by the type of cabinet used. There are single-door units, a variety of refrigerator/freezers, and two types of freezers.

Single-door refrigerators are generally small. They have a small freezing compartment across the top of the fresh food storage area where frozen foods can be stored for a short period of time. Cold air from the evaporator near the top of the cabinet naturally flows down through the refrigerated space. Beneath the evaporator is an insulated baffle. By limiting the amount of air that passes around it, the baffle enables the evaporator to provide the low temperatures needed for the frozen food section and temperatures above freezing for the fresh-food compartment.

Full-sized refrigerator/freezers feature a large freezer compartment capable of maintaining colder temperatures (Fig. 11-1). Usually the refrigerator and freezer each have their own exterior door. Some of the largest models have a side-by-side arrangement.

Refrigerator/freezers often use motor-driven fans and various ducts to carry air from the evaporator to its different compartments and areas (Fig. 11-2). Air circulated in this manner offers lower temperatures to areas reached first and higher temperatures to those areas reached last. When a single evaporator is used, it is usually located behind the freezer compartment. Some models have one or two additional evaporators located at the back or side of the refrigerator compartment.

Freezer-only units are available in chest and upright configurations. An advantage of the chest is that there is little air change when the

Figure 11-1 Full-size refrigerator/freezer combination. *(Sears, Roebuck and Co.)*

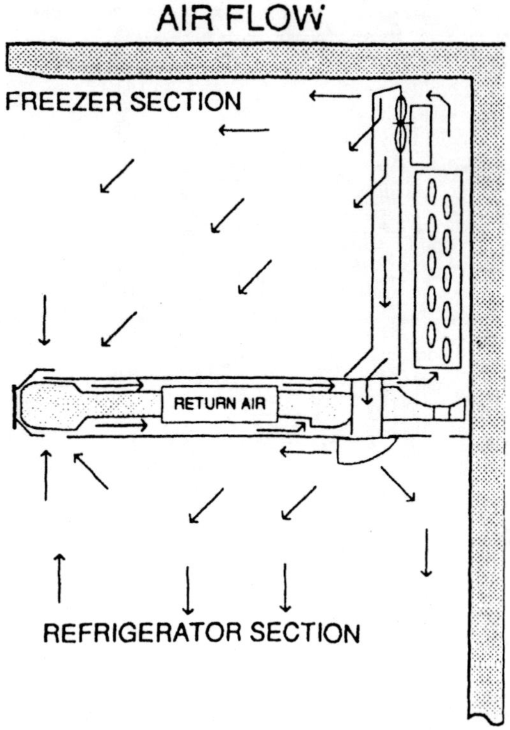

Figure 11-2 Air flow patterns in a typical domestic refrigerator/freezer.

cabinet is opened. The very cold heavy air does not spill out when the lid is opened. Consequently, it is more efficient and economical. Little moisture enters the cabinet, so the freezer only needs to be defrosted once or twice a year. The evaporator surrounds the inner liner. Another advantage of the chest-type is its availability in an extensive range of capacities. While their height and width are quite uniform, chests vary in length according to capacity. In order to make food stored at the bottom accessible, the chest is usually equipped with baskets that can be lifted out. In the storage and removal of food, however, the upright freezer offers greater convenience and the incorporation of automatic defrost systems in these freezers has made them practical. The evaporator in the upright might be located at the bottom or near the upper one third of the cabinet. Fans are used to circulate the cool air from the evaporator through the compartment.

Special-purpose compartments

Most refrigerators have compartments especially designed to store particular kinds of food (Fig. 11-3). For instance, nearly all have high-

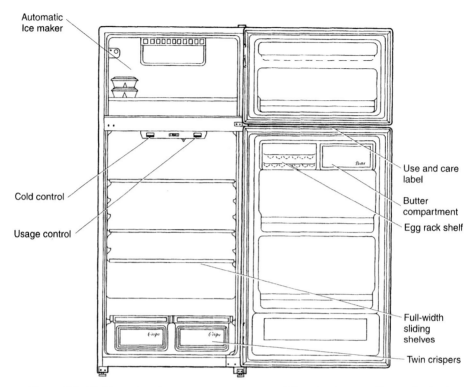

Figure 11-3 Refrigerator storage compartments and interior layout. *(Sears, Roebuck and Co.)*

humidity compartments for the storage of fresh fruits and leafy vegetables. These compartments seal off vulnerable foods in tight-fitting drawers from the effects of circulating dry air. They are usually located below the bottom shelf of the refrigerator. Butter is often stored in a slightly warmer compartment in the door so that it is suitable for spreading. This compartment has little insulating it from the room temperature outside. A small electric heater is sometimes used to warm this compartment too. Some refrigerators have a meat storage compartment that might even include an independent temperature adjustment. These bins maintain storage temperatures just above freezing. A less-common feature is a compartment that maintains a temperature of approximately 30°F for the preservation of fish. Each of these special-purpose compartments is closed off from the general storage area.

Structure and materials

A cabinet's external shell is most often a single fabricated steel structure with welded seams and a baked enamel finish (Fig. 11-4). It holds

Domestic Refrigerators and Freezers   295

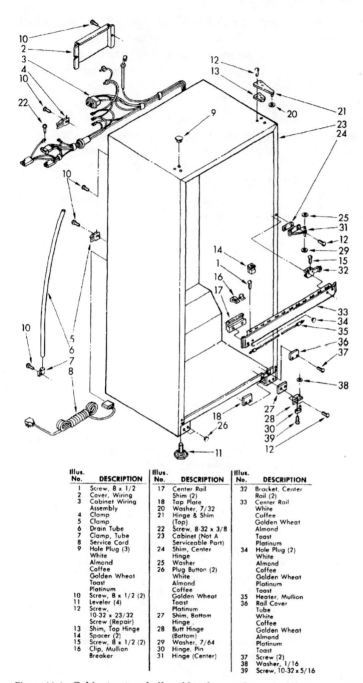

Illus. No.	DESCRIPTION	Illus. No.	DESCRIPTION	Illus. No.	DESCRIPTION
1	Screw, 8 x 1/2	17	Center Rail	32	Bracket, Center Rail (2)
2	Cover, Wiring		Shim (2)	33	Center Rail
3	Cabinet Wiring Assembly	18	Tap Plate		White
4	Clamp	20	Washer, 7/32		Coffee
5	Clamp	21	Hinge & Shim (Top)		Golden Wheat
6	Drain Tube	22	Screw, 8-32 x 3/8		Almond
7	Clamp, Tube	23	Cabinet (Not A Serviceable Part)		Toast
8	Service Cord				Platinum
9	Hole Plug (3)	24	Shim, Center Hinge	34	Hole Plug (2)
	White	25	Washer		White
	Almond	26	Plug Button (2)		Almond
	Coffee		White		Coffee
	Golden Wheat		Almond		Golden Wheat
	Toast		Coffee		Platinum
	Platinum		Golden Wheat		Toast
10	Screw, 8 x 1/2 (2)		Toast	35	Heater, Mullion
11	Leveler (4)		Platinum	36	Rail Cover Tube
12	Screw, 10-32 x 23/32 Screw (Repair)	27	Shim, Bottom Hinge		White
13	Shim, Top Hinge	28	Butt Hinge (Bottom)		Coffee
14	Spacer (2)	29	Washer, 7/64		Golden Wheat
15	Screw, 8 x 1/2 (2)	30	Hinge, Pin		Almond
16	Clip, Mullion Breaker	31	Hinge (Center)		Platinum
					Toast
				37	Screw (2)
				38	Washer, 1/16
				39	Screw, 10-32 x 5/16

Figure 11-4 Cabinet outer shell and hardware of a typical domestic refrigerator. *(Sears, Roebuck and Co.)*

the refrigeration system and the inner lining, and it supports the door(s). Some two-door cabinets have two separate inner linings housed in one outer shell. In others, a single liner with an insulated partition is used (Fig. 11-5). The partition might be an integral part of the liner or it might be installed as a separate piece.

Inner linings provide brackets for mounting things like shelves, lights, temperature controls, and thermostats. They can be steel, aluminum, or plastic. While an organic finish is used on some steel liners, an acid-resistant porcelain enamel is more common. The aluminum ones have an anodized or organic finish. Of course, plastic does not require a finish. A separate plastic breaker strip is sometimes used at the outside edge of the inner lining. Foam slabs or foamed-in-place insulation fills the space between the inner and outer walls of the cabinet. In some cases, the inner lining is not bonded to the foam, so it can be removed if required. It should be noted, however, that the lining is part of the cabinet's structural support.

The doors cover the whole front of the cabinet. Plastic door liners covering the inside of the doors are usually formed to provide shelves and racks. No separate breaker strips are used.

Hardware

Most refrigerator cabinets have adjustable feet. They should be adjusted so that the shelves are level and so that the door tends to swing closed on its own weight from any open position. For the sake of convenience in moving, cleaning, and servicing, some units are mounted on rollers. Once the refrigerator is rolled into place, the front rollers usually can be adjusted to level it and a thumbscrew lowered to touch the floor and to hold the unit in place.

Usually cabinet hinges are adjustable too. In this way, the door can be properly fitted to the cabinet. The hinges might use nylon, acetal, or ball bearings. These seldom, if ever, need lubrication. The hinge mechanisms on chest cabinets often include springs that act as counterbalances. These prevent the entire weight of the closed lid from compressing the gasket and also eliminate the need to hold the lid open. Like the hinges on an upright, these are adjusted so that the door closes by its own weight. Hinges, when they fail, ought to be replaced instead of repaired.

Most refrigerators use magnets to hold the door shut because a federal law in 1956 prohibited the shipment in interstate commerce of household refrigerators that could not be opened from the inside. These magnets might be in the form of small latch and striker plates or they might be built into the door gaskets.

Domestic Refrigerators and Freezers 297

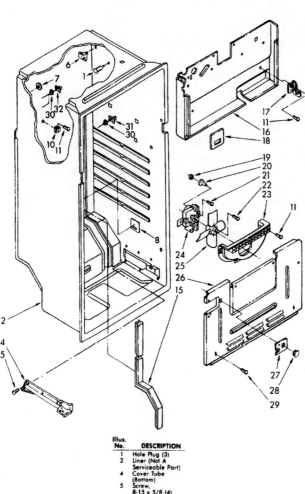

Illus. No.	DESCRIPTION
1	Hole Plug (3)
2	Liner (Not A Serviceable Part)
4	Cover Tube (Bottom)
5	Screw, 8-15 x 5/8 (4)
6	Support Stud (2)
7	Hole Plug
8	Shield Wiring
10	Bracket, Wire Receptacles
11	Screw, 8 x 1/2
15	Cover Tube (Rear)
16	Heat Shield
17	Evaporator Support (2)
18	Cover, Air Duct
19	Grommet (2)
20	Spacer, Fan Motor (2)
21	Screw (2)
22	Screw, 3-48 x 1/4 (2)
23	Fan Scroll
24	Fan Motor
25	Fan Blade
26	Evaporator Cover
27	Hole Cover
28	Hole Plug
29	Screw, 8-15 x 1/2 (7)
30	Support Stud
31	Stud
32	Saddle Stud (R.S.) Saddle Stud (L.S.)

Figure 11-5 Liner components of a typical domestic refrigerator. *(Sears, Roebuck and Co.)*

Gaskets

Soft compression gaskets made of flexible vinyl or rubber are used around the edge of the doors. As already mentioned, magnets are often built into these gaskets. As a door gasket deteriorates with age, it might harden, crack, or break. Defective gaskets should be replaced.

The effectiveness of the gasket seal can be checked in two ways. A 0.003-inch-thick plastic gauge (or a strip of paper) can be inserted at several places around the edge of the slightly opened door. When the door is closed, it should take a little pull to remove the gauge from under a properly fitted gasket. Another way to check the seal is to note the effort needed to open the door. Hold it halfway open for about 10 seconds, then close it. After it has been closed for about 15 seconds, open the door again and note the effort needed. It should take more to open the door the second time than the first. Due to the contraction of the warm air that entered when the door was open, pressure inside should be slightly lower than pressure outside.

An improper seal might result when hinges are not adjusted properly or when the door is warped. A warped door can be repaired by twisting it so that it is straight.

Ice service

The use of automatic ice makers in household refrigerators has increased. Usually, if an automatic defrost refrigerator does not already have a factory-installed ice maker, it is equipped for a field installation. Many of the newer models have solid-state controls. The mechanism, located in the freezer section, requires an attachment to a water line. A solenoid-controlled valve controls the flow of water to the ice mold. Refrigerated air passing over the ice mold freezes the water. Depending on the model, ejection is initiated either when the ice mold reaches a certain temperature or after a certain amount of time has passed. Some designs free the ice from the mold with an electric heater and a mechanism then pushes the cubes into a storage container. In others, the ice is freed and ejected into the storage container by twisting and rotating the tray. A gear motor drives the mechanism. A feeler-type ice level control or a weight control stops the production of ice once the storage container is full.

For greater convenience, some models dispense ice through the door. Either the ice is ejected into a compartment accessible from outside the freezer door or an auger mechanism in the storage container delivers the ice through a chute in the freezer door when a push-button switch is activated. Additional features on some models include crushed ice, chilled water, and juice dispensers.

Evaporators

A variety of evaporators are used in household refrigerators and freezers. Basically, the method of defrosting desired in a unit determines what kind is used. In a manual-defrost unit, the evaporator is usually shaped like a box. Three or four of its sides are refrigerated. In some units, the walls are constructed from double sheets of metal brazed or bonded together with integral passages for the refrigerant. Tubing brazed to the walls of the box carries the refrigerant in others.

There are two types of cycle defrost evaporators. One is a lightweight vertical plate usually made from bonded sheets with integral passages for the refrigerant. The other is a serpentine coil that could have fins.

The automatic defrost evaporator is usually a fin-and-tube arrangement (Fig. 11-6). The coil is arranged for forced air to flow parallel to the long dimension of the fins. This design minimizes the effect of frost accumulation. Sometimes the fins are more widely spaced at the air inlet. This minimizes the air restriction caused by frost buildup.

In chest freezers, the evaporator consists of tubing in thermal contact with the exterior of the inner lining. The tubing is more concentrated near the top of the lining than at the bottom. This spacing of the tubing takes advantage of the natural circulation of the air inside. Some upright freezers have a refrigerated surface at the top of the food compartment and refrigerated shelves connected in series with an accumulator at the end. Automatic defrost freezers have a fin-and-tube evaporator like the one explained earlier.

Condensers

The condensers used in household refrigerators and freezers can be cooled by natural draft or by fan. The natural draft condenser is located on the back of the cabinet (see Fig. 11-6). It is cooled by the natural flow of air under the cabinet and up the back. Its construction usually consists of flat serpentine steel tubing with steel cross wires welded on one or both sides. The wires run perpendicular to the tubing. Another version of the natural draft condenser consists of the tubing welded to metal sheeting.

The hot wall condenser, used on freezers for the most part, is also cooled by natural draft. In this design, the condenser tubing is attached to the inside surface of the cabinet shell. As a result, the entire shell helps to dissipate heat from the system and the problem of external sweating is largely eliminated.

The forced-draft condenser does not require any clearance space at the top or sides of the cabinet for circulation. A fan circulates room air over the condenser through a grill at the bottom of the cabinet. The

300   Chapter Eleven

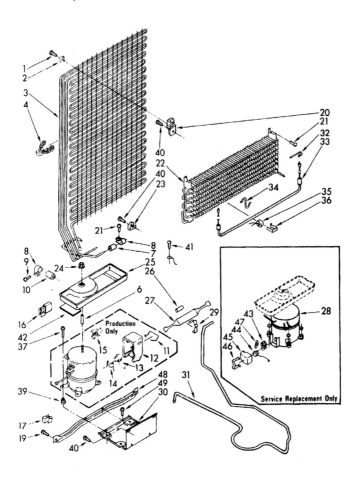

Illus. No.	DESCRIPTION	Illus. No.	DESCRIPTION	Illus. No.	DESCRIPTION
1	Screw, 8 x 7/8 (2)	24	Nut	43	Relay
2	Bumper (2)	25	Evaporator Tray	44	Overload
3	Condenser & Precooler Assembly	26	Restrictor Sleeve	45	Cover
				46	Strap
				47	Overload Spring
4	Clip Tube (2)	27	Drier	48	Unit Mounting Rail
6	Spacer	28	Compressor (Also Order 482801 Mounting Kit & 876765 Tube Kit)	49	Screw
7	Dampener				
8	Tube Clamp				
9	Screw, 8-32 x 1/2				
10	Dampener				
11	Relay Case Spring	29	Drier Clamp		
12	Relay Case	30	Unit Mounting Plate		
13	Overload	31	Heat Exchanger		
14	Overload Spring	32	Heater Clip		
15	Relay	33	Defrost Heater		
16	Housing Plug (Overload Only)	34	Heater Support		
		35	Defrost Bi-Metal		
17	Clip, Wire (5)	36	Clamp		
19	Screw, 5/16-18 x 5/8 (2)	37	Screw, 1/4-28 x 1		
		39	Grommet (4)		
20	Bracket, Condenser (2)	40	Screw, 8 x 1/2		
21	Screw, 8 x 1/2	41	Screw, 8-32 x 5/16		
22	Evaporator	42	Pad, Sound		
23	Clip, Condenser Mounting (2)				

Figure 11-6 Mechanical refrigeration system components used in a typical domestic refrigerator. *(Sears, Roebuck and Co.)*

construction of these condensers varies. They might be fin and tube, banks of tube and wire, or tube and sheet.

In some designs, a portion of the condenser tubing is routed in contact with the inside surface of the cabinet shell, particularly around the door openings. This prevents condensation from forming and eliminates the need for antisweat heaters. Another design feature on some condensers is a section for compressor cooling. From this section partially condensed refrigerant proceeds to an oil-cooling loop in the compressor. Here the liquid portion of the refrigerant absorbs heat and is re-evaporated. At this point, the vapor is routed through the balance of the condenser.

Compressors

Household refrigerators and freezers use either a rotary or a reciprocating piston compressor in which the whole motor-compressor assembly is hermetically sealed in a welded steel case (Fig. 11-7). Usually a starting relay is mounted on the body of the motor-compressor. The relay also provides overload protection for the motor. The compressor is located in the base of the cabinet.

Capillary tubes

The capillary tube is the refrigerant metering device usually used in domestic units. It connects the outlet of the condenser to the inlet of the evaporator (Fig. 11-8) and works on the principle that a given weight of gas passes through the tube less readily than the same weight of liquid at the same pressure. So if refrigerant vapor enters the tube, the mass flowing through the tube decreases. As a result, the refrigerant has

Figure 11-7 Hermetic compressor mounted in the rear of a domestic unit.

REFRIGERANT CYCLE

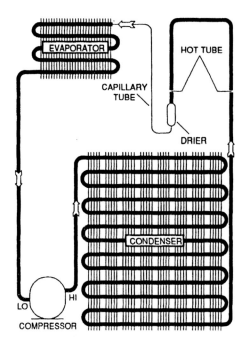

Figure 11-8 The location of capillary-tube flow control used in domestic systems.

more time to cool in the condenser. The capillary tube is often soldered to the suction line tube. This prevents the cold suction line from sweating and increases the refrigerating capacity and efficiency. Because the capillary tube is the narrowest passage in the whole system and low temperatures first occur here, a strainer-drier is often positioned just ahead of it. This prevents ice or any other foreign material from clogging the capillary tube.

Defrost systems

The temperature of the evaporator in a domestic refrigerator or freezer is below freezing. As the air inside one of these units passes over the surfaces of the evaporator, the moisture in that air condenses and forms ice on the evaporator. Over time this ice builds up and reduces the operating efficiency of the refrigeration system. For this reason it is frequently necessary to defrost the evaporator. The way in which defrost is performed depends on the design of the unit.

Cycle defrosting

Some domestic refrigerators and freezers use a cycle defrosting system. It is so called because the evaporator defrosts during each off-cycle of the compressor. In this design, the evaporator might be located either in an air duct remote from the fresh-food space or near the top of the compartment. When the evaporator is located in a remote air duct, a small motor-driven fan draws air over the cooling surface during the running part of the cycle and circulates it through the food compartment(s). In either case, heat leakage into the cabinet during the off cycle provides the energy required for defrosting. Also, the evaporator in both setups is arranged for good water drainage during the defrost cycle. As the frost melts off the evaporator, the water is conducted through tubes to a collecting surface called an *evaporating pan*. The pan is located directly over the compressor or the condenser. Here heat from either component evaporates the water so that it returns to the room's atmosphere. A cold control senses the temperature of the evaporator and cycles the compressor on when its surface is above 32°F (0°C).

In a refrigerator/freezer with two separate evaporators, this system is sometimes called *partial automatic defrost* when the evaporator in the fresh-food compartment is cycle defrosted, but the freezer evaporator must be manually defrosted at infrequent intervals.

Automatic defrosting

Two basic systems for automatic defrosting exist. One system uses 300 to 1000 W electric heaters to melt the frost off the evaporator. Some models of this system use radiative heating elements arranged to radiate heat towards the evaporator. In others, metal-sheathed heating elements stand in thermal contact with fins on the evaporator. The other automatic defrosting system uses a solenoid valve to divert hot refrigerant gas from the compressor discharge line through the evaporator(s). Because this hot gas is applied directly to the inside surfaces of the evaporator(s), it does not take very long to defrost. No matter which system is used, all surfaces of the evaporator must be heated adequately during the defrost mode to ensure complete defrosting. Automatic defrosting systems use the same drainage system described for cycle defrost systems.

Both cycle- and automatic-defrost systems initiate the defrost period by means of a timer that is activated at intervals of up to 24 hours (Fig. 11-9). Often the timer measures only compressor running time. One example of this type is driven by a self-starting electric motor geared to turn one revolution for every interval of the specified compressor running time. When the time period has elapsed in the electric heater system, the timer stops the compressor and activates the electric heaters.

In the hot gas system, the solenoid valve is opened to the evaporator at this time. The defrost timer and solenoid valve is located in the cabinet base. In some models, manufacturers have introduced microprocessor-based control systems that include an adaptive

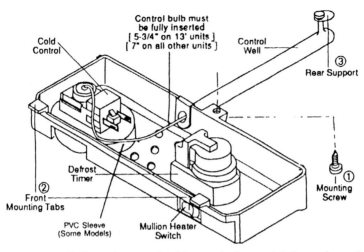

Figure 11-9 Defrost timer and cold-control thermostat for a domestic refrigerator/freezer. This unit mounts in the refrigerator compartment. *(Sears, Roebuck and Co.)*

defrost function in their software. These systems monitor certain parameters and vary the period between defrosts according to those actual measured conditions of use. The end results are an improvement in food preservation and reduced energy consumption.

The duration of the defrost period is kept short in order to limit any rises in food temperatures. In many cases its length is controlled by a timer, but in some electric heater systems it is governed by a thermostat called a *defrost terminator*. This thermostat is attached to the end plate of the evaporator. When it reaches a temperature of approximately 50°F (10°C), it opens the heater circuit. Usually the compressor is not restarted immediately when the heater circuit is opened. Instead, a timer, using the start of the defrost period as a reference point, reactivates the compressor and the air-circulating fan after a designated interval. This allows the evaporator time to drain and enables system pressures to subside.

Electrical circuits

The components and complexity of the electrical circuits used in household refrigerators and freezers varies greatly from model to model. A sample of that variety is presented here.

A relatively uncomplicated circuit for a manual defrost refrigerator/freezer is shown in Fig. 11-10. There are three separate circuits leading out of the panel disconnect.

One circuit supplies current to an electrical resistance heat wire, also called a *perimeter drier*, located in the trim of the freezer door.

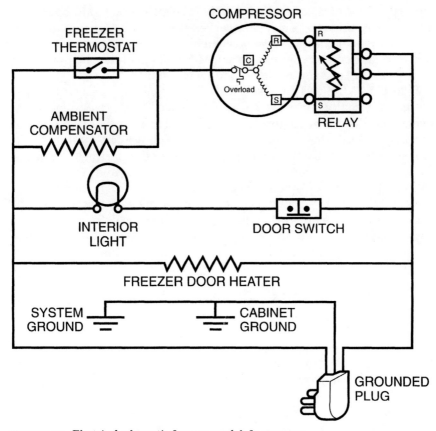

Figure 11-10 Electrical schematic for a manual defrost system.

This drier, which stops condensation from appearing, operates all the time.

The second circuit has a cabinet light and a door switch in series. The light comes on when the refrigerator door is opened. In the third circuit, the thermostat and motor-compressor are in series. The thermostat closes the circuit to the motor's starting relay when the temperature in the cabinet rises to a set temperature. The starting relay passes current to both the starting winding and the running winding of the motor. When the motor reaches approximately 75 percent of running speed, the starting relay disconnects the starting winding. The refrigeration system starts to cycle when the compressor begins to run. Once the temperature in the cabinet drops to the minimum desired temperature, the thermostat opens the compressor circuit, which stops the refrigeration cycle. An ambient compensator is connected in parallel across the thermostat. It operates only when the thermostat contacts are open. The ambient compensator is an electrical resistance heat wire that provides a small flow of heat into the food compartment. Its purpose is to cause the refrigerator to cycle even when the ambient temperature of the room is lower than the setting of the thermostat.

Figure 11-11 shows the electrical circuit and wiring diagram for a hot-gas automatic-defrost refrigerator/freezer. The defrost timer is connected parallel to the compressor so it operates whenever the compressor does. Until the period of measured compressor running time elapses, the timer directs current to the freezer compartment fan. Once the period elapses, the timer disconnects the fan and activates the water defrost solenoid and the drain sump heater. This heater prevents condensation from freezing as it moves from the evaporator down the drain tube to the evaporating pan. At the end of the defrost period, the timer disconnects from the solenoid and the heater and reconnects to the fan. The circuit has a receptacle for an optional automatic ice maker and an outlet on the panel disconnect for the water valve.

The circuit for an electric automatic defrost is shown in Fig. 11-12. This model has two heaters used to keep condensation from forming on the outside of the cabinet. (Some models have many more heaters. Besides the mullion heater and the freezer flange heater used in this model, they might include butter cavity heaters, water tank heaters, ice-chute heaters, freezer control duct heaters, and auxiliary heaters, which can be connected when the originals fail.) These heaters are in series with a power-saver switch so that they can be turned off if the refrigerator is being used in an area with low humidity. The switch is located inside the cabinet.

The motor-compressor, condenser fan motor, and evaporator fan motor are all connected in parallel so they always run at the same time. These components are in series with the defrost timer and the thermostat. The defrost timer switch must be connected to the motor circuit and the thermostat closed for these motors to operate. Once the measured period of compressor running time elapses, the defrost timer switches off the motor circuit and connects the defrost heater circuit. The defrost circuit shuts off when the defrost terminator detects that the evaporator has reached the specified temperature, usually about 50°F (10°C). However, the defrost timer does not reconnect the motor circuit until after a specified time from the start of the defrost cycle has elapsed. So there is a brief period of time after defrosting when neither the motor circuit nor the defrost circuit is operating.

In some refrigerators, the electromechanical thermostat has been replaced by a microprocessor-based control system that receives electronic signals from thermistor-sensing devices. In addition to replacing

Domestic Refrigerators and Freezers 307

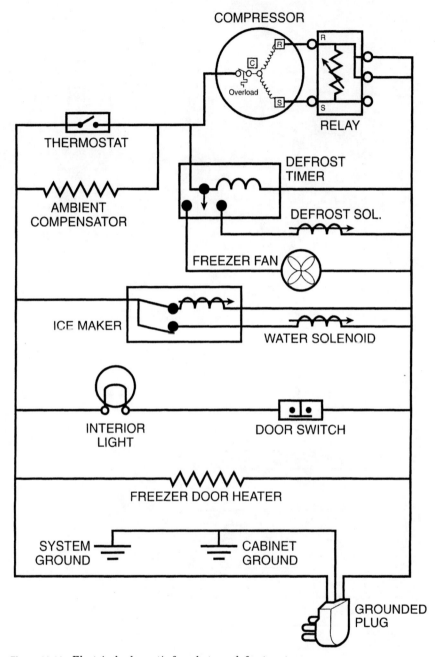

Figure 11-11 Electrical schematic for a hot gas defrost system.

308 Chapter Eleven

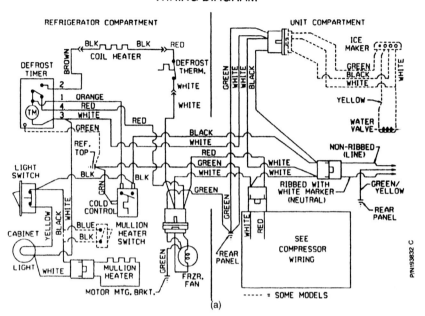

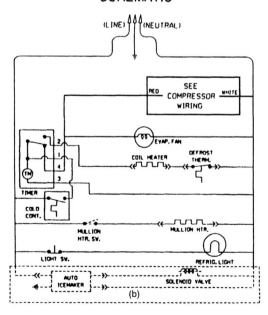

Figure 11-12 Electrical schematic and wiring diagram for an electric defrost system. *(Sears, Roebuck and Co.)*

compartment controls, the microprocessor might also control ice and water dispensers or any other accessories, sound an alarm if a door is ajar, and provide diagnostic codes. Some refrigerators even have a kitchen entertainment center.

Freezers might have signal lights to indicate when power is on and when cabinet temperature is above normal. Otherwise, freezer circuits are similar to refrigerator circuits, though they are generally less complex.

Installing units

A refrigerator/freezer should not be located in direct sunlight, or near a radiator, heat register, or stove. If it must be placed near a stove or radiator, position aluminum foil between the unit and the heat source. Also make sure that there is adequate ventilation for the type of refrigerator being used.

Use a voltmeter to see if the wall outlet provides the correct electrical supply and to see that it is properly grounded. If not grounded, attach a wire to a metal part of the cabinet and a good ground such as a copper water pipe.

Once the unit is in place, check it with a level. Adjust the front supports under the cabinet to level it, or if necessary use wood spacers.

General Troubleshooting

An unsuitable temperature or operating condition is the usual symptom of any defect in the refrigeration system. Table 11-1 lists many of the common problems encountered in domestic refrigerator/freezer service along with their likely causes and solutions. It should be used as a general guideline. It is neither exhaustive in its scope nor universally applicable.

Through good observation, the use of basic electrical tests, and pressure and temperature checks, a good technician can methodically track down the cause of any trouble in a domestic unit. Good troubleshooting involves systematically eliminating possible causes of a problem through the collection of data. In the attempt to narrow down the possible causes of a problem, external checks should be performed before internal ones. Internal checks are those that require delving inside the hermetic system or removing a part from it.

Often a simple visual inspection provides important clues indicating the cause of a problem. For instance, a lack of refrigerant might be revealed by an evaporator heavily frosted in one area and lightly frosted on the rest. The technician can check for the position of control settings and switches, bad or frozen door gaskets, frost or sweat on suc-

TABLE 11-1   Troubleshooting Domestic Refrigerators

Problem	Cause	Solution
Refrigerator does not run.	Blown fuse	Replace fuse.
	Low voltage	Check outlet with a voltmeter, which should read 115 V plus or minus 10 percent. If the circuit is overloaded, reduce the load or have a separate circuit installed. If no other solution can be found, install an autotransformer.
	Defective motor or temperature control	Connect a jumper across the terminals of the control. If unit runs and connections are all tight, replace the control.
	Faulty relay	Check relay, replacing as needed.
	Defective overload	Check overload, replacing as needed.
	Faulty compressor	Check compressor, replacing as needed.
	Defective service cord	Check with test light at outlet; if no circuit or current is detected, replace or repair.
	Break in lead to compressors, timer, or cold control	Repair or replace broken leads as needed.
Refrigerator section too warm.	Door is opened too frequently	Remind the owner to open door less often.
	Overloaded shelves are restricting normal air circulation inside the unit	Tell the owner to keep less food in unit.
	Food is too warm or hot when placed inside the unit	Inform the owner that food should cool to room temperature before it is placed in the unit.
	Door does not seal properly	Level the cabinet, then adjust the door seal.
	Interior light stays on constantly	Check light switch and replace as needed.
	Cold control knob is positioned at a warmer setting; the unit cannot operate often enough to keep food cool	Reset the knob to a colder position.
	Refrigerator recently disconnected	Allow 24 hours for complete cooldown.
	Freezer fan is not running correctly	Check the fan blades, switch, and wiring and replace as needed.
	Defective intake valve	Replace motor compressor.
	Air duct seal is not properly positioned	Reposition, making sure that it is sealed.

TABLE 11-1 Troubleshooting Domestic Refrigerators (Continued)

Problem	Cause	Solution
Refrigerator section is too cold.	Air flow knob position is set too cold	Turn knob to a warmer setting.
	Air flow control will not close	Check for any obstructions and remove them.
	Faulty airflow control	Replace.
	Defective airflow heater	Replace.
Freezer and refrigerator sections are too warm.	Fan motor broken	Check and replace as needed.
	Cold control set too warm	Reset. If still not working, replace it.
	Finned evaporator blocked with ice	Check defrost thermostat or timer.
	Shortage of refrigerant	Check for leaks and repair by system evacuation and recharging.
	Insufficient air circulation around the unit	Relocate the unit to increase clearance and allow for better circulation.
	Condenser is dirty; condenser ducts are obstructed	Clean condenser and ducts thoroughly.
	Door is not sealing properly	Level the cabinet; adjust door seal.
	Door is opened too frequently	Remind the owner to open the door less often.
Freezer section is too cold.	Cold control knob is set incorrectly	Reset knob to a warmer position.
	Cold control capillary is not fully clamped to the evaporator	Reposition and retighten the clamp.
	Faulty cold control	Check control and replace if needed.
Unit runs constantly.	Insufficient air circulation around the unit or air circulation is obstructed	Relocate cabinet to increase clearance; remove any obstructions.
	Door seals improperly	Check and make any needed adjustments.
	Too much ice being made in the freezer; too much food in the refrigerator area	Inform the owner that heavy use will increase running time.
	Refrigerant is undercharged or overcharged	Check, evacuate the system, and recharge with the correct amount of refrigerant.
	Room temperature is too warm	Increase the ventilation or lower the room temperature.
	Cold control not operating properly	If the control is not regulating the unit, check and replace if needed.

TABLE 11-1 Troubleshooting Domestic Refrigerators (Continued)

Problem	Cause	Solution
	Defective light switch	If light will not go out, replace the switch.
	Door opened too frequently	Tell the owner to open the door less often.
Noisy operation.	Uneven or weak flooring	Fix flooring; brace the unit if needed.
	Tubing is in contact with the unit or with other tubing	Reposition tubing.
	Unit is not level	Level the unit.
	Drip tray is vibrating	Move the tray and/or place on styrofoam padding
	Fan is striking liner or mechanically grounding	Move the fan away from any obstructions.
	Compressor is mechanically grounded	Replace compressor mounts.
Unit cycles on overload.	Faulty relay	Replace relay.
	Weak overload protector	Replace protector with stronger model.
	Low voltage	Check the outlet with a voltmeter; underloaded voltage should be 115 V plus or minus 10 percent. Check that there are not too many appliances on one circuit; check that any extension cords are the proper size and length.
	Compressor not operating	Check with test cord; also check for ground and replace if needed.
Stuck motor compressor.	Defective valve	Replace the compressor.
	Insufficient oil	Add oil; if the compressor will not operate, replace it.
	Overheating	Replace as needed.
Ice maker not making ice.	Wire signal arm in OFF position	Move wire arm to ON.
	Water line off or blocked	Turn on line or clear.
	Freezer not cold enough	See section of table on page 311.
Unit runs constantly; temperature is normal.	Icy buildup on the evaporator	Check door gaskets and replace as needed.
	Thermostat control bulb does not contact the evaporator surface	Move the bulb so that it does contact the surface.
Freezer runs constantly; too cold temperature.	Faulty thermostat	Check thermostat and replace as needed.

TABLE 11-1 Troubleshooting Domestic Refrigerators (Continued)

Problem	Cause	Solution
Freezer runs constantly; too warm temperature.	Icy buildup in the insulation	Remove breaker strips. Turn off the unit, melt the ice, and dry the insulation. Seal all outer shell leaks and joints, then reassemble.
Rapid ice buildup on the evaporator.	Leak in the door gasket	Adjust the door hinges; replace the gasket if cracked, brittle, or worn.
Door on freezer compartment freezes shut.	Defective electric gasket heater Faulty gasket seal	Use an alternate gas heater or install a new one. Inspect the gasket. If worn, cracked, or hardened, replace it.
Freezer works, then warms up.	Excessive moisture in the refrigerator	Install drier in liquid line.
Loss of freezing capacity.	Wax buildup in the capillary tube	Clean the capillary tube with a tube cleaning tool or replace it.

tion lines and on driers, ice on the evaporator, and dirt collecting around the condenser. A careful eye can pick up other hints as well.

Other elementary checks include listening for rattles, hisses, and hums, and checking the temperatures of the different system components. Electrical parts and wiring can be checked relatively quickly with the proper equipment too.

Once all external checks have been performed or have eliminated as possible causes any external problem, internal checks should be made.

Normal operating conditions

Diagnosing refrigerating units successfully requires knowing how each unit is supposed to perform. One source of this type of information is the identification plate mounted on the compressor motor or inside the refrigerator door. The I.D. plate lists important information such as the refrigerant used, the correct refrigerant charge (in ounces), compressor horsepower, rpm, and electrical data including running amps, voltage, and phase.

Also, schematics of the electrical circuits are often attached to the back of units, or taped to the underside near the compressor. Refer to these drawings and return them to their original location when service is completed. You might need them again in several years.

The manufacturer's service literature provides additional information about normal operating conditions. Table 11-2 lists typical operating parameters for a domestic refrigerator/freezer. Always check these

TABLE 11-2 Domestic Refrigerator Midpoint Control Settings (No Load or Door Openings)

Type A: with run capacitor Type B: without capacitor	65°F ambient		90°F ambient	
	Type A	Type B	Type A	Type B
Operating time	30 to 45%	35 to 45%	54 to 65%	55 to 67%
Freezer temperature	2 to 8°F	2 to 6°F	2 to 6°F	2 to 6°F
Refrigerator temperature	35 to 40°F	35 to 40°F	35 to 40°F	35 to 40°F
Low-side pressure, cut in	8 to 16#	8 to 12#	8 to 16#	8 to 12#
Low-side pressure, cut out	1 to 4#	1 to 4#	1 to 6#	1 to 4#
High-side pressure (last 1/3 cycle)	110 to 120#	110 to 120#	150 to 175#	150 to 175#
Wattage (last 1/3 cycle)	150 to 200	190 to 200	150 to 200	190 to 220
Amps, running	1.3 to 1.8	2.2 to 2.6	1.3 to 1.8	2.2 to 2.6

sources for the appropriate information before starting any troubleshooting procedure.

Knowing the normal temperature-pressure conditions during the operating cycle of the refrigeration system is crucial to performing certain diagnostic tests. The technician should know the normal temperature in the evaporator and in the condenser and also the normal pressure on the high- and low-pressure sides. These conditions vary depending on the refrigerant used. While HFC-12 has been the most popular refrigerant for domestic units for decades, new refrigerators are charged with R-134a. Because domestic refrigerators can operate

TABLE 11-3 Approximate Temperature-Pressure Conditions in Refrigerator/Freezer

	Ambient Room Temperature 72°F.			
HFC-12 Refrigerant	Evaporator Temperature*	Evaporator Pressure*	Condenser Temperature	Condenser Pressure
Beginning of Cycle	15°F	12 psig	75°F	75
Midpoint of Cycle	5°F	9 psig	100°F	125
End of Cycle	0°F	5 psig	100°F	125
	Ambient Room Temperature 90°F.			
HFC-12 Refrigerant	Evaporator Temperature*	Evaporator Pressure*	Condenser Temperature	Condenser Pressure
Beginning of Cycle	15°F	12 psig	90°F	85
Midpoint of Cycle	5°F	9 psig	115°F	160
End of Cycle	0°F	5 psig	115°F	160

* Readings in Fresh-Food Compartment

for decades without a major problem, you will encounter systems charged with R-12 well into the next century.

Table 11-3 lists the average temperature-pressure conditions for an evaporator and a condenser of a typical refrigerator/freezer charged with HFC-12.

Troubleshooting electrical circuits

Troubleshooting electrical systems involves more than measuring volts and ohms. Once again, simple observation is a powerful tool not to be neglected. Look for electrical connections discolored to a blue or green tint by overheating or corrosion. Obviously, charred insulation indicates overheating as well. Connections overheat when they are loose or dirty. A frozen door gasket might indicate that the heater in the door is defective.

Some electrical tests do not require a meter. A test light is all that is needed to check for open circuits and grounded electrical wires. Motors can be checked with a manual start test cord. Some manufacturers have developed their own test equipment that tests multiple circuits at once. The technician just connects the tester to a multiple circuit connection incorporated into the circuits. Manufacturer's instructions should be followed when using this equipment.

When dealing with a motor-compressor that either short-cycles or will not start, it is important to check all of the electrical components involved before judging the unit to be defective. The thermostat, capacitor (if used), and relay are easier to check when they are removed from the circuit. The parts themselves can be tested directly or they can be temporarily replaced with an appropriate test part to see if that solves the problem. All wiring, connections, and terminals from the outlet to the motor should be tested too. Use a voltmeter to measure the open circuit voltage at the wall outlet and then the voltage at the outlet when the unit is running. A voltage drop of 10 volts or more indicates an overload, a problem in the motor windings, or poor wiring to the wall outlet.

A manual test cord makes it easier to determine whether the problem is in the compressor-motor or in the external circuit. Just make sure that the appropriate test cord is available for the type of compressor motor being checked. Different setups are required for different motors (Fig. 11-13). When either a capacitor-start, induction-run motor, or a capacitor-start, capacitor-run motor is tested, new test capacitors with the same ratings as those in the system being tested should be used. To avoid shocks, remember to discharge the test capacitors after testing the motor.

Once the motor is disconnected from all wiring in the circuit and the test leads are correctly attached, press the manual switch button and hold it in for one or two seconds. Then release the button enough to

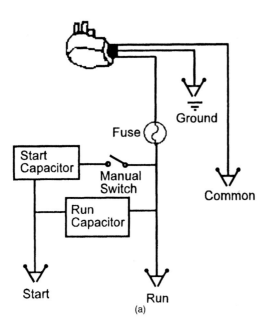
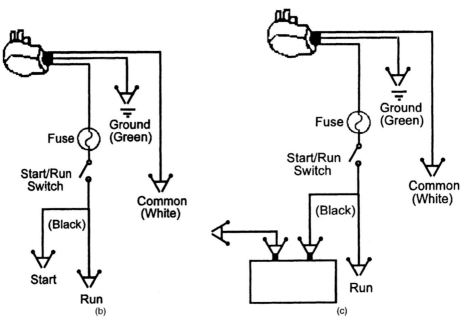

Figure 11-13 Test cord sets for domestic refrigerators (a) cord for hermetic motors, (b) cord for capacitor-start, induction-run motors, and (c) cord for capacitor-start, capacitor-run motors.

open the starting circuit. If the motor works, the problem is somewhere in the electrical system up to the compressor. If it does not work, see Chapter 10 for additional information about troubleshooting compressor-motors.

Internal troubleshooting and service

Internal troubleshooting and service requires attaching gauges and servicing devices, such as vacuum pumps and refrigerant cylinders, to the system. These devices must be connected in such a way that no air, moisture, or dirt is introduced to the system. There are several ways to do this. Some systems have adapters to which service valves can be attached. In systems without these adapters, either piercing valves or core valves must be mounted on the suction line tubing, the discharge line tubing, or the process tube. (The process tube is used during the manufacturing process to evacuate, test, and charge the new unit.) In some cases, the technician might install these valves on both the suction and discharge lines.

Service valves and adapters

The service valve adapter is welded or brazed onto the dome of the compressor. Inside the adaptor, there is a needle that can be screwed opened and closed by the service valve when it is fastened to the adaptor. The service valve screws onto the adaptor. A copper or synthetic gasket seals the joint between the adaptor and the valve attachment. This attachment has a valve stem that is inserted like a screwdriver into a screwhead on the adaptor needle. A handwheel on the other end of the drive can be turned to open and shut the needle. A service valve might have one or two openings. On valves with two openings, a pressure gauge can be attached while service operations are being performed.

Before a service valve is attached, the adaptor should be cleaned and the gasket inspected. Once the valve is threaded onto the adaptor, the packing unit surrounding the valve stem should be tightened into the top of the valve body before the valve is purged and opened.

Piercing valves

As explained in Chapter 2, there are basically two types of piercing valves: those that are bolted and those that are brazed onto the line. When attaching a bolt-type, check to see that there is enough room to connect the attachment valve and service line(s). Also make sure the tubing is straight and round. Then carefully clean it and apply a small amount of clean refrigerant oil on the tubing. Make sure the bushing gasket is in place and the needle valve stem is completely retracted.

Mount the valve on the tubing and then tighten the clamping screws evenly.

When attaching a braze-mounted valve, make sure that there are no flammables in the area, that the tubing is round and straight at the brazing site, and that there are no soft-soldered joints near it. Also, check to see that there is enough room to connect the attachment valve and service line(s). Remove the needle valve stem and the gasket from the valve assembly. Use clean sandpaper or steel wool to clean the saddle surface of the valve assembly and the tubing surface. Then apply clean brazing flux to the saddle surface (unless a phosphorous-copper brazing filler rod is used for brazing). Mount the saddle of the valve assembly on the tubing. If necessary, use a small C-clamp to keep the assembly from moving during the brazing procedure. Remember to use safety goggles. Heat the saddle and tubing until filler flows around the saddle. Be careful to avoid overheating because it can cause the tubing to burst. Once the brazed joint has cooled, return the needle valve stem and gasket to the valve assembly. After installing either type of piercing valve, service valve attachments can be installed.

Core valves

In many systems, a core valve is used to gain access to the hermetic system. The adaptor for these valves can be threaded to the system or either brazed or clamped to the tubing. When a service line fitting or service valve is mounted on the adaptor, a pin in the fitting or service valve depresses the core valve stem in the adaptor and allows gas to flow.

Leak detection and repair

Poor refrigeration is sometimes caused by a shortage of refrigerant in the system, and a lack of refrigerant indicates a leak. Any leak must be located and repaired. After a leak has been found and repaired, the entire system should be checked again to make sure that the repair is successful and that there are no other leaks.

A common method used for detecting leaks is the bubble method. A soap and water solution, or some similar patented solution, is brushed over an area of the system. Any gas leaking from the system causes the solution to bubble. The solution is easy to use and low in cost. Another advantage is that it can be used around urethane insulation whereas a halide torch and electronic leak detectors cannot.

The method used to repair a leak depends on the metal or the combination of metals used at the leaky point. Copper and steel can be brazed. Aluminum can be brazed, aluminum soldered, or repaired with epoxy cement. Aluminum and copper or steel connections can be resistance welded.

Most leaks occur at tubing connections. When a leak is found at a brazed or silver-soldered connection, the repair is relatively simple. Just clean the connection, coat it with flux and reheat it. However, the best way to repair a leak at a flared connection is to use new flare fittings and to make a new flare.

During leak repairs, a metal sheet should be inserted between the repair site and the cabinet to reflect heat away from the cabinet. It is also important to remember that a drier should never be heated because the applied heat will drive moisture back into the system.

Undercharged hermetic systems

A shortage of refrigerant might be indicated by a refrigerator that is running too much, a warm or partially frosted evaporator, or a low pressure on either the high-side or low-side. If one or more of these conditions exists, the system might need to be charged with refrigerant.

Any leak in the system should be repaired before charging, and the system should always be charged through a low-side valve with the refrigerant in vapor form. See Chapter 8 for details on purging, evacuation, and charging systems.

Locating compressor faults

Often a motor-compressor in good condition is replaced when an electrical problem is not discovered or is misdiagnosed. As stated earlier, it is important to rule out any external electrical problems before labeling the motor-compressor faulty.

If external circuits and the electrical motor are working properly, the compressor can be checked by making a volt-amp reading at normal low- and high-side pressures. A volt-amp reading below the motor rating means that the pump might be worn out.

A test can be performed to check the pumping ability of the compressor. A gauge manifold is installed on a low-side valve and the suction line is pinched off between the valve and the evaporator. Then the motor is turned on to see how much vacuum it can pull. Once the pressure reaches about 25- to 28-in. Hg (7 to 17 kPa), turn off the motor. If the pressure starts to drop toward 10-in. Hg (68 kPa), the exhaust valves are leaking and the compressor must be replaced or overhauled.

If there is an offensive odor when a piercing valve is opened, it indicates that the compressor motor is burned out. The heat from the motor breaks down the refrigerant, and if moisture is present, hydrochloric and hydrofluoric acid are formed. A system with a burnt motor should be repaired within two or three days to prevent these corroding chemicals from damaging the system. It should be flushed with R-11 or the refrigerant used in the system. Also, use goggles, rubber gloves, and extreme caution when handling contaminated oil. It can cause a severe burn. The oil should be collected in a glass container.

There are a number of things that can cause a motor to burn out: low voltage, a short, a ground, an overload, a faulty circuit protection device, a stiff compressor, high head pressure, or a lack of refrigerant that could lead to inadequate cooling. It is important that any external cause of trouble is located and corrected before a new unit is installed.

Checking for restrictions

If a compressor starts but then short cycles on the overload protection, or a liquid line is frosted or sweating, the problem could be a restriction in the system. A restriction can also cause the compressor to run continuously or excessively long.

The technician can check for restrictions, after the system has run for a few minutes, by listening for a hissing sound where the capillary tube enters the evaporator. The absence of a hissing sound indicates a restriction. In that case, the next step is to heat the end of the capillary tube that feeds into the evaporator with warm water and a rag. If ice is restricting the line, the warmth will melt the ice and a hissing sound will result.

Another test involves measuring the pressure on the low side. A vacuum indicates that the system has a restriction or it is low in refrigerant. To determine which is the problem, open a valve at the condenser outlet just a little bit and allow some refrigerant to escape. (It might be necessary to install the piercing valve first.) If nothing escapes or if gas escapes, the system needs to be charged. If liquid refrigerant escapes, there is a restriction either in the capillary tube or the filter-drier. In that case, it must be determined which component is at fault. First, discharge the system and then clean, flux, and heat the connection between the capillary tube and filter-drier to separate the two components. Charge a small amount of refrigerant into the low side of the system. No gas will flow out of whichever component is clogged.

A capillary tube might not need to be replaced if it can be repaired with a capillary tube cleaner.

Installing filter-driers

Filter-driers keep a system dry and clean inside. A new filter-drier should be installed in the liquid line just before the refrigerant control. In this position, it is far away from the heat of the condenser, which would cause the drier to give up the moisture it had absorbed. Also, any heat that reaches the drier in this position would also reach the refrigerant control, so the chance of ice forming in the control is minimized. The sealing caps on the drier should not be removed until just before installation. Make sure that the filter-drier is installed so that the arrow on its body points in the direction that refrigerant should flow. Either blazed or flared connections can be used for installation.

A filter-drier should be replaced whenever it is clogged or a new motor-compressor is installed. Often two filters are installed when a system is repaired. The second one is placed between the evaporator and the compressor. This helps to ensure that any moisture or contaminants that entered the system during servicing are removed. There are two drying agents used in these devices: liquid and solid. If a second drier is added to a system, it should be the same type as the first. Otherwise the solid drier will cause the liquid one to release its moisture.

Hot gas defrost problems

A heavily frosted evaporator might indicate a broken timer or a problem with the solenoid of the hot gas valve. The electric coil of the solenoid and the timer can be checked electrically. If there is no electrical problem, the valve stem of the solenoid might be stuck. Tapping the valve body while the timer switch is closed might break loose the valve stem. If it does, a surge of hot gas can be heard and the line between the solenoid and the evaporator will become warm to the touch. A faulty valve should be replaced.

If the evaporator and the line between it and the solenoid are warm, the valve stem might be stuck in an open position. If tapping frees the valve stem, low-side pressure will begin to drop and the evaporator will begin to cool.

Dismantling a system

Certain preparatory steps should be taken before any component in a hermetic system is replaced. First the electrical circuit should be disconnected. Then carefully clean all system surfaces to reduce the chance of contamination when the system is opened. Next, purge the system of refrigerant with a recovery/recycling unit. Finally, cut the tubing and remove the defective component. Remember to wear goggles whenever a system is being dismantled.

Installing a motor-compressor

When replacing a defective hermetic compressor, make sure the replacement is an exact match with the same capacity. Carefully clean the ends of the oil lines, suction line, and discharge line, along with the connections for these lines on the motor-compressor. If a piercing valve has not already been installed in the system, one should be installed now. Then, after attaching a gauge manifold to the valve, connect the lines to the compressor. It might be necessary to use adaptor fittings, lengths of tube, or an expander. Before brazing the connections, take precautions to protect nearby items from the heat. Position sheet metal to keep the heat away from the cabinet, and any plastic parts or wires. Cloth or heat-absorbing compounds can be used to shield other brazed joints in the vicinity.

Next, install two filter-driers in the system. Installing both reduces the chance of the motor-compressor burning out again soon due to dirt or moisture in the system.

Once the filters are installed, leak-check the system. If no leaks are found, evacuate the system and then charge the lines with the refrigerant used in the system so that system pressure is equal to or slightly above atmospheric pressure. Then reconnect the electrical circuit and charge the system to specifications.

Once the installation is complete, the operation of the system should be checked. Place a recording thermometer in the refrigerator cabinet for at least 24 hours.

Repairing condensers and evaporators

Sometimes domestic refrigerator condenser problems are as easy to solve as removing dirt and lint from the outside. At other times, when a leak cannot be reached for repair, the only solution is to replace it. For those leaks within reach, see the section on leak repair for the appropriate method of repair.

A stainless-steel evaporator can be brazed or welded with an inert gas tungsten arc welding system. Aluminum evaporators can be aluminum soldered (no flux is used), aluminum welded with an inert gas tungsten arc welding system, or sealed with an epoxy. When an epoxy is used, the surface around the leak should be cleaned immediately before application because oxidation occurs quickly. Also, the system should be open to the atmosphere at some other opening when the epoxy is applied.

Always remove all refrigerant from the system before repairing a leak in either the condenser or evaporator. Once the leak is repaired, leak test, evacuate, charge, and check the unit for proper operation.

Testing a Rebuilt System

The final step involved in repairing a domestic refrigerator or freezer is testing. The system should be monitored for at least 24 hours with a recording thermometer placed in the refrigerator as it runs under thermostatic control. Dome and line temperatures should be checked as well. Also check to see if the evaporator is frosting evenly. The electrical section can be checked with a voltmeter and ammeter or with a combination hermetic compressor analyzer and electrical tester. The efficiency of a system can be checked by installing a thermometer on one of the last coils of the evaporator and then covering the evaporator with insulation. After about 30 minutes, the thermometer should read as low as 25°F in a room with an air temperature of about 70° to 75°F.

Shutting Down Units

When shutting down a unit, allow several hours for it to thoroughly defrost. Then wash the inside of the cabinet with a solution of baking soda and water. Finally, make sure that the cabinet is completely dried out. Place a portable heater inside the cabinet to expedite the process. The doors should be ajar slightly so that air can circulate during shutdown. Following these steps will help to prevent odors and rusting.

Federal law, passed to protect children from suffocating in unused units, states that cabinet doors must be removed from refrigerators or freezers being taken out of service.

Chapter

12

Commercial Reach-In and Walk-In Units

This chapter discusses the operation and troubleshooting of reach-in and walk-in type refrigerators and freezers used in commercial applications. It is important to remember that much of the information found in this chapter pertaining to control systems, timer clocks, and defrost systems also can be applied to systems described in Chapters 13 and 14. Similarly, Chapters 13 and 14 contain additional information on compressors, condensing systems and other items that is useful in servicing reach-in and walk-in units.

Reach-In Refrigerators

As a service person, the reach-in refrigerator is the most common type of unit you will be called on to service and troubleshoot. They are used in virtually all types of food service and retail applications.

Commercial reach-in refrigeration units are designed for the short term storage of perishable items in a temperature range of 35 to 40°F. Characteristics of commercial reach-in refrigerators include stainless steel or aluminum construction with a minimum of plastic parts. Strong construction in the door and frame components enable them to be opened and closed literally hundreds of times per day. Large-size compressor units can handle the higher cooling loads caused by constant door openings. Reach-in refrigerators can be further classified as either display or storage units.

Display units. Display or merchandising units feature glass doors and rack and shelf configurations gear toward customer self-service (Fig. 12-1). The use of glass doors reduces the insulating efficiency of the unit. Almost constant customer accessing also increases heat leakage into these units. As a result they are often equipped with slightly oversized condenser units.

Storage units. Reach-in storage units feature all stainless steel or aluminum construction with no glass panels (Fig. 12-2). They are used in restaurants, cafeterias, and the food-service industry.

Construction

Reach-in refrigerators can have their condensing unit integrated into the cabinet or located at a remote location. When the condenser unit is integrated into the cabinet design, it can be located either above or below the refrigerated space.

Bottom-mounted condenser units are located where the coolest air in the room will pass over the condenser coils. This is often important in kitchen applications where the ambient temperature can become quite warm. However, being near the floor also exposes the condenser to higher levels of dirt and dust, so regular cleaning is a must for bottom-mount units.

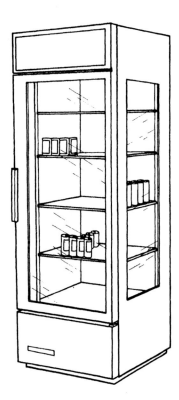

Figure 12-1 Reach-in refrigerated display unit.

Commercial Reach-In and Walk-In Units 327

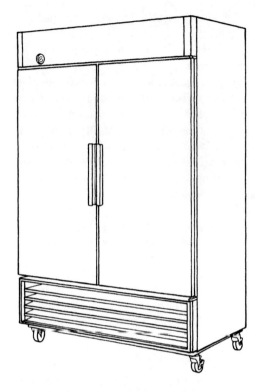

Figure 12-2 Stainless-steel storage freezer.

Top-mounted condenser units help maximize the amount of refrigerated space accessible to the user. They also stay cleaner longer. However, their one major drawback is being exposed to high temperatures near the ceiling of the room.

Locating the condensing unit remote from a reach-in refrigerator maximizes refrigerated space. Easily relocating the unit becomes a problem. The temperature and cleanliness of the remote location can also range from acceptable to detrimental.

Most reach-in refrigeration units use air-cooled condensing units. Compressors of ⅓ to 1 horsepower are typical depending on the unit's size. Electrical power is single-phase 115 V.

Evaporators are normally mounted inside the storage compartment and use either draw or blow-through fans to create air flow across the coils. Evaporator fans used on commercial units are designed to run constantly. If you see one that is not turning, its a sure sign of electrical or motor problems. Gravity coils that use natural convection to create air movement over the coils are not widely used.

In some cases the evaporator can be mounted on the outside of the storage cabinet. Air to be cooled is drawn out of the compartment, passed over the evaporator coils, and returned to the refrigerated space.

Refrigerant flow controls can be either capillary tubes or thermostatic expansion valves. Depending on the evaporator size, the TEV can be either internally or externally equalized. See page 338 for details.

Reach-in refrigerators require a control that maintains compartment temperature within the design requirements. In most reach-in refrigerators, a temperature control monitors evaporator temperature and cycles the compressor on and off as required. The sensing bulb of the thermostat mounts to the evaporator coil and measures coil temperature rather than air space temperature. The cut-in temperature is fixed and the cut-out temperature is adjusted to raise or lower the temperature maintained in the refrigerated compartment. The cut-in temperature must be set high enough (about 37 or 38°F) to provide for self-defrosting of the evaporator coils. A low pressure control is normally used to provide suction pressure sensing that will shut down the compressor if low-side pressure drops below a certain level. This saves the compressor from damage if refrigerant charge is lost.

Some units are also equipped with door anticondensate (defogger) circuits. These are heater circuits wired around all door openings. In hot or humid ambient conditions the switch-controlled circuits are turned on to prevent moisture from forming around the door openings.

Operation

During normal operation, a typical commercial refrigerator continuously circulates cabinet air that is slightly above freezing temperature (36–38°F) through the evaporator coil. The compressor's off cycle provides the time needed to melt any frost that accumulates on the coils during compressor operation. The water from melting frost or ice is directed to a drip pan. In some systems, such as the top-mounted system illustrated in Fig. 12-3, the hot gas line from the compressor to the condenser is routed through the pan to provide heat for evaporating the collected water. In other systems, the drip pan contains electric heating circuits. In still others, the hot air from the condenser fan is directed across the drip pan. The use of high capacity evaporating coils can also help eliminate frost from forming during the cooling cycle. With standard holding refrigerators, high relative humidity is also maintained to prevent dehydration of the products stored in the unit.

Electrical system

The electrical system of a commercial reach-in refrigerator is normally quite simple (see Fig. 9-16). As mentioned earlier, the evaporator fans run whenever the unit is plugged in. The condensing unit (compressor and condenser fans) operate via a relay controlled by the temperature controller. Both the interior light circuit and defogger circuit are switch

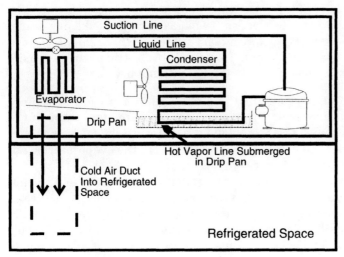

Figure 12-3 Commercial refrigeration unit with top-mounted condensing unit and hot gas defrost.

controlled, with the light switch being activated on the door opening. The defogger switch is manually operated.

Reach-In Freezers

Reach-in freezers maintain case temperatures in the −20 to 0°F temperature range. To generate these temperatures, evaporator coil temperatures in the range of −8 to −20° must be generated. Refrigerants such as R502, R12, and their alternatives are found in these systems. Freezer units must be equipped with a compressor designed for low-temperature operation.

The refrigerant flow control is normally a thermal expansion valve. Both the evaporator and condenser are the forced-air type. The evaporator coils can use either electric or hot gas defrost systems.

On systems with a remote condensing unit, a low-pressure controller monitors suction line pressure to provide for pumpdown. A high-pressure control is used to stop the compressor if high-side pressures exceed a preset limit (usually in the range of 350 psi). Refer to Chapter 5 for details on the various types of temperature and pressure controllers.

Operation and defrost cycle

During normal operation (compressor on) the freezer unit continuously circulates air at temperatures below freezing (0°) through the evaporator coil. Because the air temperature in a freezer unit never rises above freezing, an outside source of heat is needed to defrost the evaporator coil of a freezer. Two types of defrost methods are used in reach-in freezer units: hot gas defrost and electrical heating element. The defrost cycle, whether it be electric or hot gas, is controlled using a time clock. The time clock is typically factory preset for three equally spaced defrost cycles every 24 hours. However, this setting can be easily changed to suit various climates and operational needs.

Hot gas defrost. Hot gas defrost is performed by temporarily routing hot refrigerant vapor from the compressor discharge line through the evaporator coils (Fig. 12-4). A solenoid valve controls vapor flow into this by-pass line. The hot gas bypass line enters the evaporator downstream of the refrigerant flow control.

At the start of the defrost cycle, the solenoid valve opens and the hot gas follows the path of least resistance back to the evaporator. The compressor continues to run during the hot gas defrost cycle. As the hot refrigerant vapor flows through the evaporator coils, the ice on the outside of the coils melts away. At the end of the defrost cycle the solenoid valve is closed.

Hot gas defrost systems can generate higher-than-desirable suction line pressure that can place a strain on the compressor. If the discharge line is exposed to cool or cold ambient temperatures, the refrigerant vapor might cool to the point where it is not very effective at melting the ice off the evaporator coils. The result is extended defrost cycles or incomplete defrosting. Another potential problem is refrigerant vapor condensing in the evaporator. As has been stressed throughout this book, any liquid that reaches the compressor can cause major damage.

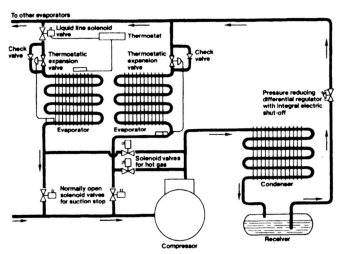

Hot Gas Defrost System

Figure 12-4 Layout of a typical hot gas defrost system. *(Courtesy of Parker Hannifin Corporation)*

Because of these drawbacks, hot gas defrost is not the preferred method of defrosting the relatively small evaporator coils of reach-in freezers. It is normally used on larger walk-in freezers or multiplexed systems. Hot gas defrost is covered in more detail in Chapter 13.

Electric defrost. The most common defrost system for reach-in freezer units is electric defrost. In this system, electric heating elements are mounted to (or built into) the body of the evaporator coils. In a typical freezer unit, both the compressor and evaporator fans are off at the beginning of the freezer defrost cycle. The electronic defrost heater attached to the evaporator coil is energized. If the unit is equipped with a condensate drain pan heater, this heater is also energized. A temperature control located at the evaporator is used to terminate the defrost cycle. This temperature control is normally set to energize the compressor when the coils have reached a temperature that ensures all frost and ice have melted. This temperature is normally about 50°F. When the evaporator coils reach this temperature, compressor operation is resumed and the defrost and drain pan heaters are de-energized. Coil fans are delayed from starting at the end of the defrost cycle by a separate heat sensing control affixed to the coil. The fans will not turn on until the control senses a coil temperature of 25°F. This ensures the fans will not blow warm air into the freezer's storage space.

As the freezer defrosts, heat is confined to the enclosed coil area to prevent any rise in air temperature in the food zone. The heat sensor delays the fans at the end of the cycle for two reasons. First, it prevents any warm air from blowing into the food storage area. Second, it does not permit any condensate on the defrosted coil to be blown into the food storage area.

Normally, defrosting will be ended by temperature sensors after 15 to 20 minutes. Most systems are designed with an absolute maximum time limit for the defrost cycle. This fail-safe measure is normally factory set into the time clock. For example, if the maximum time limit for defrost is set at 25 minutes, the clock will activate the compressor 25 minutes into the defrost cycle regardless of the temperature control reading. Do not tamper with the 25-minute setting on the clock without consulting with the factory first.

Drain line and pan. As the ice accumulation on low-temperature evaporator coils can be substantial, provisions must be made to dispose of the water produced during defrost. Water is directed into a drain or drip pan that can be routed to a facility drain line. Drain lines passing through the low-temperature refrigerated space must be kept as short as possible to prevent freezing in the lines. No traps should be used in drain lines because these hold water that will quickly freeze and cause backups into the freezer unit. As mentioned earlier, the drain pan might be equipped with is own defrost system, either electric or hot gas.

When facility drain lines are not conveniently located, a small condesate pump can be installed to pump the water to a remote drain line.

Tips on Installing a Reach-In Unit

Once the refrigerator or freezer unit has been placed in a suitable location and connected to a power source, turn on the power and check the following:

- Is the compressor running?
- Are the coil fans running if the unit is in a refrigerator? (On freezers, there will be a delay on starting.)
- When the door is opened, are the lights working?

When the unit has been running for about three hours, check the temperature readout to be sure that the unit has reached normal operating temperature. For a freezer, the typical temperature range should be −5°F to 0°F, while the refrigerator cycles at 38°F to 42°F. Adjust the temperature control up or down one number at a time as needed to ensure that the temperature falls in the proper range.

The digital thermometer in some refrigerators and freezers is battery operated. Be sure batteries are installed, and remind the owner that replacement will be needed in the future.

If equipped, set the defogger switch to the appropriate setting to match ambient conditions. In hot or humid conditions, the switch should be left in the on position, while in less humid conditions, the switch can be turned off. Used properly, the defogger switch will reduce the electrical cost of operating the unit.

Product placement. Items should be arranged on the shelves so that air can freely circulate. Packing the items tightly can cut off air circulation. In addition, leave one inch between shelf items and interior walls so that air can circulate throughout the cabinet's interior space. Shelves are often located on pins that are set into mounting holes in the cabinet frame. Interior wall covers might need to be removed to access the pins.

Maintenance

Periodic maintenance and cleaning will extend the life of a refrigerator or freezer. In addition, proper equipment care will help to keep the food well preserved and in excellent condition.

Cleaning. Advise customers to use a mild detergent for cleaning, followed by a sanitizing rinse solution. For interior cabinets, first remove all food and disconnect the power. Remove the shelves from their support pins. Clean the shelves using a detergent with a nylon-bristle brush. Rinse them first in clear water, then in a sanitizing solution. Allow the shelving to air dry.

Using a soft nylon-bristled brush, remove loose food particles from the interior walls, floor, and ceiling of the cabinet. Scrub these areas with a warm detergent solution. Be sure to clean the interior door panel and gasket as well. Rinse first with clear water, then with a sanitizing solution.

After all surfaces have air dried, set the shelves back onto their support pins. Reconnect the refrigerator to the power source. When the unit has cooled to the proper temperature, return the food to the shelves. Repeat this entire process at least once a week, as needed.

Wipe all exterior surfaces using a soft cloth or a sponge soaked in mild cleaning solution. Never use strong soaps or gritty abrasive cleansers, as they will only mar the surface. For exterior hardware, use warm water, a mild cleaning solution, and a soft cloth. Dry all areas with a clean cloth. Do the above cleaning on a daily basis.

Compressor housing/condensing unit. The condensing unit should be cleaned periodically to keep the fins free of lint and dust. Once a month is the recommended minimum cleaning interval. Clean the unit more often if conditions warrant it. A clean condenser will use less current, increasing the cabinet's efficiency.

To clean the condenser, first disconnect the electrical power to the cabinet. Then thoroughly vacuum or brush away any dust or lint from the finned condenser coil, compressor, and other cooling system components. For a heavily soiled condenser, the dirt can be blown out of the fins with compressed air.

Service and Troubleshooting

Service and repair of reach-in refrigerators and freezers can involve a number of items.

Doors and gaskets

Door hinges rarely require replacement on quality units. But if a hinge must be replaced, check to see if the hinge houses a microswitch for the interior light. Be sure to replace the hinge in its original position so the interior light works properly (Fig. 12-5).

Over time a door gasket can become brittle, loose, or damaged. When replacing a door gasket, allow the gasket to reach room temperature

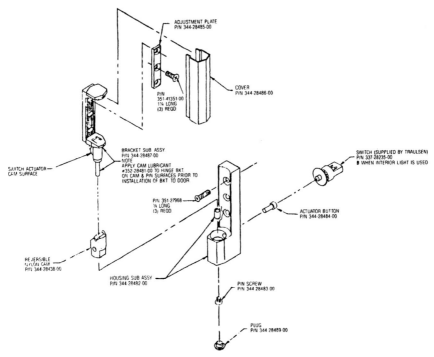

Figure 12-5 Hinge equipped with an interior light switch. *(Courtesy of Traulsen & Co., Inc.)*

before installing it. Using a rubber mallet with a block of wood, gently tap in the four corners first, working slowly toward the center from both ends. In some cases the correct replacement gasket might appear too large. However, if installed properly it will fit into place.

Replacing the door defogger circuit heater wire

If the door defogger circuit is not functioning, locate the terminal connections for the heater wires in the units junction box. Continuity for heater wires can be checked using an ohmmeter connected between the white and black wires and the white trace wires of the appropriate numbered door located in the junction box. If there is continuity, this is a clear indication that the heater wire is working. No continuity indicates that the heater wire is defective and should be replaced. (The heater wire should have good contact with the metal door frame.)

Cabinet door heater wires are commonly located under a breaker strip or are face mounted (Fig. 12-6). Face-mounted wires are located inside the foam. Replace a heater wire as follows: Remove the four breaker strips from around the door frame by unscrewing the sheet

Commercial Reach-In and Walk-In Units

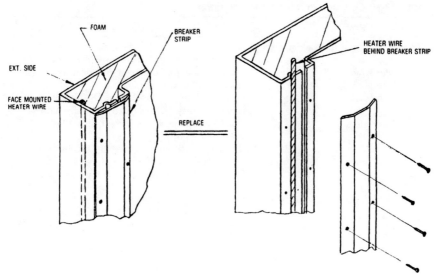

Figure 12-6 Location of cabinet door "defogger" heater wiring. *(Courtesy of Traulsen & Co., Inc.)*

metal screws. If there is a heater wire behind the breaker strip, remove it. Place the new heater wire in the groove around the door frame and tape it into place. Reinstall the breaker strips.

Troubleshooting

Table 12-1 lists the major problems associated with reach-in refrigerators and freezer units.

Walk-In Refrigerators and Freezers

Walk-in refrigerators and freezers are simply refrigerated rooms used for long-term storage of perishable foods. Temperatures in walk-in refrigeration units are kept in the 33 to 36°F range. Freezer units are kept at around 0°F.

Construction

Older walk-in units normally use standard wood or masonry walls with fiberglass or cork insulation. Modern walk-in units are constructed using foamed-in-place urethane panels. The panel's outer shell can be constructed of galvanized steel, stainless steel, or aluminum. The panel frame can be high-density urethane or kiln-dried wood. Panel thickness ranges from three to six inches (Fig. 12-7).

TABLE 12-1 Common Problems and Solutions

Problem	Possible solution
The condensing unit does not start.	Make sure that the line has not been disconnected. Check the fuse for burnout, replacing as needed.
The condensing unit is operating for extremely long intervals, or fails to shut off.	Check to see if amount of refrigerant is low. Add more if needed.
	Look for evaporator coil icing. Open the time clock door and reset it to activate the defrost cycle. Reset the timer to the correct time when the cycle is finished.
	Check the refrigerant lines for any obstructions. If condensers or filters are dirty, clean them.
	Check that all doors are properly closed and sealed.
The food compartment is too warm.	See that the gaskets on the doors are sealed. If the control setting is too high, lower it.
	If a large amount of food was recently put in the unit, or the doors have been open awhile, wait for the unit to cool by itself.
	Check for any possible restrictions to air circulation.
The food compartment is too cold.	Reset the temperature control to a warmer setting.
There is condensation on exterior surfaces.	When the unit is located in a very humid area, this condition is not unusual. To remedy, set the defogger switch to the on position.
	Check for a tight seal on all gaskets, also that doors are aligned correctly.

Walls and ceiling assemble in sections using tongue-and-groove or butted frame construction (Fig. 12-8). The structural integrity of all panel joints is important, particularly in low-temperature (freezer) applications. Any air leakage into the refrigerated space will cause ice to form on the inside walls.

The floor can be a concrete slab placed over rigid insulation, or it can consist of special rigid panels similar to the wall and ceilings.

Doors offer the same lightweight urethane foam construction. Hinged and sliding door designs are used with the former being the most popular. On low-temperature storage units, mullion heaters are often installed around the door frame to prevent the buildup of ice.

Storage/display units. In convenience stores and supermarkets, many walk-in refrigerators and freezers serve a dual purpose of storage and display (Fig. 12-9). The front of the unit is equipped with either sliding or hinged glass doors and shelving so the customer can access the stored product. These door sections are often prefabricated and set into an opening within the front of the unit. Allowance must be made in computing the refrigeration load for the extra service load.

The storage area of the unit is accessed through a separate door. Workers can restock the display shelves and racks from the rear.

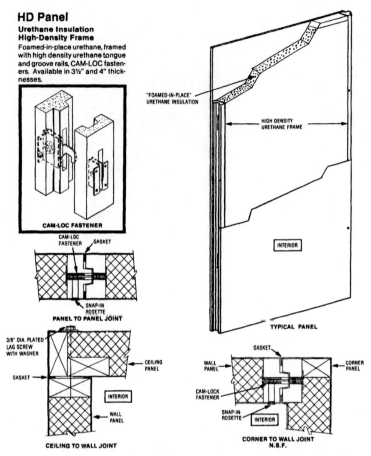

Figure 12-7 Foamed-in-place urethane panel construction with tongue-and-groove fastener detail. *(Courtesy of Three Star Refrigeration Engineering, Inc.)*

Condensing units and compressors

While small walk-in units might use 120-V single-phase condensing units, 230-V single-phase and 208 three-phase units are more common in larger installations. The larger size compressor motors (and fan motors) require contactors or motor starters. The temperature and pressure control units in larger walk-in units are part of 208- or 24-V control circuits. Compressors are usually semihermetic models, although full hermetic compressors are finding increased usage as higher horsepower units are introduced to the market.

Lube oil protection control. The larger size compressors found on walk-in units will likely have a lube oil-protection control wired in series with the compressor contactor or motor starter.

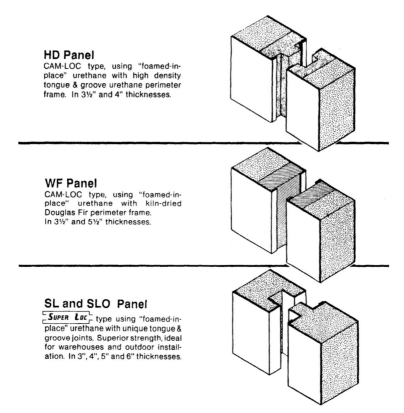

Figure 12-8 Modular panel construction of modern walk-in unit. *(Courtesy of Three Star Refrigeration Engineering, Inc.)*

The evaporator coil fans of a walk-in refrigerator run constantly. The condenser fan motors are energized by contactors or motor starters whenever the compressor is started.

Refrigerant flow control

The refrigerant flow control in most walk-in units is a thermal expansion valve that is externally equalized. An externally equalized TEV is used to eliminate the adverse effects that will occur if there is a substantial pressure drop (5 psi or more) across the evaporator coil. Such a pressure drop is common in large or multicircuited evaporates commonly used on high-capacity, low- to medium-temperature refrigerators and freezers.

Internal versus external operation. As discussed in Chapter 4, the operation of a TEV is determined by the relationship between three basic pressures: sensing bulb pressure, spring pressure, and evaporator pres-

Figure 12-9 Combination storage/display walk-in unit commonly found in convenience stores and supermarkets. *(Courtesy of Ram Freezers and Coolers Manufacturing Inc.)*

sure (see page 70 and Fig. 4-10). Evaporator pressure is transmitted to the underside of the valve in one of two ways. In an internally equalized valve, the evaporator pressure at the valve outlet is transmitted to the diaphragm via a passageway within the valve body or through a clearance around the pushrods. If the valve is externally equalized, the underside of the valve diaphragm is isolated from the valve outlet pressure. Evaporator pressure is transmitted to the diaphragm by a tube connecting the suction line near the evaporator outlet to an external fitting on the valve. The external fitting is connected to a passageway that leads to the underside of the diaphragm (Fig. 12-10).

Caution: When an externally equalized TEV is used, the equalizer connection on the TEV must be connected to the outlet of the evaporator and not capped.

Figure 12-11 illustrates an internally equalized TEV on a system having no pressure drop across the evaporator coil. Because an internally equalized valve senses evaporator pressure at the valve outlet, the total pressure in the closing direction is 52 psig plus the 12 psi spring pressure or 64 psig. A bulb pressure of 64 psig is required for proper valve regulation. This translates to a bulb temperature of 37°F if HCFC-22 refrigerant is used in both the refrigeration system and the bulb charge. The saturated temperature corresponding to the outlet pressure of 52

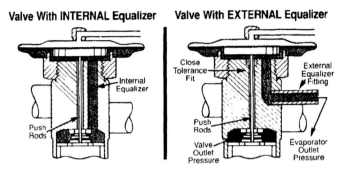

Figure 12-10 Details of an externally equalized TEV. *(Courtesy of Sporlan Valve Company)*

psig is 28°F for HCFC-22 refrigerant. So superheat which is bulb temperature minus saturated evaporator temperature is 37 – 28 or 9°F.

Now let's consider what happens to the same system when there is a pressure drop across the evaporator of 6 psig (Fig. 12-12) Because an internally equalized valve senses evaporator pressure at the valve outlet, the total pressure in the closing direction becomes 58 psig plus the 12 psi spring pressure or 70 psig. A bulb pressure of 70 psig is now required for proper valve regulation. This translates to a bulb tempera-

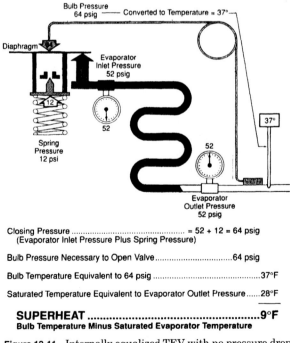

Closing Pressure .... = 52 + 12 = 64 psig
(Evaporator Inlet Pressure Plus Spring Pressure)

Bulb Pressure Necessary to Open Valve .... 64 psig

Bulb Temperature Equivalent to 64 psig .... 37°F

Saturated Temperature Equivalent to Evaporator Outlet Pressure ...... 28°F

SUPERHEAT .... **9°F**
Bulb Temperature Minus Saturated Evaporator Temperature

Figure 12-11 Internally equalized TEV with no pressure drop across the evaporator. *(Courtesy of Sporlan Valve Company)*

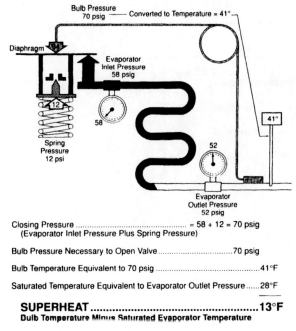

Figure 12-12 Internally equalized TEV with 6 psi pressure drop across the evaporator. *(Courtesy of Sporlan Valve Company)*

ture of 41°F. So superheat now becomes 41 minus 28 or 13°. This increase in superheat means less refrigerant is entering the evaporator and cooling capacity is lost. The temperature of the refrigerant vapor in the suction line to the compressor also increases. This increases the strain on the system compressor.

Figure 12-13 shows how an externally equalized TEV eliminates the adverse affects of pressure drop by sensing evaporator pressure at the evaporator outlet. As a result, the TEV senses the correct pressure and controls the proper superheat. Always follow manufacturer's recommendations for maximum pressure drops that can be safely tolerated by internally equalized TEVs.

Temperature control

Walk-in refrigerators often use space-temperature sensing to control the temperature of the refrigerated space.

Condensate disposal

Evaporator coil condensate disposal is similar to reach-in refrigerators and freezers covered earlier in this chapter. On freezer units with long condensate disposal runs, a heater ribbon is often installed in the drain line to keep it from freezing.

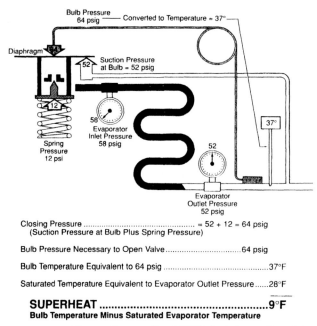

Closing Pressure .... = 52 + 12 = 64 psig
(Suction Pressure at Bulb Plus Spring Pressure)

Bulb Pressure Necessary to Open Valve .... 64 psig

Bulb Temperature Equivalent to 64 psig .... 37°F

Saturated Temperature Equivalent to Evaporator Outlet Pressure ...... 28°F

SUPERHEAT .... **9°F**
Bulb Temperature Minus Saturated Evaporator Temperature

Figure 12-13 Externally equalized TEV that is unaffected by evaporator coil pressure drop. *(Courtesy of Sporlan Valve Company)*

Defrost systems

Walk-in units can use either electric or hot gas defrost systems. The number of defrost cycles needed during a 24-hour period will vary according to the overall usage of the unit and the ambient conditions surrounding the unit. For example, a combination display/storage walk-in unit installed in a convenience store might be open and shut hundreds of times a day. If the store is not air-conditioned, the increased humidity of the summer months might also increase the frost buildup on the coils. For this reason, an additional defrost cycle can be programmed into the 24-hour clock timer during the summer months.

Liquid-suction heat exchanger

To form a liquid-suction heat exchanger, hot liquid refrigerant from the liquid line is passed through a coil of small diameter tubing that is tightly wound around the system's suction line (Fig. 12-14). The cool vapor in the suction line absorbs heat from the hot liquid refrigerant, creating subcooling in the liquid line. This subcooling decreases the amount of flash gas generated at the flow control and increases the overall efficiency of the system.

Commercial Reach-In and Walk-In Units   343

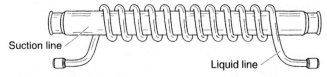

Figure 12-14 Liquid-suction heat exchanger.

As the temperature of the suction line vapor increases, its superheat will also increase. This rise in suction line temperature helps ensure that all refrigerant in the suction line is vaporized before it reaches the compressor.

On low-temperature applications, the temperature of the vapor exiting the evaporator might be below freezing. This can cause frost or ice build-up on the suction line or on the compressor itself. The use of a liquid-suction heat exchanger often raises the temperature of the suction line above freezing, eliminating this problem.

However, you must be careful of avoiding too much heat transfer to the suction line. If the temperature of the suction line vapor is raised substantially, it will not provide the needed cooling effect in the compressor. As a result the compressor might overheat. A tripped compressor overload might be an indication that this situation exists.

Crankcase pressure regulator

Compressors used on low-temperature freezer units are designed to handle only low-pressure (10–20 psi) refrigerant vapor in the suction line. During normal operation, these compressors draw current near or at their maximum rating.

If pressure in the suction line increases, the compressor must work harder and draw more current. There are several cases in which the low-pressure compressor might be subjected to overload conditions. The first is during initial start-up of the unit or when the unit is started after being out of service. Temperatures inside the refrigerated space can be quite high. This results in increased evaporator pressures and the possibility of an electrical overload. The second case occurs during a hot gas defrost cycle or immediately after an electric defrost cycle. In both instances evaporator temperature is raised, resulting in increased pressure in the suction line to the compressor.

To prevent high vapor pressure from reaching the compressor, a crankcase pressure regulator is installed in the suction line (Fig. 12-15). This is a spring-loaded mechanical regulator that is normally open at low pressure but closes down as pressure on the upstream side of the regulator increases. This throttles suction line pressure down to

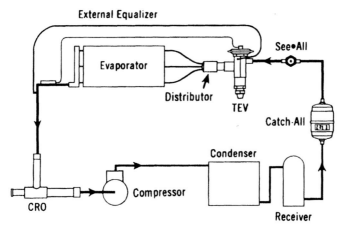

Figure 12-15 Location of crankcase pressure regulating valve. *(Courtesy of Sporlan Valve Company)*

safe levels. So while pressures inside the evaporate might exceed 50 psi during high load conditions, only 10 psi or so is allowed to reach the compressor at all times.

Tips on Controller Installation

The accuracy of a temperature or pressure controller can be adversely affected by incorrect installation of the unit. Always follow manufacturer's specific instructions when installing a controller. The following sections outline some general good installation practices.

- Avoid rough handling or dropping the controller. It is a precision instrument and can be damaged or lose its accuracy.
- Do not locate controllers in areas of high moisture or dusty or dirty areas. Exposure to moisture or dirt can shorten controller life.
- Never install the controller in an explosive or corrosive environment. Sparking inside the controller can lead to an explosion.
- Mount the controller on brackets designed for this purpose. Avoid turning screws into the controller body because they can damage internal components.
- If brackets are not used, mount the controller to a flat surface. If it is twisted, the controller might not operate properly.
- Be sure electrical wiring to all controllers conforms to local codes and good wiring practices.
- Provide fused disconnects to protect motors and controllers.
- Keep the total electrical load within the control's limits.

- Attach all electrical leads to the control switch terminals using terminal screws.
- Allow slight slack in electrical leads to provide for expansion and contraction due to changes in temperature.

Temperature controllers

Temperature controllers are available with a variety of different sensing elements all of which affect how the unit is used and installed (Fig. 12-16). Air-coil sensing elements are used when the temperature controller is mounted in the space to be controlled, such as a walk-in cooler. Mount the controller away from door openings and evaporator discharge fans that can cause false readings. Locate the sensing element in an area when air can move freely around it. The preferred location is near the evaporator return air. Use a bracket to mount the controller out from the wall to prevent a heat sink effect.

Vapor filled temperature control sensing elements can be a capillary tube or a capillary tube with a sensing bulb on the end.

If the controller is monitoring the temperature of a coil, make certain the bulb or a minimum of six inches of the capillary tube comes in direct contact with the coil. If a vapor-filled sensing element is used to sense air temperature, a coil of at least 18 inches of tubing should be exposed to the air stream. On applications where the element is submerged in a liquid, at least four inches of tubing must be submerged. If the capillary tube is inserted into a well, at least 10 inches of tubing should be formed into a bulb that makes direct contact with the sides of the well. If the sensing point is in a vertical position, form an "S" or "U" shape with the tubing, as shown in Fig. 12-17.

On elements with a sensing bulb, the bulb must be securely clamped in place with contact along its entire length. Avoid crushing or deforming the sensing bulb. To improve the accuracy of the sensing bulb, insulate the bulb, clamp, and refrigerant line after installing the bulb in place. If the bulb is mounted in a well, be certain the bulb is in good contact with the well wall and seal the well to prevent water from collecting on the bulb.

With vapor-filled elements, always remember that the controller will respond to the coldest temperature along the entire length of the capillary tube. If part of the tube is exposed to colder temperatures than the controller setting, a cross ambient condition occurs and inaccurate temperature control will be the result. If cross ambient conditions exist, use a special cross-ambient sensing element. Follow these tips as well.

- Do not expose the sensing element of temperature controllers to temperatures above 150°F or 50° above the maximum setting of the controller, whichever is less.

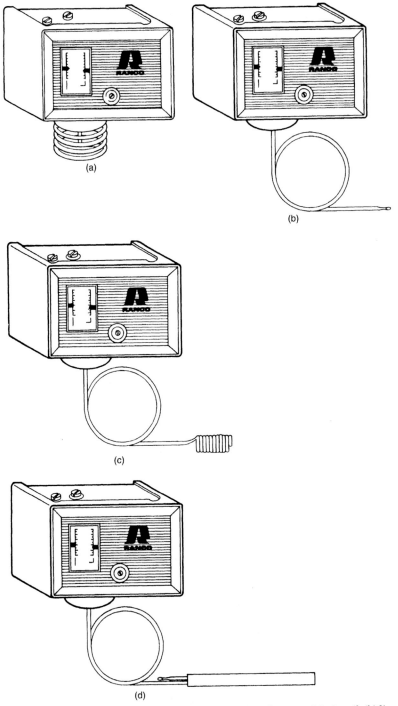

Figure 12-16 Various types of temperature controller elements: (a) air-coil, (b) limited vapor-fill capillary tube, (c) limited vapor-fill remote bulb, and (d) cross-ambient sensing element. *(Courtesy of Ranco Incoporated)*

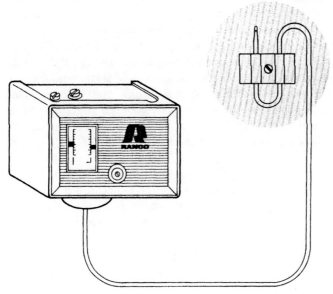

Figure 12-17 Proper installation of limited-vapor capillary sensing element. *(Courtesy of Ranco Incoporated)*

- Route the capillary sensing element away from extremes in temperature. For example, avoid contacting the systems liquid or suction lines.
- Do not mount the capillary tube to vibrating surfaces as this will cause metal fatigue and eventual failure.
- When passing a capillary-sensing element through a wall or partition, insert a soft rubber grommet into the hole to prevent damage to the tube from rough edges.
- Coil excessive tubing in a minimum three-inch diameter loop and prevent it from rubbing against itself or other surfaces.
- Remember that altitude will have a slight effect on temperature controller settings. Higher altitudes will lower the setting of the controller about 1°F per 1000 feet.

Pressure and oil lube controllers

Correct installation of pressure controls is very important for accurate readings and long service life. When the controller will be located near the sensing point, it is equipped with a ¼-inch diameter capillary tube, which is fitted into the refrigerant line. If the controller must be located remotely from the sensing point, it is supplied with a ¼-inch male fitting. A properly sized length of ¼-inch copper tubing is then installed between the controller fitting and a ¼-inch fitting at the sensing point.

For proper operation, the pressure in the controller's bellows head must be the same as the pressure at the point being sensed. The most common cause of improper pressure sensing is oil logging. This condition results when lubricating oil in the circulating refrigerant accumulates in the pressure controller's capillary tube and bellows. This refrigerant oil acts as a restrictor and prevents the controller from responding to changes in system pressure. Oil logging is further aggravated by cold temperatures that can turn the oil to sludge that further clogs the capillary.

Problems with oil logging are most common on systems using pumpdown cycles. During pumpdown, pressure in the suction line changes rapidly. An oil-logged controller cannot respond quickly too these changes and suction line pressure might drop well below the cut out setpoint before the controller shuts the compressor down.

To reduce oil logging problems, large bore capillaries should be used for all low-pressure controllers. Proper installation is also very important. Follow these guidelines.

- Make pressure connections self-draining as shown in Fig. 12-18. This is particularly important on suction line connections.
- Mount the control box above the pressure sensing connection whenever possible. This will allow any oil that enters the capillary to drain out of it.
- On systems with outdoor condensing units exposed to cold temperatures, use ¼-inch copper tubing for all pressure line runs regardless of their length.

Work-hardening due to system vibration is a major cause of pressure control capillary tube breakage. To minimize the problems associated with work-hardening follow these guidelines:

- Avoid bending the capillary or tubing multiple times during installation. Plan ahead and bend the tubing once.
- If the capillary is mounted to the compressor, secure it firmly so the capillary and compressor vibrate together (Fig. 12-19). If the capillary is vibrating out of sync with the compressing, work-hardening and tubing failure will result.
- Do not make capillary tubing runs too tight. Allow for a slight play in the run.
- Coil excess tubing and keep individual coils from vibrating or rubbing against one another.

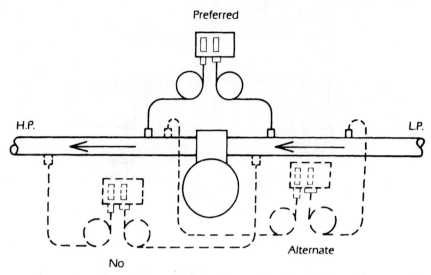

Figure 12-18 Installing the pressure controller so that the capillary tubes are self-draining. *(Courtesy of Ranco Incoporated)*

- When the controller is mounted remote from the compressor, but the pressure connection is on the compressor, form a vibration coil between the compressor and the condensing unit base (Fig. 12-20). Secure the capillary between the pressure connection and the coil to the compressor and the capillary between the coil and the controller to the condenser base.

Post installation checks

The following are some post installation check out tips:

- Check a high-pressure control cut-out setting by disconnecting the condenser fan or blocking air flow across the condenser coils. On water cooled condensers, temporarily turn off the water flow and ensure the controller cuts out.
- A low-pressure control can be checked or adjusted using a manifold gauge set connected to the suction line service valve. Close the service valve in the liquid line and run the compressor to pump the refrigerant form the evaporator and suction line into the condenser and receiver. Note the pressure at which the controller cuts out and stops the compressor.

Next, open the liquid line service valve to allow refrigerant to flow into the evaporator. As the suction line pressure increases, the controller should start the compressor at the cut-in setting.

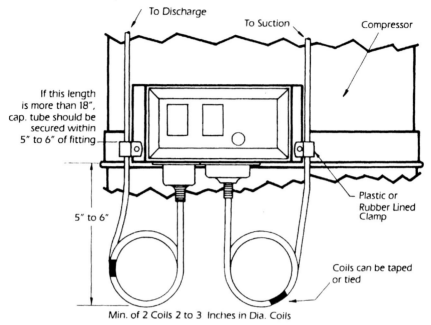

Figure 12-19 Proper mounting of pressure controller to compressor to minimize work-hardening due to vibration. *(Courtesy of Ranco Incoporated)*

Setting pressure controls

In a typical close high/open, low-temperature, or pressure controller, adjust the range screw until the scale pointer indicates the desired cut-in setting.

Next, adjust the differential screw until the scale pointer indicates the desired differential setting. The controller's cut-out setting is equal to the cut-in setting minus the differential. Fig. 12-21a illustrates a low-pressure controller with a cut-in setting of 36 psi and a differential of 20. This gives the controller a cut-out setting of 16 psi.

An open high/close low-high pressure controller can be set in the same way. In some cases, a high pressure side controller is designed with no differential. Only the cut-out pressure is set (Fig. 12-21b). The controller will cut out at any reading above this setting.

Troubleshooting Refrigeration System Controls

The following section covers common troubleshooting techniques for temperature and pressure controllers used in commercial refrigeration system.

Commercial Reach-In and Walk-In Units 351

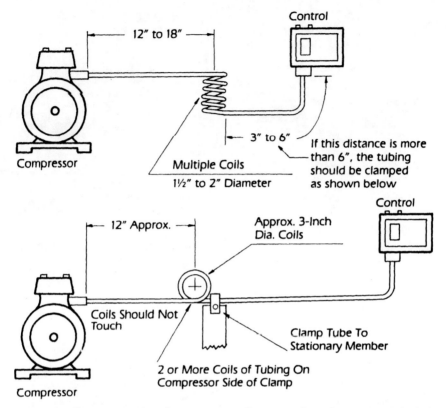

Figure 12-20 Proper mounting of pressure controller remote from the compressor including the use of a vibration coil. *(Courtesy of Ranco Incoporated)*

Checking operation

To check the operation of a pressure or thermostat control toggle the bellows lever using a screwdriver. When toggling, a "crisp" snap should be heard when the control cuts in and out. The bellows lever should engage and disengage the pickup arm at the proper level with the proper clearance (Fig. 12-22).

If the control does not have proper pick up, it might operate erratically, or it might not have differential adjustment capability. Because the pick up is critical to proper control operation, always keep screwdrivers out of the switch area. Toggling the control in the switch area can affect the control operation because the pressure of a screwdriver can change the relationship of the bellows lever to the switch operating points and destroy pick up. The only point that the control should be manually toggled is the bellows lever. Controls that have been improperly toggled frequently have a small dent on the edge of the scaleplate where the screwdriver has been rested to pry the switch lever closed.

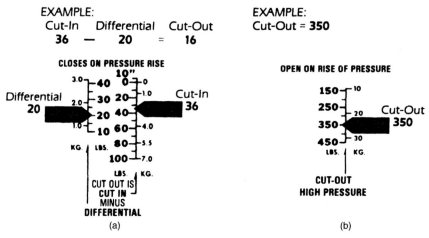

Figure 12-21 (a) Setting of a low-pressure controller cut-in and differential. (b) High-pressure controller with cut-out pressure setting only. *(Courtesy of Ranco Incoporated)*

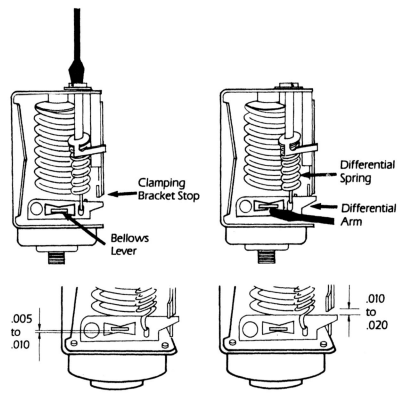

Figure 12-22 Pick-up adjustment check of pressure or thermostatic controller. *(Courtesy of Ranco Incoporated)*

Oil logging of controllers

As mentioned earlier, oil logging of the control capillary and bellows is a major cause of erratic control of operation. Controls that have been oil logged can be cleared of oil by hanging the control with the capillary extended down. Attach a vacuum pump to the capillary extension tube and pull a vacuum for 10 to 15 minutes (Fig. 12-23).

Lube oil protection controls

Hermetic and smaller semihermetic compressors used on reach-in refrigerator and freezer units use internal oil pumps or splash lubrication to provide compressor lubrication. These systems do not require the use of controls to monitor the compressor for proper lubrication.

Larger semihermetic compressors often use an external oil pump. These compressors often use lube oil protection controls to monitor the compressor oil pressure. These controls will shut down the compressor if oil pressure drops below a certain level. The normally closed control switch is wired with a thermal bimetal time delay switch. This thermal time delay allows the oil pressure to build up to proper operating pressure at start-up and prevents nuisance compressor shutdowns due to momentary drop in oil pressure during normal operation. It can be set for a 30-, 60-, or 120-second delay based on compressor manufacturer's suggestions.

Timing delay values are affected by control circuit voltage. Low voltage will extend timing, while high voltage will shorten timing.

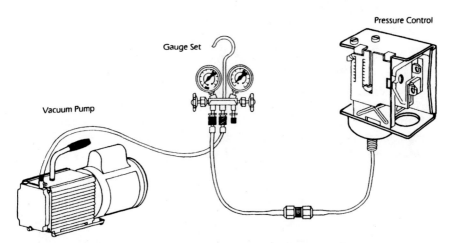

Figure 12-23 Pulling a vacuum to clear the capillary tube of oil. (*Courtesy of Ranco Incorporated*)

Lockout on run cycle

When the compressor is running normally and the oil pressure drops below the proper lubricating point, the pressure switch closes and energizes the heater timing circuit. If the oil pressure does not build up within the time specified, the heater circuit will open and shut the compressor down. This is called *lock out*. The control has done its job of protecting the compressor. You should next find out why the oil pressure dropped, causing the control to lock out. Correct the problem, then push the manual reset button to restart the compressor.

Short cycling of the compressor on the operating or limiting controls can also cause the lube oil protection control to lock out. This occurs when the timer module accumulates heat during the repeated start-ups with insufficient time for cooling down.

On compressors with internal line-break motor protectors, when the compressor stops due to the opening of the motor protector, the lack of oil pressure will cause the timer module to begin timing, because it is still receiving electrical power. When the timer reaches the end of its cycle, it will open the P30 contacts. Even though the motor protector will automatically reset, the P30 control must be manually reset before the compressor can be restarted.

If this condition is not considered desirable, a current-sensing relay can be wired in series with the S30 timer-delay circuit. Then when the motor protector opens, the current-sensing relay will de-energize the timer-delay circuit.

Causes of low oil pressure

Causes of low oil pressure include:

- Insufficient oil in the compressor crankcase.
- Oil logging.
- Migration of liquid refrigeration to the compressor crankcase during an off-cycle or a floodback condition that occurs at start-up. The refrigerant floodback to the compressor crankcase causes the oil to foam on compressor start-up. The oil pump can't pump foam and the refrigerant can't lubricate the compressor. The lube oil control reads this as low oil pressure then locks out.

Chapter

13

Supermarket Refrigeration Equipment

Supermarkets represent the largest and most complex refrigeration systems encountered by most service technicians. In addition to the reach-in and walk-in equipment described in Chapter 12, supermarkets contain a wide variety of specialized low- to medium-temperature display units. These normally include:

- Open-well frozen-food cases
- Island frozen food and ice cream cases
- Back to back island frozen food and ice cream cases
- Low-temperature glass-top spot merchandisers
- Frozen meat display cases
- Glass-front frozen meat display cases
- Multishelf frozen food cases
- Low-temperature service merchandisers
- Medium- or low-temperature glass-door merchandisers
- Ice cream glass door merchandisers
- Top-display produce and meat case
- Deli and food service cases
- Dairy/deli cases

Supermarket Display Case Design

Virtually all supermarkets use banks of refrigeration units and freezers installed end to end to provide a continuous display of products as the customer moves down an aisle (Fig. 13-1). Special end units are used for the first and last units in a continuous display. Methods of joining self-service refrigerators vary, but they are usually bolted or cam-locked together.

Transition zones

Open-display refrigerators are constructed with surface zones of transition between the refrigerated area and the room atmosphere. Thermal breaks of various designs separate the zones. These breaks reduce the surface area on the refrigerator that is below the *dew point* (the point where condensation and/or ice can form).

On open-display cases, the surfaces in front of discharge air nozzles, whether on the nose of the shelving or at the refrigerator's front rails or center flue, can be below the dew point. In these locations, electric heater coils are used to raise the exterior surface temperature above the dew point to prevent accumulation of condensation (Fig. 13-2).

These heaters can be controlled by programmed cycled timers. Electronic PID heaters controllers can also be used to vary heater output to match the stores ambient temperature. This reduces overall energy consumption.

Figure 13-1 Continuous display refrigeration unit used in supermarket installations. *(Courtesy of Hussmann Corporation)*

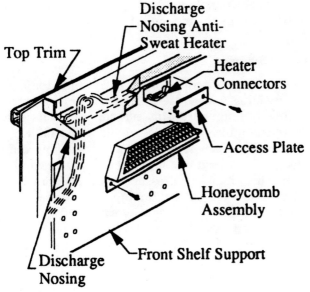

Figure 13-2 Electric heaters used in display cases to eliminate unwanted condensation.

Evaporators

Evaporators and air-distribution systems for specialized supermarket display refrigerators are custom engineered to fit precisely into a particular unit. As a result, they are inherent in the fixture and are not standard independent evaporators. The design of the air-circuit system, the evaporator, and the means of defrosting are the result of extensive testing to produce the particular display results desired.

Cooling requirements

Cooling requirements for continuous-run displays are often stated as a refrigeration load per unit length. Manufacturers of refrigeration cases and displays will sometimes state a lower load per unit length on systems driven by parallel compressors. This is because peak loads are less when a programmed defrost is used to defrost portions of the system, rather than the entire system at one time. On a parallel system, temperature recovery after a defrost period is less of a strain than it would be to a single compressor system.

Published refrigerator load requirements allow for extra capacity for temperature pull-down of the refrigerated space once the defrost cycle is complete. The refrigeration industry states standard store conditions to be 75°F and 55 percent relative humidity. These store conditions should be maintained with air-conditioning. Open display cases will affect ambient air temperature, and store heat removed by them must be accounted for when sizing the air-conditioning system.

Multiplexed Refrigeration Systems

A large supermarket can have literally dozens of refrigerator and freezer display units. Providing a separate condensing unit for each display unit would be highly impractical. Instead supermarkets and large convenience stores use multiplexed systems in which multiple refrigeration units (evaporators) are served by one condensing unit (Fig. 13-3). The heat load requirement of a multiplexed system can be matched to a large-capacity single-compressor condensing unit or divided into manageable circuits for parallel compressor systems.

The piping for individual units running on the same refrigeration system can be run from one unit to another via the end frame saddles.

Caution: Never run refrigerant lines through units that are not part of the same refrigeration system. Doing so could result in poor refrigeration control, with possible compressor failure.

If receiver cool gas defrost is used, the liquid line inside the unit must be two sizes larger to ensure even drainage from the evaporators during defrost.

Single-compressor units

Single-compressor units make up half of the supermarket compressor equipment in use today (Fig. 13-4). As covered in Chapter 4, these single-compressor units can use either air-cooled or water-cooled condensers that are normally remotely located from the compressor.

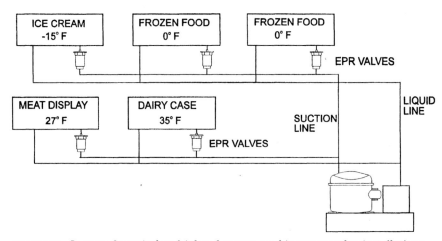

Figure 13-3 Layout of a typical multiplexed system used in supermarket installations.

Supermarket Refrigeration Equipment   359

Figure 13-4 Single compressor condenser unit used in supermarket installations. Note the use of an electronic control system. *(Courtesy of Hussmann Corporation)*

A solid-state pressure control for single units can help control excess capacity when the ambient temperature drops. The control senses the pressure and adjusts the cutout point to eliminate short cycling during low-load conditions. The pressure control also saves energy by maintaining a higher suction pressure than would otherwise be possible and by reducing overall running time.

Parallel compressor systems

A modern large-capacity condensing unit might use six or more compressors operating in parallel to provide a wide range of cooling capacities (Fig. 13-5). These rack systems usually have a remote condenser and frequently recover heat from a secondary condenser coil in the air handler (Fig. 13-6).

The rack compressors are typically provided with high- and low-pressure controls, an oil pressure safety control, primary overload control, and compressor cooling fans on low-temperature applications. A single large volume receiver or smaller dual receivers can be used. Factory-supplied piping might include suction, discharge and liquid headers, an oil separator and return system, dual receiver tanks, and a suction filter on each compressor. A liquid line filter drier, sight glass, and level indicator are usually standard items.

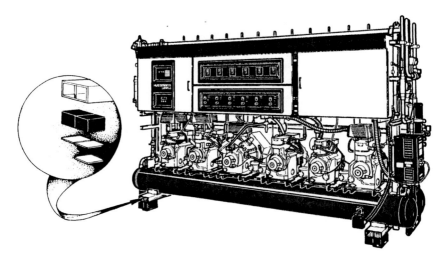

Figure 13-5 Large capacity condensing unit rack with multiple compressors operating in parallel. *(Courtesy of Hussmann Corporation)*

The rack system provides a factory-wired control panel with a prewired distribution power block and circuit breakers and contactors for individual components. Compressor time delays are wired into the system and wires are typically color coded for easy identification and traceability.

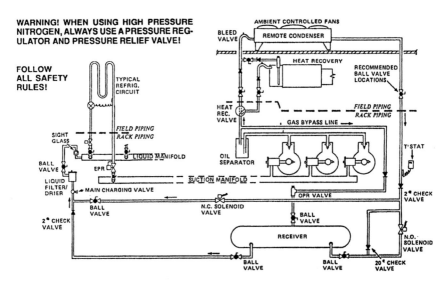

Figure 13-6 Parallel rack system with an air-cooled condenser and heat-recovery coil. *(Courtesy of Tyler Refrigeration)*

Rack systems are normally set up in the supermarket's machine room. The location must provide sufficient ventilation to cool the compressors. Duct work might be needed to provide this cooling for larger units. Floor drains are required to dispose of condensate. Clearances around all equipment must meet NEC and all local codes.

Typically supermarket refrigeration falls into low or medium temperature ranges. An average low-temperature rack maintains a suction temperature of $-25°F$ and has a low-end satellite operating at $-33°F$. A typical medium-temperature rack operates at $+16°F$ with a low-end satellite at $+7°F$. High-end satellites are often applied to preparation room cooling.

Satellite compressors

A separate satellite compressor for ice cream refrigerators on a low-temperature system or for meat refrigerators on a medium-temperature system can be physically mounted on the rack and piped so that only the heat removed from the lower temperature refrigerators is supplied at the less efficient rate.

Parallel Compressor Operation

All larger markets include one or more medium-temperature parallel systems for meat, deli, dairy, and produce refrigerators and the medium-temperature walk-in coolers. The system might have a separate compressor for some refrigerators or all units might operate off a single compressor.

Low-temperature refrigerators and coolers are grouped on one or more parallel systems. Ice cream refrigerators normally operate using a satellite compressor or a single independent compressor. Supermarket cutting and preparation rooms are also normally cooled using a single independent unit.

Refrigeration cycle

Beginning with the parallel compressors, vapor refrigerant is compressed into the discharge manifold and flows to the oil separator (Fig. 13-7). Oil is removed from the refrigerant vapor and returned to the compressor. A three-way heat reclaim valve directs the refrigerant to either the condenser or a heat reclaim coil. When the valve's solenoid is de-energized, the valve directs the refrigerant to the condenser.

A flooding valve maintains head pressure in low-ambient conditions by reducing the available condensing area. Restricting liquid refrigerant flow from the condenser, this flooding valve prevents the liquid refrigerant from leaving the condenser as fast as it is forming, so the

362 Chapter Thirteen

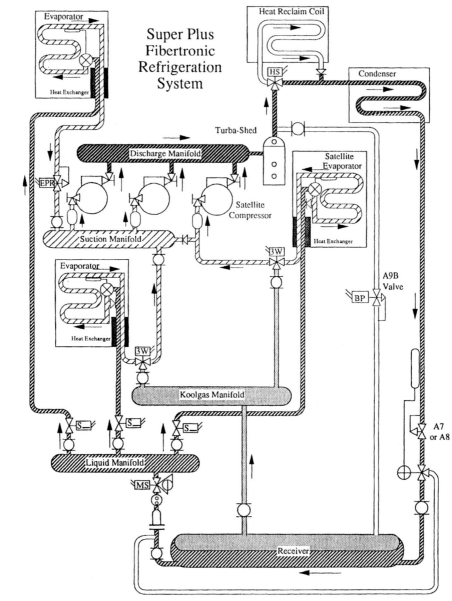

Figure 13-7 Operation of large capacity rack system. *(Courtesy of Hussmann Corporation)*

condenser floods with its own condensate. The twin receivers act as a vapor trap and supply the main liquid solenoid with quality liquid refrigerant. Sufficient liquid, vapor, and pressure are critical to the operation of the system. The system maintains these balances in the twin receivers for use as needed.

The main liquid pressure differential valve functions during the receiver cool gas defrost cycle to reduce pressure to the liquid manifold. This solenoid-operated valve never closes completely but restricts the liquid line, reducing the pressure in the liquid manifold. The reduced pressure allows reverse flow of refrigerant necessary for defrost.

The branch liquid line solenoid valve closes off refrigerant supply to the evaporator, yet allows back flow of refrigerant into the liquid manifold.

The TEV, located in the supermarket display case or merchandiser, meters the liquid refrigerant into the evaporator. A CDA valve can be used to control the evaporator temperature by preventing the evaporator pressure from dipping below a set point. Service valves or ball valves at critical points along the refrigerant lines allow isolation of components.

Capacity control of parallel systems

Supermarket refrigeration systems must be designed to maintain proper temperatures even under peak summer load. Until recently, most supermarket refrigeration systems were operated at 90°F or greater condensing conditions. This high-condensing temperature maintained sufficient high-side pressure to properly feed the system's thermal expansion valve.

Modern technology now enables engineers to design systems in which the condensing temperature tracks the ambient temperature down to about 70°F or less. As long as the liquid-line is correctly sized and installed, the TEV will continue to feed the evaporators sufficient refrigerant at these lower condensing temperatures. This means that at partial load, the refrigeration system will have sufficient capacity to maintain the desired product temperature.

Multiple compressors can be controlled or staged, depending on a drop in system suction pressure. With same-size compressors, a mechanical device can turn off one compressor at a time until only one is running. The suction pressure will be about 5 psi or more below optimum. Newer, solid-state control devices can cycle units "on" and "off" while the system remains at one economical pressure range. The run time for each compressor motor can be equalized. Satellite compressors can also be controlled with one control that also monitors other functions, such as oil pressure, alarm functions, and more. Compressor cylinder unloading can also be used to vary rack capacity.

Staging unequally sized compressors

Unequally sized compressors can be staged to obtain more steps of capacity than the same number of equally sized compressors. Improved unloaders on multi-bank compressors, and the use of variable speed compressors, provide for excellent capacity control. See page 382 for more information on how compressors can be staged in a typical computer-controlled rack system.

Two-stage compressor systems

Parallel operation is also applied in two-stage compressor systems that are used to provide more efficient, reliable refrigeration in low-temperature applications using HCFC-22 refrigerant. HCFC-22 refrigerant has chemical properties that make it reach very high temperatures rapidly in high-compression applications. CFC-502 refrigerant, which has been used for many years in low-temperature applications, does not reach these high temperatures as rapidly as HCFC-22, and does not usually require two-stage compression.

Two-stage compressors should be used when the compression ratio is greater than 10 to 1. The compression ratio is found by dividing the absolute condensing pressure (head pressure) by the absolute suction or low-side pressure.

$$\text{Compression ratio} = \frac{Head\ pressure}{Suction\ pressure}$$

As the compression ratio increases, the number of BTUs the compressor can pump drops, while the heat added to the refrigerant from the compressor increases. Simply stated, the greater the difference between the suction and head temperatures, the more work must be done on the refrigerant by the compressor. Work done by the compressor results in heat added to the refrigerant. With HCFC-22 refrigerant, temperatures inside the compressor cylinders can reach 330°F. This temperature is hot enough to evaporate most refrigerant lubricants, resulting in extreme ring and piston wear. Above 350°F, the oil will begin to carbonize.

As its name implies, two-stage compression is done in two steps or stages. Each of the stages performs about half of the required compression and lowers the compression ratio of each of the stages. (The overall compression ratio remains constant).

Some special compressors have cylinders that operate at two different pressures, but in most cases two-stage compression is accomplished using two separate compressors. The first stage compressor pumps the vapor up to a midpoint on the compression curve. At this point the vapor is cooled, but does not condense. The second compressor then compresses the cooled, intermediate vapor to the final pressure-temperature condition.

Parallel compressors might be equal or unequal in size and can have unloaders or variable speed drives.

Remote air-cooled condenser and heat-recovery coil

Remote air-cooled condensers are popular for use with parallel compressors. The condenser coil TD is in the range of 10°F for low-temperature and 15°F for medium-temperature applications. For energy conservation, generous sizing of the condenser with a lower TD is recommended.

Evaporative condensers

Evaporative condensers used in rack and supermarket applications have a coil in which refrigerant is condensed, or closed-loop circulating water is cooled, and has a means to supply air and water over its external surfaces. Heat is transferred from the condensing refrigerant inside the coil or the water in the loop to the external wetted surface and then into the moving airstream chiefly by evaporation. In areas where the wet-bulb temperature is about 30°F below the dry-bulb temperature, the condensing temperature can run from 10 to 30°F above the wet-bulb temperature. This lower condensing temperature saves energy. One evaporative condenser can be installed for an entire store.

Water-cooled condensers

Large commercial refrigeration systems can be designed with water-cooled condensers. A water-cooled condenser is commonly a shell with coils or straight runs of tubing installed inside the shell. Hot refrigerant vapor enters the shell and is cooled by water running through the tubes or coils. The vapor condenses into liquid and collects in the bottom of the shell housing.

A tube-within-a-tube condenser is constructed out of a small diameter tube inside of a larger diameter tube. Cooling water is pumped through the inner tube and refrigerant vapor is passed through the larger tube.

Advantages/disadvantages. Water transfers heat much more efficiently than air, so water-cooled condensers cool refrigerant quicker and at a more consistent rate. Water-cooled condensers retain their cooling ability in warmer weather, so the high-side pressure, and thus the operating efficiency of the refrigeration system, does not decrease during the spring and summer months.

The major disadvantages with water-cooled condensers involve handling and disposing of the large volume of water needed for operation. A single three-horsepower compressor might require a water-cooled condenser that uses upwards of 5000 gallons of water in a 24-hour period. Water is no longer an inexpensive commodity, and restrictions on using city water for cooling and/or dumping of warm condenser water into municipal sewage systems can cause design problems and increase operating costs.

Operation. In an air-cooled condenser, the condensing temperature is approximately 25 to 35°F higher than the ambient air temperature. So as outside air temperature rises, the high-side temperature and consequently the high-side pressure also rises. The compressor must work harder at higher pressures and operating efficiency is lost.

On the other hand, the condensing temperature of water-cooled condensers is generally only 15 to 25 degrees higher than the cooling water temperature. This translates into lower high-side temperatures and pressures, and increased operating efficiency.

With the use of cooling towers, the cooling water temperature rarely exceeds 85°F, and is often much cooler. In fact, water-cooled condensers require some type of water flow control to ensure the condensing temperature does not drop too low during winter months. A water-regulating valve is used to meter the proper amount of water to the condenser for the amount of cooling needed during winter months. The valve is designed to react to changes in high-side pressure. As high-side pressure decreases, the valve closes to reduce the amount of cooling water entering the condenser. When the system cycles off, the regulating valve closes completely.

Cooling towers. Cooling towers are often used to help reduce the amount of water consumed by water-cooled condensers. The tower is mounted on the facility roof. Cooling water from the condenser is pumped to the top of the tower and allowed to cascade down over a series of wooden slats. A fan can be used to create air flow across the slats or the system might rely solely on natural air flow.

As the water cascades down, roughly 10 to 20 percent of the water evaporates, creating natural cooling. The remaining 80 percent of the water piped to the tower is cooled and recirculated back to the condenser.

Unfortunately, as the water cascades through the tower, it picks up dirt and sulfur dioxide from the air. Strainers and filters installed in the water lines require regular service. Algae buildup on the slats and tower body can also be a problem that requires treatment similar to that used on swimming pools.

The sulfur dioxide combines with water to form a mild acid that corrodes the condenser tubes and damages the cooling tower over time. Impurities in the city water supply also lead to the buildup of mineral deposits on the inside surfaces of the cooling tube piping. The water system of the cooling tower requires chemical treatment to maintain proper pH levels and deter the build up of deposits that decrease heat transfer. Some cooling tower systems feed a small amount of fresh "bleed" water into the lines to help deter mineral buildup.

Defrost Methods

In supermarket equipment, not only must the evaporator be defrosted, especially in low-temperature equipment, but also frost in the flues and around the fan blades in various areas of the air-distribution system must be melted and completely drained. Refrigeration systems used in supermarket applications use any one of a number of defrost methods. Electric defrost, reverse air defrost, and off-cycle defrost can be used on both parallel and single-unit systems.

Hot gas defrost is a popular choice for parallel compressor refrigeration systems. Hot gas directly from the compressor discharge or cooler gas from the top of the receiver is routed to the evaporator manifold.

Receiver or "cool" gas defrost

Receiver or "cool" gas defrost is normally used on continuous display multiplexed systems served by parallel compressors. Cool gas defrost is performed by using the relatively cool refrigerant gas that forms at the top of the refrigerant receiver. This cool gas is discharged down the suction line to the evaporator. At the evaporator, the gas condenses as it gives up its heat to melt the ice and frost on the evaporator coils (Fig. 13-8). Because the receiver defrost gas is at a relatively cool temperature at the start of the defrost cycle, it reduces thermal stress on the piping and the possibility of eventual leaks and line breakage. Cool temperature gas is continuously supplied from the compressor discharge line throughout the defrost cycle.

The compressor discharge gas is injected into the receiver through a port at a 90-degree angle to the gas exit port. As the discharge gas passes over the liquid in the receiver, it is desuperheated. This helps generate the positive pressure needed to maintain the flow of liquid to the refrigerated fixtures during defrost. On a multiplexed system, roughly 25 percent of the total load can be defrosted at any one time. The remaining 75 percent of the load is needed as a heat source to generate the defrost gas.

Cool gas defrost can be started using either mechanical or electronic multicircuited time clocks that provide the proper sequence of defrosts. Only one circuit can be defrosted at a given time or the system operation might be adversely affected.

Systems operating in cold (below 30°F) ambient temperatures require a discharge differential pressure regulator in the discharge line to ensure there is sufficient gas flow during the defrost cycle. This valve should be set for a 10–20 psi differential across the valve.

In the typical control strategy for a receiver gas defrost system, the remote condenser fans are controlled by a pressure control set to the minimum target pressure that corresponds to an 88 to 89°F saturation temperature. This ensures sufficient defrost during colder ambient con-

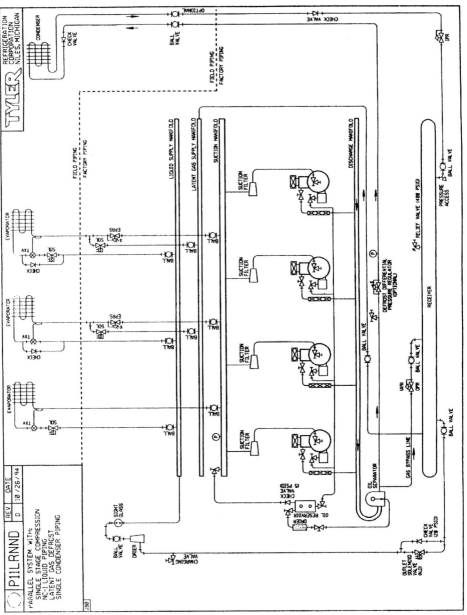

Figure 13-8 Typical receiver or cool gas defrost system. *(Courtesy of Tyler Refrigeration)*

ditions. The outlet pressure regulator (OPR) valve is set for the target pressure that corresponds to an 86 to 87°F saturation temperature. The inlet pressure regulator (IPR) valve is set for a target pressure that corresponds to a 94 to 95°F saturation pressure. The main solenoid valve is closed during defrost and the 20 psi check valve becomes operational.

Defrost control

A variety of mechanical and electronic clocks, often part of a compressor controller system, usually controls defrosting. Electronic systems often have communication capabilities from telephones outside the store. Regardless of the controls, the manufacturer's recommendations should be followed.

Liquid and/or suction line solenoid valves control the circuits for defrosting. Often, a suction stop is combined with the temperature controlling Evaporator Pressure Regulator (EPR). Temperature control on the branch circuits is also achieved by refrigerator thermostats actuating liquid line solenoid valves. Shutoff valves isolate each circuit for service convenience.

Setting mechanical defrost timers

Mechanical defrost clocks are used on many supermarket refrigeration systems as well as walk-in and reach-in boxes described in previous chapters (Fig. 13-9). Knowing how these units work and how to set them is essential. There are several major manufacturers of mechanical time clocks, but the basic operating principles are similar for all types.

A typical mechanical defrost clock is designed to handle defrost periods of two hours or less. Units that require longer defrost cycles will need additional time clocks. General instructions for setting time clocks are presented here.

There are three main components that must be set up in order to program the defrost cycle for a refrigeration system:

- Program timer dials
- Cycle timer dials
- Time setting dial

Program timer dials are marked in one-hour slotted increments and rotate once every 24 hours. A tripper is installed for each time that the defrost cycle should start within a 24-hour period.

Cycle timer dials rotate through each defrost cycle, then stop when the cycle ends. The length of a cycle might be set for a period of 2 to 120 minutes.

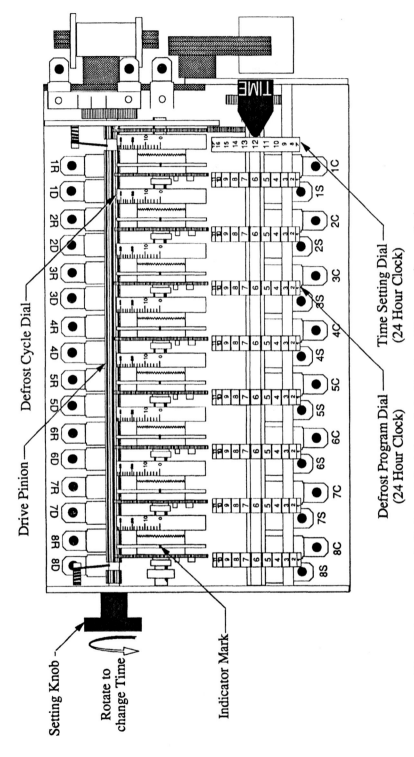

Figure 13-9 Timer dials on a mechanically driven defrost timer clock. *(Courtesy of Hussmann Corporation)*

The time setting dial rotates once every 24 hours and appears to be seven hours ahead of the program timer. This time difference is meant to compensate for the location of the defrost switches and the pointer marked TIME. The time setting dial should be set according to the time of day.

Setting the defrost timer

To set the defrost start time, rotate the program timer dials by turning the setting knob located at the end of the timer opposite the motors. As the dial slots marking the start of the defrost cycle appear, install a tripper for each time that the beginning of a cycle is desired. The slots for the tripper time are found directly above its number on the dial face.

To install a tripper, push straight in until it snaps over its holding detent (Fig. 13-10). When properly placed, the tripper's shoulders should extend $\frac{1}{32}$ inch above the dial face, in square alignment to it. Placement is important as just one misaligned tripper could jam the timer.

To set the length of defrost, turn the setting knob until the indicator mark faces forward and the cycle timer dial stops.

Caution: The indicator dial must be visible before the cycle timer dial is adjusted, or damage will result to the clock.

Insert the time setting tool and pull the dials apart (Fig. 13-11). Reset the teeth once the number indicating the desired length of defrost has been positioned opposite of the indicator mark.

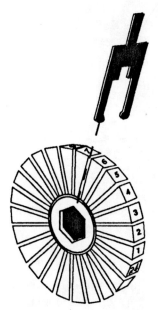

Figure 13-10 Installing a timer tripper. *(Courtesy of Hussmann Corporation)*

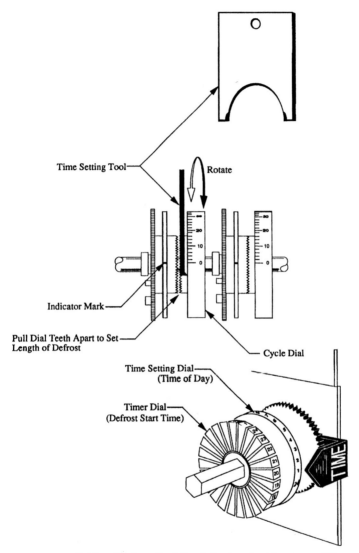

Figure 13-11 Setting the length of the defrost cycle. *(Courtesy of Hussmann Corporation)*

To set the time of day, turn the setting knob until the correct number on the dial indicating the time of day is opposite the arrow time marker. The time on the time setting dial should not correspond to the numbers for the program timer dials.

Alarm switch

The timer shown in Fig. 13-9 consists of two motors and an alarm switch, which is normally open. Both motors operate continuously. If one motor fails, then the other one will still drive the timer. The alarm switch closes to indicate that the timer needs to be serviced.

To reset the alarm switch, first push gear A toward the motor so that lever B does not snap (Fig. 13-12). Using a nonmetallic "stick," depress the plastic cam until it snaps into position. The switch should now be held in an open position.

Modular defrost timers

The defrost timer shown in Fig. 13-13 has the following three components: the frame, the drive motor module, and the program modules.

The frame holds one drive motor module, and the correct number of program modules needed for a given defrost system. The drive motor module is mounted on the frame end. It powers the program modules through the main drive gear. The drive motor and the defrost circuits need not be the same voltage.

A program module contains two dials. The time-of-day dial rotates once every 24 hours, and the minute dial rotates once for every defrost cycle. The time-of-day dial is notched so that it can accept a defrost tripper for an "even hour" or an "odd hour" in its 24-hour rotation. These modules cannot be substituted—one type cannot be converted to another.

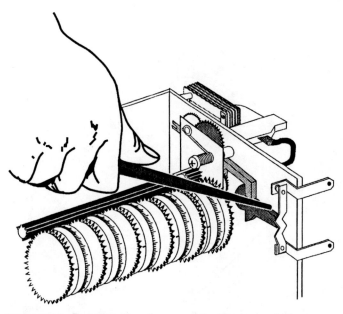

Figure 13-12 Resetting the alarm switch. *(Courtesy of Hussmann Corporation)*

Setting a modular defrost timer

To set a defrost start time, insert one black tripper in the desired white program module notch. (Afternoon or PM hours are indicated by numbers 13 through 24. Thirteen is 1 PM, 14 is 2 PM, and so on.) Remove any extra trippers before starting the cycle.

To set the length of defrost, rotate the copper termination lever of the minute dial until the desired minutes are shown. Take care not to bend the lever any more than is needed to disengage it from the dial teeth.

Caution: Do not move the red tab.

To set the time of day, rotate the main drive gear on the motor module. Push upward with a finger or thumb so that the desired hour on the black time-of-day dial lines up with the pointed alignment mark on the modules.

Program module removal

First, turn off the power to the control panel and meter check the Paragon as a safety measure. Disconnect and mark the wires from the switch found at the top rear of the module. Switch terminals are labeled C, NC, and NO.

Adjust the timer so that all the red tabs on the minute dials are directly facing the front of the timer. At the bottom rear of the module, pull down on the plastic latching lever and out on the bottom of the module (Fig. 13-14).

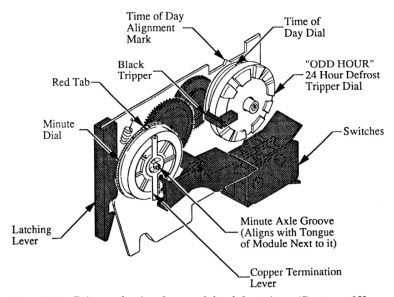

Figure 13-13 Drive mechanism for a modular defrost timer. *(Courtesy of Hussmann Corporation)*

Supermarket Refrigeration Equipment 375

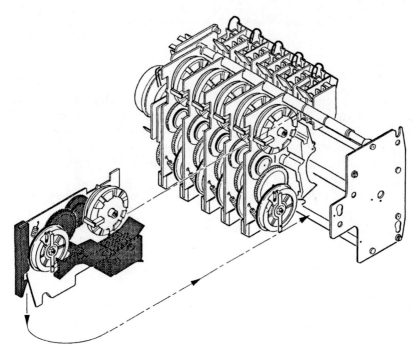

Figure 13-14 Program module removal and installation. *(Courtesy of Hussmann Corporation)*

Module installation

Check to be sure that the power is off. Set all modules, including the one to be installed, at the same hour on the dials. Set all red tabs on the minute dials (including the new one) in a front most position.

Slide the module into the slotted rod of the frame top. Mate the minute dial axle into the axles on both sides of it. Check that all red tabs are aligned. Replace the wires on the switch terminals. Reset the time of day and restore power to the control circuit.

Program motor removal

Begin by turning off the power to the control panel. Use lock-out, tag-out procedures and use a voltmeter to check the unit as a safety measure. Disconnect and mark all wires that are attached to the motor. Next, rotate the main drive gear until the axle tongues and grooves are vertical—the red tabs should lie directly above the axle. Loosen the hex head bolt next to the motor, sliding the module upward until all three locator studs clear the key slots.

To install the new motor, begin by checking the new motor voltage application as well. Next rotate the main drive gear until the axle tongues and grooves are vertical—the red tabs will be found directly above the axle. Place the locator studs in the key slots and slide them down. Tighten the hex head bolt, and reconnect all wires (Fig. 13-15).

Subcooling Liquid Refrigerant

By lowering the temperature of the liquid supplied to the TEV, the efficiency of the evaporator is increased because less flash gas is produced. Many larger supermarket refrigeration systems with remote condensers are designed to allow refrigerant to *naturally subcool* in cool weather as it returns from the remote condenser. This saves energy. The most common method to naturally subcool refrigerant is to flood the condenser and allow the liquid refrigerant to cool close to the ambient temperature.

Subcooling can also be done mechanically by installing a subcooling compressor on a parallel system (Fig. 13-16). Because mechanical subcooling uses a direct expansion device, it is not limited by ambient temperature. A liquid line solenoid valve and a TEV control refrigerant to the subcooler. An EPR prevents the subcooler temperature from dropping below the desired liquid temperature. Electrically, a thermostat responding to main liquid line temperature controls a solenoid valve on the liquid supply line. The mechanical subcooling system would be set to operate when the ambient temperature raises the refrigerant temperature above the desired subcooled temperature. The mechanical subcooling compressor can run at twice the efficiency of the main sys-

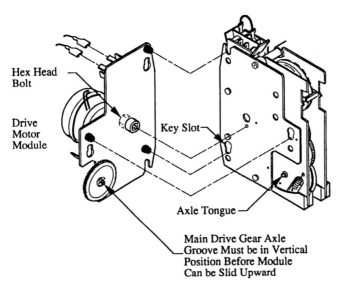

Figure 13-15 Motor removal and installation. *(Courtesy of Hussmann Corporation)*

tem, thereby saving energy through year-round liquid temperature control.

Heat-Recovery Systems

Heat recovery can be an important feature of virtually every compressor system, parallel or single. A heat-recovery coil is simply a second condenser coil placed in the store's air handler. If the store needs heat, this coil is energized and is run in series with the regular condenser (Fig. 13-6). The heat-recovery coil can be sized for a 30 to 50°F TD depending on the capacity in cool weather. Lower head pressures in parallel systems permit little heat recovery. However, when heat is required in the store, simple controls create a higher head pressure for heat recovery. When compared with the cost of auxiliary gas or electric

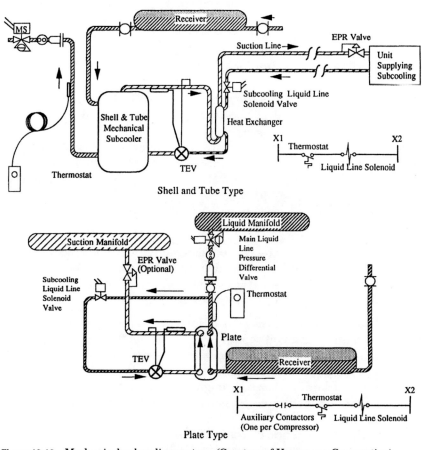

Figure 13-16 Mechanical subcooling system. *(Courtesy of Hussmann Corporation)*

heat, the higher energy consumption might be compensated for by the value of the heat gained.

Water heating. Recovered heat can also be used to heat water. Water can be heated by a desuperheater on one large, single compressor unit. The most common system for heating water is by the interstage desuperheater on two-stage or compound parallel systems used in low-temperature HCFC-22 applications.

Store environment control

Store environmental control is the key factor of energy management when using recovered heat. Store temperature and humidity must be properly controlled for full-energy management. If store humidity is high (as it usually is during the spring, summer, and fall in uncontrolled stores), the refrigerated or freezer display cases will undertake the job of removing excess moisture from the air. This moisture is removed as extra frost on the case evaporator coils. Consequently, extra defrost time and heat are required to melt the excess ice off the coils.

High temperature and uncontrolled humidity add to the case refrigeration load. Cool, dry conditions subtract from the load. The key is to avoid high refrigeration loads and, thereby, lower overall operating cost. When ideal conditions are maintained (70°F, 35 percent RH), the load is typically reduced 20 to 40 percent. The reduced load reflects directly in lower energy usage.

Proper store environmental control takes heat from case condensing units, supplemental heaters, and air-conditioning capacity to maintain the store environment, creating optimal case operating cost and customer and employee comfort (Fig. 13-17).

Remote air-cooled systems

Parallel refrigeration systems are ideally suited for heat recovery. The necessary heat recovery valve is commonly installed on the rack. The valve is activated as a stage or part of a stage of heat recovery by the store's environmental control system.

Normal operation (heat recovery off). Hot gas from the compressor discharge goes through an oil separator and is piped directly to its section in the remote condenser. The condenser is usually on the roof and should not be more than 50 equivalent piping feet from the parallel. The condensed liquid passes through the winter control valve, which helps to maintain a suitable pressure in the receiver, and then returns to the parallel receiver.

An O.P.R. valve in a line direct from the discharge provides a positive means of maintaining pressure in the liquid receiver in very cold

Supermarket Refrigeration Equipment

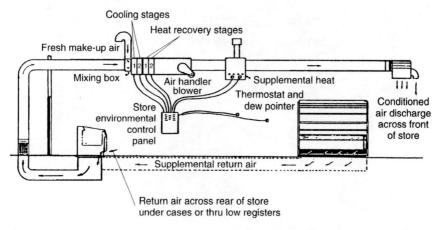

Figure 13-17 Details of in-store heat recovery and environmental control system. *(Courtesy of Tyler Refrigeration)*

weather. This is necessary to ensure proper flow of liquid refrigerant to the fixture expansion valves (Fig. 13-18).

Heat-recovery operation. When heat is called for by the store environmental controller, the heat-recovery valve switches the flow of hot gas to the heat recovery coil in the store air handler. After heat has been transferred to the flow of recirculating store air, the refrigerant goes to the remote condenser. If proper condensation does not take place in the heat-recovery coil, the remote condenser is there to finish the job. Piping should not exceed 100 equivalent feet from the parallel to the heat-recovery coil in the air handler.

Heat reclaim sizing

To calculate the heat available for heat reclaim usage, start with the design evaporator loads and reduce them to simulate a winter condition. Next, increase this value to add the heat of compression from the compressor. The combined values result in the heat available for heat reclaim.

Medium-temp. case load $\times$ 0.70 $\times$ HCF = heat available for reclaim

Low-temp. case load $\times$ 0.80 $\times$ HCF = heat available for reclaim

Walk-in cooler load $\times$ 0.50 $\times$ HCF = $\dfrac{\text{heat available for reclaim}}{\text{Total heat reclaim available}}$

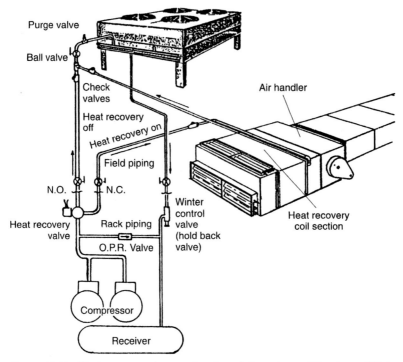

Figure 13-18 Heat recovery in remote air-cooled systems. *(Courtesy of Tyler Refrigeration)*

Table 13-1 lists the heat of compression factors (HCF) for various suction temperatures at a condensing temperature of 90°F. Ninety degrees is the normal heat recovery condensing temperature. Higher temperatures only produce heating at the same cost as electric heat. Heat-recovery coils are typically sized for 30 to 40 degrees TD. This anticipates a 50- to 60-degree mixed air temperature entering the coil.

TABLE 13-1 Heat of Compression Factor*

Suction temperature (°F)	Compression factor
+35	1.20
+25	1.24
+15	1.28
+5	1.33
0	1.36
−5	1.39
−15	1.46
−25	1.53
−35	1.61

* Condensing temperature is 90°F.

Air-conditioning credits

Specific air-conditioning credits vary substantially from case style to case style. Case loads are made up of internal electric loads, radiant heat from exterior lighting (very substantial on top display cases), wall conduction and the infiltration load, which is a combination of sensible and latent heat. Manufacturers also usually add some percentage for safety to account for loading techniques, traffic, warm product, and times of poor ambient control. It is not practical to try to calculate precisely the credit for each style of case and its application. A general rule takes into account the mix of cases in a typical store because the air conditioning calculation itself is partly an educated guess, especially when considering outside air influences and person loads.

Consider only open cases for this calculation. Do not use walk-in cooler and freezers and glass-door merchandisers. The following credit is then usually acceptable. Calculate using the total published open case load.

- Sensible credit = 40 percent
- Latent credit = 30 percent

The importance of good air distribution

The effect and efficiency of heating, heat recovery, and air-conditioning in a supermarket is only as good as the air distribution system that is to deliver conditioned air and return the store air for reconditioning. The unique construction of food stores (in that they are usually made up of many rows of both refrigerated and nonrefrigerated displays) makes air distribution difficult.

The open refrigerated aisles are a particular problem because substantial amounts of heat are removed from these areas daily, posing two concerns for efficiency and comfort. The lost heat must be replaced for comfort and the cold air that is removed should be drawn back into the system for air conditioning. Both of these concerns can be dealt with by properly removing the cold air from the aisle and letting the natural flow of upper-level warmer air to replace it. Cold air is heavier than warmer air and will pocket in aisles that do not have positive air movement. This, of course, must be done in a way that does not disturb the internal air flows of the display cases.

A number of methods to accomplish optimal air flow patterns exist, and the proper one for a given application should be thoroughly discussed with the HVAC contractor. The case manufacturers can usually offer recommendations as well. Good air distribution is always more expensive in a supermarket than other types of buildings, but without proper consideration, poor air distribution will result in cold aisles, which will increase operating costs for both the refrigeration and the HVAC system.

Air circulation

Fresh air introduced through the air handler should operate 100 percent of the time. It should be done at the rate required by local building codes plus allowance for exhaust fans. Shutting off the fresh air does not lead to economy of operation. Normal customer and service door openings, wide impingements, and exhaust fans will always cause an exchange of indoor and outdoor air. By properly introducing the fresh air through the air handler, uncontrolled air introduction is minimized or eliminated, and it is conditioned for comfort and economy.

Contrary to the statement above, the use of economizer cycles are not recommended for supermarkets because the times that cooler air is available for building cooling is usually at the same time that the moisture content of this air is high. Introduction of this humid air will have adverse effects on the performance of open refrigerated cases.

Overall design considerations

The high refrigeration loads that produce cooling, and the high lighting loads that produce heating of supermarkets make them unique in the design and application of the HVAC systems. Finished versus open ceilings and the mix of refrigerated and nonrefrigerated areas of large stores make these designs even more challenging. Because cold air is heavy, duct and grill systems must be provided that will flush the cold air from refrigerated aisles without producing drafts that can wipe the air out of the display cases. This is particularly critical for vertical cases.

The accepted techniques of discharging air at the front of the store and returning under cases, and at the rear of the store are valid, and perform well in flushing a store and providing even and comfortable conditions as well as efficient operation. Consultation with architects and contractors should result in methods that ensure warm aisles.

Modern Control Systems

Electronic control systems are a fact of life in modern technology. You will find them on your car, your microwave oven, your TV and VCR, and on most rack-type refrigeration systems installed in the past five to seven years. Not all electronic control systems provide the same level of control and sophistication.

It is important to remember that electronic controls do not change the way the compression refrigeration cycle works. They simply allow us to monitor and operate it more precisely. In fact, if a problems occurs with an electronic controller, most quality systems provide for switchback to mechanical control. This means the electronic controller turns over control of the system to standard mechanical and pressure switches and controls. Electronic control systems also provide battery backup and redundant control of critical operations such as defrost control.

Just as manufacturers do not make the exact same rack systems, they do not design their control systems to be exactly alike. The following sections outline several examples of control systems available today. There are many other excellent control systems available. Learn as much as you can about the control systems on the equipment you service. Read manufacturer manuals, view their videos, and attend specialized training classes when available.

The system illustrated in Fig. 13-19 routes all operating signals to a diagnostic center/alarm panel. While only one compressor is illustrated, the system can be used to pinpoint an individual problem area in any one of the multi-compressor circuitry systems. This makes the complex circuitry of the system manageable.

The internal pilot circuitry throughout employs 220 volts, which readily permits simple replacement of components. The alarm systems can signal the following types of malfunctions:

- low oil pressure
- high discharge pressure
- compressor power overload
- low liquid level (with 20-minute time delay)
- low suction pressure

All signal devices, such as high-pressure switches, differential pressure switches, and liquid level indicators are circuited through an isolating relay that allows for individual signaling between warning devices. Signals and alarms can also be routed via telephone lines to a remote location.

On systems equipped with electronic controls, a transducer must be disconnected or isolated during system evacuation or charging to prevent internal damage.

The electronic device illustrated in Fig. 13-20 monitors system status with respect to compressor motor operation and current draw. The status is indicated using green, amber, and red status lights. When the green light is flashing, the motor is in acceleration; when it is solid green, a successful start cycle has been completed. When the amber light flashes, the system is in its defrost cycle. When the amber light is solid, the compressor is cycled off.

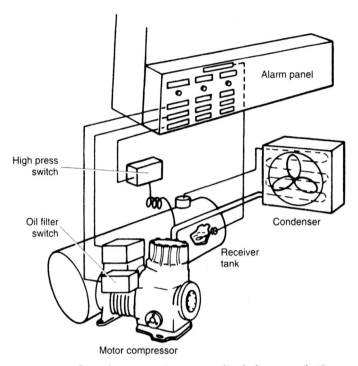

Figure 13-19 Control system using a centralized alarm panel. *(Courtesy of Tyler Refrigeration)*

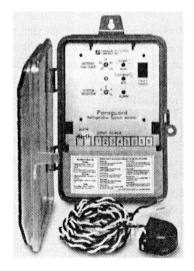

Figure 13-20 Electronic controller with fault code capabilities. *(Courtesy of Paragon Electronics)*

When the red light is illuminated, it indicates a fault condition has occurred. Faults are indicated by a numerical code, which is displayed on a LED on the monitor. The code number corresponds to specific fault conditions that are listed in Table 13-2.

TABLE 13-2 System Fault Code and Alarm Indicator Troubleshooting Guide

FAULT CODE 1 The compressor has cycled too often with a preset time period.	Probable causes: Condenser fan motor failure Blocked condenser Machine room ventilation failure Refrigerant overcharge/noncondensibles (air) in system Incorrect pressure control settings Intermittent electrical connections in the control circuit Compressor overloading or overheating (motor protector trip) Sticking or partially restricted flow-control devices Refrigerant loss Extreme motor overheating
FAULT CODE 2 There is a repeated high current draw during the start or stabilizing period.	Probable causes: Large amount of liquid refrigerant/oil is returning to the compressor Lack of lubrication Seizing bearings (before lock-up) Intermittent load voltage supply Intermittent control power (contact chatter)
FAULT CODE 3 There is repeated high current draw during the run portion of the cycle.	Probable causes: Excessive amounts of refrigerant/oil returning to the compressor Lack of lubrication Seizing bearings (before lock-up) Intermittent voltage supply Intermittent control power (contact chatter)
FAULT CODE 4 There is a continuous high current draw during the start or stabilizing period.	Probable causes: Single phasing Starter contact failure Rotor or stator problems, i.e., dead spots in the windings Seized bearings Extreme motor overloading Low voltage conditions Faulty starting components (relay, capacitors) Loose electrical connections
FAULT CODE 5 There is a sustained high current condition during the run portion of the cycle.	Probable causes: Single phasing Seized bearings Extreme motor overloading Low voltage conditions Loose electrical connections

TABLE 13-2 System Fault Code and Alarm Indicator Troubleshooting Guide (Continued)

FAULT CODE 6 There is a loss of mechanical performance of the compressor or a loss of fixture load during the start portion of the cycle.	Probable causes: Broken rods or crankshaft Valve plate failure Extreme ambient temperature changes Partly restricted flow-control devices Low voltage conditions (seasonal)
FAULT CODE 7 There is a loss of mechanical performance of the compressor or a loss of system capacity during the run portion of the cycle.	Probable causes: Partly restricted flow-control devices Extremely low ambient temperature Refrigerant loss Evaporator fans off or frozen
FAULT CODE 8 The compressor has not restarted within an acceptable period.	Probable causes: Loss of control voltage Open safety or manual reset controls Extremely low ambient temperature
FAULT CODE 9 The compressor has not restarted following a regularly scheduled defrost period.	Probable cause: Defrost circuit problem

Computer-Controlled Systems

A true electronic control system is driven by a microprocessor. That is, it uses a computer to process information about how the refrigeration system is operating. This data is used to make changes to system settings to increase operating efficiency.

The most sophisticated systems control virtually all aspects of operation and defrost cycling of multiple compressor rack systems. Operating parameters can be programmed into the control system and the system can provide detailed self-diagnostic and troubleshooting information. By monitoring the refrigeration system over extended periods of time, some systems can actually adjust themselves automatically to more closely match operating conditions.

Communications between the controller and its sensors and actuating devices is done using low-voltage signals provided from a 24-Vac transformer.

Operators or service technicians interface with such systems through a computer monitor, flat screen display panel, or a hand-held monitor that plugs into a communications port on the system central processing unit (CPU).

The technician has the ability to test individual compressors, valves, and other components that are under system control. This makes troubleshooting and isolating problems much simpler. Trouble codes or specific alarm messages can be logged and/or displayed when the control system sees a particular problem occur. Multiple rack systems can be

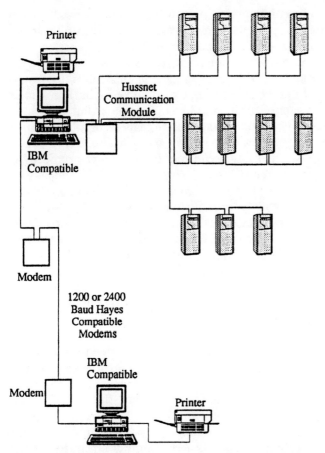

Figure 13-21 Remote monitoring center for multiple rack systems. Communications are via modems and phone lines. *(Courtesy of Hussmann Corporation)*

tied into a central monitoring station through the use of a communications module, modem connections, and telephone communication lines (Fig. 13-21). A computer-controlled refrigeration system for rack applications typically contains a control module or computer and a number of interface boards. These interface boards might include an input board to route signals from sensors and valves to the controller, relay boards to control signal outputs from the controller that turn compressors and condenser fans on and off, and a defrost control board to handle specific defrost control functions. Relay and input boards make checking voltages to and from sensors and components easy and convenient.

These interface boards are located outside of the controller and connect to ports on the controller via signal cables. The compressor relay board can handle up to eight separate compressors, two of which can be satellite compressors. Data is supplied to the input board from a number of sources including the suction pressure transducer, discharge or head pressure transducer, and split rack pressure transducer. Temperature inputs can be provided from sensors in the liquid and suction lines.

The controller monitors suction pressure to decide when to cycle the compressors. Electronic control systems can do this more accurately because there is no differential between cut-in and cut-out. The system can control a single-suction unit, a split suction unit that allows for independent control of two separate suction pressures, or a low temperature compound or two-stage rack, which also operates using two separate suction pressures. The system can handle up to eight compressor unloaders or two variable speed compressors instead of the unloaders.

The condenser fan board can be configured to split the fan operation. Only one-half of the fans will operate during cold weather, cutting operating cost without lowering system efficiency. A "floating head" feature changes the discharge pressure and liquid set point as measured by a probe on the drop leg and as ambient temperature changes. This is done to optimize performance and efficiency. The defrost management system can control up to 32 separate branches and features a 24-hour programming memory so information on defrost cycling is not lost if power to the unit is lost. The discharge pressure is also used to control the condenser fan operation. The controller matches the compressor to the load and then provides the amount of condensing fans needed to support the compressors. The controller uses two methods to control head pressure: a fixed head and a floating head. A fixed head operates the compressor on a fixed set point. With the floating head option, the controller can lower the head pressure to take advantage of operating efficiencies through subcooled liquid from the condenser. It adjusts liquid temperature and discharges pressure set points accordingly. It optimizes system efficiency by taking advantage of lower head pressures.

When satellite compressors are used, case temperature determines when to cycle the compressors on and off. This is because satellites do not operate from the same suction pressure as the rack. They have an independent suction line from the case line-up. The controller uses suction and discharge pressures to troubleshoot compressor operation and safety. The discharge pressure and ambient and liquid line temperatures aid in the control of the condenser.

Compressor unloaders allow step variation in the compressor capacity, and variable speed compressors have an infinite variation in capacity. The controller is designed to cycle unloaders or variable speed compressors faster than standard compressors to allow for close matching of capacity or demand.

There are four ways the controller can cycle the rack compressors. Sequential is a first-on, last-off method. Round-robin alternates run time between the compressors with the first compressor turned on being the one that has been turned off the longest. Swing cycling is used when compressors in a given rack do not have the same capacity. It cycles small- and large-capacity compressors to derive the greatest variety of capacities available. Mix-match is a custom-tailored cycling pattern that brings on different combinations in a prescribed pattern to meet specific capacity needs.

Software and programming. All electronic control systems that run off of a computer or CPU are driven by the manufacturer's supplied software programs. A software program is essentially a set of instructions to the computer that tells it how to operate. The software program also allows the service technician or operator to program operating values into the system and view system data. Manufacturers continually update their software programs to provide new features, enhance its speed and performance, or eliminate any problems with earlier software versions.

A software program is normally set up in a series of menus that organize the program for easy, logical use. The software program usually includes individual system status and parameter programming menus, plus menus for performing maintenance tasks and system configuration work. Menus used for programming or testing system operation are normally pass-code protected so that only authorized personnel can use them.

Installation Tips

Supermarket equipment manufacturers normally provide detailed installation and maintenance instructions with their various units. Videotapes and special training seminars are also available from major equipment suppliers. As a serious technician, you should take advantage of these materials when they are available to you. The following sections list a number of common installation tips and checks.

Piping

Make certain to use the correct connection size for the liquid line and suction lines to the unit. Once the connections have been made, seal the outlet thoroughly inside and out using an expanding polyethylene-type foam. Refer to the original legend furnished for the store when sizing refrigerant lines. If no legend can be found, refer to the refrigeration unit manufacturer's application data for guidance.

Note: If the defrost system uses "cool" gas from the system's receiver, the liquid line might need to be larger inside the merchandiser area to ensure even liquid drainage from all evaporators during defrost. Refer to the unit's installation manual for sizing details.

Be very careful when making line connections with a torch. If heat from the torch travels up the back side of the plastic discharge honeycomb, it could damage it. As a precaution remove the honeycomb section until all connections are made.

Excessive pressure drop can keep the refrigeration system from running at full capacity. To keep the pressure drop to a minimum, the refrigerant line should run as short as possible with as few elbows as possible. Remember, use long radius elbows when needed.

Line insulation

For units that do not use cool receiver gas defrost, the suction and liquid lines should be clamped or taped together. These lines should be insulated starting at least 30 feet from the unit. For units using receiver cool gas defrost, the suction and liquid lines should not contact each other. They should be insulated separately beginning at least 30 feet from the unit. Where excess moisture from condensation is a problem, insulation might be started at the beginning of the line.

Oil traps

Oil traps/P-traps must be installed at the base of all suction line vertical risers.

Typical control settings

Table 13-3 lists control settings for a single-deck island display case for ice cream or frozen foods. Some large units might use two evaporators to provide for separate temperature control at the front and rear of the unit. If this is the case, each evaporator must be adjusted independently of the other.

Single-compressor systems. On a single-compressor system, the island display case temperature is controlled by a thermostat with a 3 to 6°F differential. The thermostat is wired to control the compressor motor contactor.

When electric defrost is used, it is terminated at 52°F. The defrost thermostats for all units on the single compressor are wired in series. Fail-safe must not control defrost cycle length, especially when less than 208-V power supply is used for defrost heaters, or if frost buildup is heavy from shopping demands.

On outdoor units, the defrost timer will control a liquid line solenoid beginning a defrost pumpdown four minutes prior to defrost.

Low-pressure control settings are applicable to outdoor condenser units where ambient temperatures do not fall below 0°F. The display thermostat must control the display unit temperature.

Parallel compressor racks. In a unit equipped with parallel compressors, merchandiser temperature is controlled using a thermostat or CDA valve. The CDA sensor should be mounted in the same location as the thermostat sensing bulb. The CDA valve and control board are

TABLE 13-3 Typical Control Settings for a Frozen Food/Ice Cream Merchandiser

	Conventional Single Compressor		Parallel Compressor Rack	
	Frozen Food	Ice Cream	Frozen Food	Ice Cream
Discharge Air Temp., °F	−10	−20	−10	−20
Evaporator Temp., °F	−20	−30	−20	−30
Refrigeration Controls				
EPR	When used EPR valve setting must maintain evaporator temperature listed above			
Thermostat, Cut-Out °F	−10	−20	−10	−20
CDA valve, set point °F	—	—	−10	−20
Fan Cycling Cut-in/Cut-out °F	—	—	—	—
Defrost Controls				
Frequency	Every 24 hours		Every 24 hours	
Termination Temperature for Electric Defrost, °F	52	52	52	52
Length in Minutes for:				
Koolgas Defrost	—	—	20	24
Electric Defrost	60	60	60	60
Fail-safe for:				
Single compressor, outdoor condenser, pumpdown included	64	64	—	—
Reverse air	90	90	90	90
Low Pressure Setting (With CDA or Thermostat Temperature Control)				
Cut-in/Cut-out psig for R-502 refrigerant	32/10	15/4	—	—
Cut-in/Cut-out psig for R-22 refrigerant	25/6	10/0	—	—

mounted on the rack. If electric defrost is used, the setting and precautions are the same as those for single compressor-driven units.

When cool gas defrost is used, it is time-terminated. The defrost frequency and lengths might require adjustment to match specific store conditions such as temperature and humidity, low head pressure, long refrigerant line runs, and seasonal changes. Stagger defrost to maintain stable compressor loading and sufficient defrost gas. If possible, defrost when the store is closed.

Expansion valve superheat adjustment

Expansion valves should be adjusted to fully feed the evaporator. To prepare the evaporator before adjustments are made, do the following:

Check that the evaporator is clear or just lightly covered with frost. The fixture should be within 10°F of its normal operating temperature.

To measure the existing superheat setting, attach two temperature sensing probes (either thermocouple or thermistor) to the evaporator. Place the first probe under the clamp holding the expansion valve bulb. Securely tape the second to the coil outlet (Fig. 13-22). An alternate method of measuring evaporator superheat is to take a pressure reading on the coil outlet and use the system refrigerant pressure/temperature table to convert this pressure to a temperature. Take a temperature reading from the suction line adjacent to the expansion valve bulb. Subtract the temperature conversion from the actual valve bulb temperature to determine the superheat of the evaporator.

Some "hunting" of the expansion valves is normal. The valve should be adjusted so that the greatest difference between the temperature readings during hunting is within system specifications. Specifications will vary with the type of equipment, but are normally in the 3 to 5°F or 6 to 8°F range.

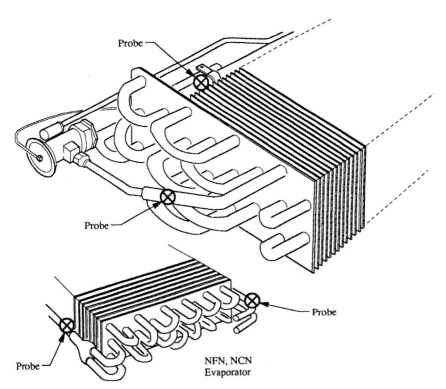

Figure 13-22 Expansion valve superheat adjustment on supermarket display unit. *(Courtesy of Hussmann Corporation)*

To adjust valve superheat, start by removing the valve stem cover. To decrease superheat, turn the valve stem counterclockwise (as viewed from the stem end of the valve). To increase superheat, turn the valve stem clockwise. Turn the valve stem no more than ¼ turn at a time, waiting at least 15 minutes before rechecking the temperature and making further adjustments.

During the normal cycle, the temperature might be less than the setpoint, and might reach 0°F. All pans, coil covers, and doors should be closed or in place while checking the cycle. If left ajar, the load on the evaporator will be increased.

All TEVs have a recommended maximum number of turns (in or out) allowed for adjustment. Depending on the valve manufacturer, the number might be as low as 8 and as high as 12; check the installation manual for specifications. Do not force the adjustment beyond these stops, or damage to the valve will result.

The following values are approximations and should only be used as a guideline:

- For R-12: One turn will result in a 3 to 4°F change in superheat.
- For R-22: One turn will result in a 2 to 2½°F change in superheat.
- For R-502: One turn will result in a 2°F change in superheat.

Refrigeration thermostat

An optional thermostat can be factory installed (Fig. 13-23). The thermostat body is typically located under the overhang at the left end of the unit. A factory-installed bulb is found in the discharge flue under the display pan, attached to the lower lip of the front shelf support. Twin temperature models should have two thermostats—one for each side of the unit.

CDA sensor

An optional CDA sensor is factory installed at the spot where the thermostat bulb would normally be found. Its leads are routed through an electrical raceway to the rack control panel. Leads are tagged in the raceway. Twin temperature models have two sensors—one for each side of the fixture.

Electrical connections

Connections are made at an electrical connection box located on the unit or in the end case. All wiring should comply with NEC and local codes. Some manufacturers mark the leads of all electrical circuits

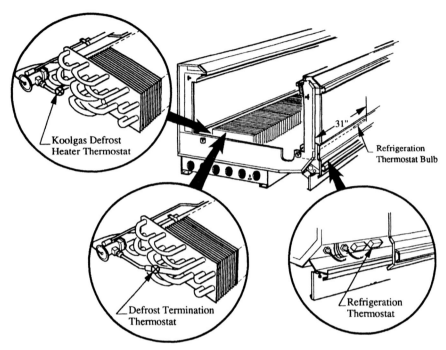

Figure 13-23 Optional thermostat control for display unit temperature control. *(Courtesy of Hussmann Corporation)*

using colored plastic bands. These bands correspond to a "color code sticker" located inside the unit's raceway. Defrost heaters, defrost termination thermostats, and refrigeration thermostats on wide island units might also be tagged to show whether they belong in the front or rear case display section.

Field wiring

Field wiring should be sized to correspond with the component ampere ratings stamped on the serial plate although the actual ampere rating could be less than specified. Field wiring should run from the refrigeration control panel to the unit where the optional defrost termination thermostats, optional refrigeration thermostats, or CDA sensors are used. When several units share the same defrost circuit, defrost termination thermostats should be wired in series. Always check the serial plate before selecting wiring.

Night covers

Ice cream models are typically supplied with night covers—frozen food products do not need them. Night covers are specially made from polished aluminum that prevents the top packages from softening due to radiant heat or defrost air. Aluminum is best; other materials that are porous or nonreflective would provide the same conditions.

Night covers should be laid flat on top of packages each night at store closing time. They should not extend above the load-limit marking because this could result in defrost air being trapped between the product and the cover.

Final Checks

Review the following questions, in addition to the standard installation procedures, to help ensure that the installation is complete.

- Check all electrical plug-in connections for a positive seal. Remove all packing material, and check that the expansion valve feeler bulb is attached securely to the suction line.

- Do not remove plugs from flare nuts or caps from threaded connections until the unit is ready for final hookup. All coils should be pressurized and have an access fitting. If pressure has been lost, check for leaks.

- The case must be level for the coils to work properly and the drain to function correctly.

- After final hook-up, let the case run through at least one defrost (preferably two or more). Check the duration and frequency of defrost to make sure the case is defrosting properly.

- Check the operating temperature, comparing it with the recommended guidelines.

- Verify that store temperature and humidity levels fall within the recommended guidelines. Use a wet and dry bulb psychrometer to check the store temperature. Also, check for outside influences. Draft from ventilator openings or radiant heat from light fixtures could adversely affect the performance of the case.

- Do not load the case with product until it has reached the proper operating temperature and fully completed a defrost cycle. Do not load the case with warm product. Do not exceed the "Load Lines" when filling the case with product.

- On remote units, has the refrigerant line entry been caulked inside and out thoroughly?

- Are all fans running? Check connections for tightness.

- Is the defrost control properly set?

- Is the defrost fail-safe set for the proper time?

- Is the discharge air temperature −10°F for frozen food and −19°F for ice cream?
- Are all the connections tight?
- Is the expansion valve superheat set within specifications?

Start-Up Procedures

Before starting up the compressor, the following three steps must be completed: leak testing, evacuation, and charging start-up. Completing these steps will help prevent any problems when unit start-up is initialized.

Leak testing

Leak testing is important because the unit cannot run properly with even the smallest leak present. For this reason, all leaks should be taken care of first before any other testing is done.

Caution: Do not start any compressor unless specifically instructed to do so. Starting a unit at the improper time could result in serious damage to the compressor.

To test for leaks, begin by checking that the pilot circuitry on-off switch at the store power distribution panel is in the off position. Also check that all compressor primary on-off switches are in the off position. All of the following valves should be open:

- Compressor discharge service valves
- Compressor suction service valves
- Liquid return on receiver (from remote condenser)
- Liquid outlet on receiver
- All field-supplied hand shut-off valves
- All liquid line manifold valves
- All suction line manifold valves
- All hot gas manifold valves
- All oil equalization system valves

Next, remove the black power wire from the multicircuit time clock motor located in the defrost control panel. This step will keep the clock from advancing until the start-up procedure is complete. Turn on the power at the store distribution panel. Reset all time clock modules so that they are on refrigeration. Set the on-off toggle switches to on, which will open all branch circuit liquid line solenoid valves.

Connect the charging lines. Locate the ⅜-inch flare connection of the main charging valve, next to the liquid drier. Backseat the receiver liquid outlet valve and connect a charging line to the valve gauge port connection. Gradually charge about 25 lbs of refrigerant into the system. Following this charge, pressurize the system with nitrogen to 162 psi.

Warning: When using high-pressure nitrogen, always use a pressure regulator and a pressure-relief valve.

Caution: If a pressure greater than 162 psi will be used, disconnect the low-pressure control lines and seal the pressure port. This will eliminate any damage to the control's bellows.

Use an electronic leak detector to carefully inspect the system for leaks, especially at the joint areas. Check the line pressure gauge at the nitrogen tank for any pressure drop, which is a good indication that a leak is present. After the initial inspection is complete, let the system stand for about 12 hours with the pressure on (Nitrogen tank off). Check for any pressure changes. If none are found, the system is leak-free. If there is a pressure drop, close off the hand valves to isolate the part of the system where the leak is present. Let the damaged section depressurize, then repair the leak immediately.

Note: When damaged lines are welded, nitrogen or carbon dioxide should flow through the lines to minimize the possibility of oxide or scale formation. If these contaminants enter the system, they could quickly clog the small ports of pilot-operated valves and other valves within the system.

Evacuation procedure

Once any leaks have been located and repaired, the system should be slowly and thoroughly evacuated using a quality vacuum pump. For practical field applications, some manufacturers recommend using a triple evacuation procedure, breaking vacuum each time to about 2 psi with the refrigerant to be used in the system. It is best to leave the system in a vacuum to aid in charging.

It might be necessary to initially purge the refrigerant used to break vacuum into a reclaimer due to refrigerant costs, availability, and CFC concerns.

Piping of heat recovery coils is also recommended. This means that it is necessary to field supply a temporary bypass between the line downstream from the inlet check valve on the heat recovery coil and the discharge line downstream of the IPR hold back valve. Failure to bypass the IPR will result in the inability to evacuate the reclaim coil. The bypass line should be removed after evacuation so it does not impinge on the proper operation of the system.

Begin the triple evacuation procedure by attaching the vacuum pump to the system, using all of the following ports: discharge line after oil separator, liquid line prior to filter, and suction manifold. Check to be sure that all of the following valves are open:

- Compressor discharge service valves
- Compressor suction service valves
- Liquid return on receiver (from remote condenser)
- Liquid outlet on receiver
- All field-supplied hand shut-off valves
- All liquid line manifold valves
- All suction line manifold valves
- All hot gas manifold valves

The most careful evacuation and purging will not clean a system that is poorly constructed or inefficiently operated. With this in mind, evacuate the system as follows:

1. Draw a 28-in. Hg (49-mm) vacuum on the system. Stop the vacuum pump. Charge the system to 100 psi with dry nitrogen, leaving the nitrogen in the system for one hour.
2. Once again, draw 28-in. Hg (49-mm) vacuum on the system. Stop the vacuum pump. Charge the system to 100 psi with dry nitrogen, leaving the nitrogen in the system for one hour.
3. Finally, draw a vacuum down to 500 microns with the vacuum pump. The system is now ready for charging.

Note: R-22 and R-502 have a higher discharge temperature and pressure than R-12.

All dirt, sludge, moisture, and air should be removed from any system in order to minimize the chance of compressor burnout. Complete evacuation as just described is an ideal way to ensure that your system is clean.

Charging and start-up

Always use the appropriate refrigerant recommended for your particular system. Lower temperature systems typically use R-502 or R-22 refrigerant, while medium systems use either R-12, R-22, or R-502. Tyler recommends the high-side charging method, as follows.

Precautions

1. Clean and purge all charging lines to be sure they are free of air and moisture.
2. Test the system for leaks and evacuate it properly before charging it with refrigerant. Always wear goggles when handling a refrigerant.

3. *Never* allow liquid refrigerant to enter the compressor, as this could cause serious damage.
4. Check that all temperature controls in all circuits are preset to the correct operating temperatures.
5. Be sure that all high and low-side pressure gauges are connected to common connection points or headers.
6. Check that all fixtures have been supplied with false loads before proceeding with start-up.

Charging and start-up

1. Follow manufacturer's recommendations when determining how much refrigerant is needed to charge the system.
2. Locate the ⅜-inch Schrader valve next to the downstream regulator. Connect the refrigerant tank with gauge and dehydrator.
3. Fill the receiver with as much refrigerant as it will hold (this should be approximately one tank).
4. Attach a refrigerant tank with gauge and dehydrator to the receiver outlet valve service port. A 16-cubic inch drier would be used on a 145-pound cylinder.
5. Close the receiver liquid outlet valve.
6. Slowly open the refrigerant tank valve and charge liquid refrigerant into the system. The vacuum should pull almost all of the refrigerant from a 145-pound tank.
7. Close all liquid line and suction manifold valves.
8. Choose a branch circuit and open both the suction and liquid service isolation valves ¼ turn.
9. Turn on the condenser fan circuit.
10. Start one of the compressors. Check and record the compressor amperage readings.
11. Slowly open the suction and liquid isolation valves ¼ turn at a time to activate the first branch circuit. Examine the branch circuit to be sure that the expansion valve feeler bulbs have started controlling refrigerant flow through the cases.
    Note: Check the refrigeration circuits while they are operating for any signs of liquid slugging at the compressor. Stop the compressor immediately if any slugging is observed.
12. Be sure that the oil equalization system is operating properly.
13. Check the oil level of the compressors, adding oil as needed so that the oil level remains ¼ to ⅓ full at the sight glass. Add oil directly

to the reservoir, not at individual compressors. If there is any foaming, run the compressors on and off until the foam settles.

14. Begin activating the branch circuits one at a time. If the pressure exceeds 15 psi, start a second compressor to prevent overloading. Keep the suction pressure at or below 20 psi at all times.

15. Adjust the EPR and TEV valves according to recommendations. Sweat or frost on the suction line indicates that the valve is working properly.

16. Continue starting refrigeration circuits until they are all on the line. Charge the circuits as needed to maintain the correct refrigerant level in the receiver. Examine the liquid line sight glass during charging. If bubbles are present, this may mean there is a low refrigerant charge. Check the liquid level indicator when in doubt.

17. On an electric defrost system, compare the defrost load amperage with the summary sheet.

18. Make sure the multicircuit time clock settings indicate proper time termination and sequence of defrosting.

19. Check the following to be sure they have been correctly sized and designed for the system: starters, heaters, contactors, and circuit breakers.

20. On the multicircuit time clock, replace the black power wire to the motor.

21. Check the ability of the compressors' motors to restart after a power failure by stopping and restarting them. Use an ammeter to determine the operation of a loaded start.

22. Record the motor amperage at normal operating pressures and temperatures.

23. Check that the remote condenser and heat recovery coil are operating properly.

24. Referring to #13, check the compressor oil level.

25. On a gas defrost system, be sure that the system operates properly during defrost.

26. Remember to replace liquid line drier and suction filters as needed. (Normally they will need to be changed in about two weeks.)

Post start-up inspection

When you have observed the system in continuous operation for a two-hour period with no complications, check for the following:

1. Case fans should be operating smoothly and rotating in the proper direction.
2. All thermostatic expansion valves should be set for the correct superheat setting.
3. The following compressor operating parameters should comply with manufacturer's specifications: head pressure, suction pressure, line voltage, and compressor amperage. If any of the readings exceed or fall below recommended values, determine the cause of the problem and correct it.
4. The sight glass for compressor oil level should be $\frac{1}{4}$ to $\frac{1}{2}$ full.

When the final inspection has been fully and properly completed, the system can now begin operating on automatic controls.

Typical Serviceable Components

Serviceable components of a supermarket display unit include antisweat heaters, defrost heaters, drip pan heaters, waste outlet heaters, honeycomb air discharge outlets, and fan blades and motors.

Replacing antisweat heaters

Antisweat heaters are normally mounted at the discharge and return air nosing (see Fig. 13-2). Once power is turned off using lock out tag procedures, the heater replacement is a simple procedure of removing the various trim, access plate, and honeycomb sections to access the heater. The heater is then disconnected at the connector (do not cut the heater wire). Install the new heater using the above instructions in reverse order.

Warning: Always disconnect the electrical power at the main disconnect when servicing or replacing any electrical component—for example, fans, heaters, thermostats, and lights.

Replacing defrost heaters

Finned defrost heaters are typically located beneath the removable display pans. The heater connections are often located in their own electrical raceway. Make sure to reconnect the heater leads to the proper condensing unit circuit.

Drip pan heaters and waste outlet heaters used on cool gas defrost systems have similar change-out procedures.

Servicing the honeycomb

The honeycomb helps distribute refrigerated air evenly throughout the case. Over time it can become blocked by dirt or damaged. Remove the honeycomb assembly by removing the mounting screws and clean or replace it as required. Reinstall the honeycomb assembly in reverse order, making sure the honeycomb is nested behind the plastic extrusion.

Replacing fan motors and blades

Evaporator fans are normally located directly beneath the display pans. When servicing the fans, always install new fan blades with the raised, embossed side toward the motor. To replace a typical fan, remove the bottom display pans, disconnect the fans from the wiring harness, and remove the fan blade. Remove the screws that hold the fan blade to the plenum, and slide the fan motor and bracket clear. Lift the bracket and motor up and out through the fan plenum. Reinstall the fan in reverse order of removal.

Repairing aluminum coils

Aluminum coils used in supermarket display units can be easily serviced in the field. The materials needed are readily available from local refrigeration wholesalers. To repair a coil, proceed as follows:

1. Locate the leak.
2. Remove/disconnect all pressure sources.
3. Brush the area under heat.
4. A torch with a number 6 tip is recommended. Use a separate set of stainless-steel brushes that are meant for use on aluminum only.
5. Tin the surface around the area. Brush the tinned surface under heat, thoroughly filling the open pores around the leak.
6. Repair the leak, taking care that the aluminum melts the solder, not the torch.
7. A repair should be thick enough to last. Do not judge a repair by looks alone. Perform a leak check.
8. Wash the area with water, then cover it with a high quality flexible sealant.

Chapter

14

Ice Makers, Water Coolers, and Food-Service Equipment

Ice-making machines, water coolers, and other self-contained specialty equipment make up a sizable portion of the equipment serviced by today's refrigeration technicians.

Ice-Making Equipment

Ice makers are a particularly important segment of refrigeration work. There are literally dozens of companies that manufacture ice-making equipment, and each product line has unique features. The ice is produced in one of two forms: cubes or flakes (nuggets). Each type necessitates changes in the way the evaporator is designed, how water is routed through the machine, and how the ice is "harvested."

The ice maker illustrated in Fig. 14-1 is modular in design. The actual ice-making unit is separate from the lower storage bin. The ice maker is stacked on top of the bin, and the two are joined using hardware supplied by the manufacturer. In general all ice-making machines can be broken down into three subsystems: refrigeration, water supply, and electrical.

Ice maker refrigeration cycle

The components of the ice-making unit are shown in Fig. 14-2. The refrigeration cycle of an ice-making machine follows the standard mechanical vapor compression cycle discussed throughout this book (Fig. 14-3). The refrigerant is compressed into a high-temperature gas. The discharge line directs this gas to the condenser. At the air- or water-cooled condenser, the gas is cooled and condenses into a liquid.

404 Chapter Fourteen

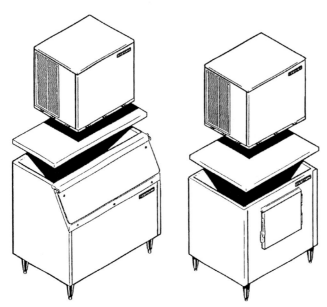

Figure 14-1 Modular ice-making machine with a separate ice-making unit and storage bin. *(Courtesy of Scotsman®, Vernon Hills, IL)*

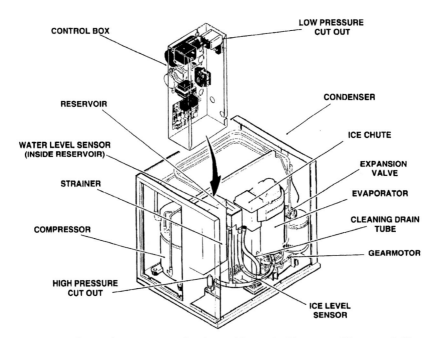

Figure 14-2 Internal components of an ice-making unit. *(Courtesy of Scotsman®, Vernon Hills, IL)*

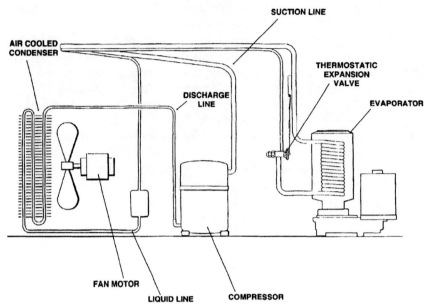

Figure 14-3 Refrigeration system for an ice maker with an auger-driven nugget system. *(Courtesy of Scotsman®, Vernon Hills, IL)*

This high-pressure liquid then goes through the liquid line to the thermostatic expansion valve or capillary tube. The thermostatic expansion valve or capillary tube meters liquid refrigerant into the evaporator. At the evaporator, the refrigerant absorbs heat from the evaporator and the water that contacts it. The temperature of the water is lowered until ice begins to form and build up on the evaporator. After the evaporator, the refrigerant, now a low-pressure vapor, goes through the suction line back to the compressor, where the cycle is repeated.

Flaker systems

In a machine that produces flaked ice or nuggets, the evaporator is a stainless-steel tube that is water filled and refrigerated. It contains a stainless-steel double spiral auger that is driven by a small motor. Water entering the unit goes to the water reservoir where its flow is controlled by a float valve. This float valve also controls the water level in the evaporator tube. Water leaving the reservoir is fed into the base of the evaporator and runs over the surface of the evaporator and auger. As ice forms on the surfaces, the turning auger pushes ice crystals upward until they exit the evaporator and fall into the ice chute (Fig. 14-4). A breaker compresses the ice, squeezing out any excess water just before the ice falls into the bin.

A control box contains electrical controls that operate the unit. A circuit board in the box contains two sensors, one for ice level and one for water level. It also contains one relay for the compressor and another relay for the auger's gear motor. The gear motor relay has a built-in time delay that clears the evaporator of ice when the unit turns off. A transformer supplies low voltage to the circuit board.

A high-pressure cutout with a manual reset switch senses the high side of refrigeration pressure. It is preset to shut off the unit at a given pressure (usually around 450 psig). A low-pressure cutout with a manual reset switch senses the low-side pressure. It is preset to shut the unit off if the low-side pressure is less than 0 to 4 psig.

An ice-level sensor senses the presence of ice in the bottom of the discharge chute. It will turn the unit on and off automatically, according to the amount of ice that it senses is in the bin.

In a flaker machine, harvest is more or less continuous, and no special harvest cycle is needed.

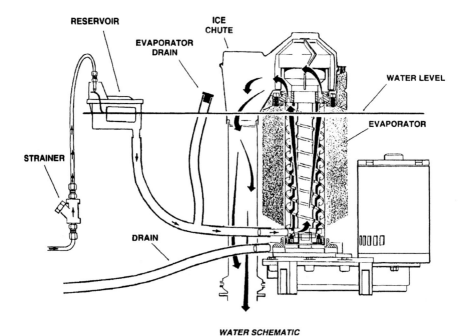

Figure 14-4 A water flow system for an auger-driven nugget-type ice maker. *(Courtesy of Scotsman®, Vernon Hills, IL)*

Cuber systems

Ice machines that produce cubes require separate freezing and harvest cycles. A typical freezing cycle operates as follows: water from a sump assembly is pumped to the water distributor at the top of each evaporator plate. From the distributor, the water cascades down the cells of the plate to the sump assembly below (Fig. 14-5).

As the freezing cycle begins, the electrical circuit is completed to the compressor and the water pump. The water pump operates continuously, through both the freezing and harvest cycles. The constant flow of water across the evaporator plate ensures the ice formed is clear, not cloudy. When the machine is equipped with a water-cooled condenser, water also flows through the condenser and out the drain.

The hot gas solenoid valve and the water solenoid valve are closed in the freezing cycle (Fig. 14-6). As the ice cubes begin to form, the cube-size control (a sensing bulb attached to the suction line) closes, connecting power to the timer motor. The timer will control the rest of the freezing cycle, keeping the unit in this cycle for a preset length of time.

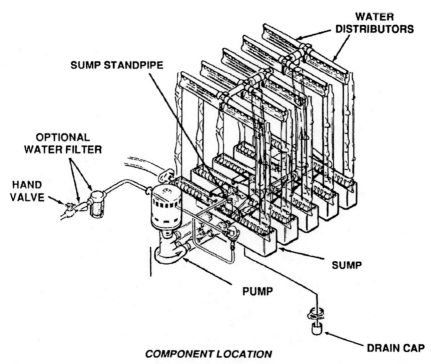

Figure 14-5 Water distribution system for a cell-type evaporator plate used to form ice cubes. *(Courtesy of Scotsman®, Vernon Hills, IL)*

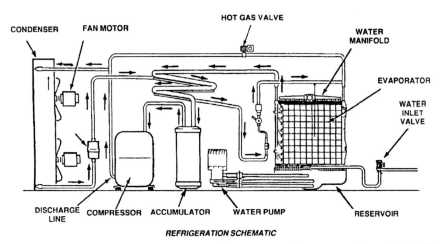

Figure 14-6 A cuber ice-machine freezing cycle. *(Courtesy of Scotsman®, Vernon Hills, IL)*

When the ice cubes are completely formed, contacts in the timer assembly microswitch will switch the unit into the harvest cycle.

When the timer switches the unit into the harvest cycle, the finish relay is de-energized and the hot gas valve and inlet water valve open (Fig. 14-7). As high-pressure, high-temperature gas leaves the compressor, it is diverted from the condenser through the hot gas solenoid valve into each evaporator plate. During the harvest cycle, the refrigerant bypasses the condenser. Both the compressor and the water pump are operating.

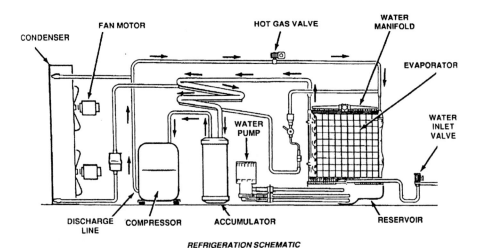

Figure 14-7 A cuber ice-machine harvest cycle. *(Courtesy of Scotsman®, Vernon Hills, IL)*

Hot gas condenses in each evaporator plate while water cascades over the ice cubes. The result is a warming effect that releases the ice from the sides of each evaporator plate. The finished cubes drop into a storage bin below.

At the end of the harvest cycle, the timer cam pushes in the microswitch actuator arm. If there is no ice close enough to the bin level control's transducer, a whole new cycle will begin. If there is ice within eight inches of the base of the machine (with the bin level control on full), the ice maker will shut off.

A number of components are unique to ice-making equipment and merit special mention.

Cube size control. A reverse-acting thermostat controls the length of the freezing cycle by sensing the temperature of the suction line. When the suction line gets cold enough, the sensor closes and starts the timer. A change in ambient air or incoming water temperature will affect the length of time it takes the evaporator to cool. This, in turn, will change the amount of time that it takes until the sensor closes.

Control box and relay. A multifunction, three-pole, double-throw relay plugs into a receptacle on the printed circuit board in the control box (Fig. 14-8). The relay is designed to bypass the bin thermostat control, preventing the ice maker from shutting off as the bin thermostat opens during the freezing cycle. The relay helps produce full-sized ice cubes

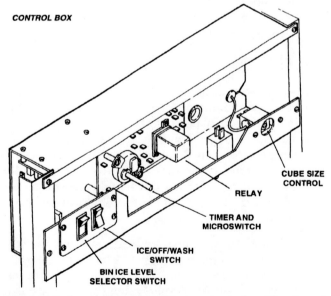

Figure 14-8 A control box for cuber-type ice maker. *(Courtesy of Scotsman®, Vernon Hills, IL)*

with each harvest cycle. It also prevents short-cycling on the bin thermostat control.

Timer switch and assembly. The timer cam is activated by the cube-size control. The inner-surface, small-diameter lobe determines the length of time for the harvest cycle. The outer-surface, large-diameter lobe determines the length of time necessary to finish freezing of the cubes.

When the microswitch button is pushed in, power connects to the relay coil, putting the unit into the freeze cycle. When the microswitch button is released, the relay shuts down, moving the unit into the harvest cycle. This microswitch is actuated by a cam assembly directly connected to the timer motor.

One complete cam rotation typically takes eight to ten minutes. The harvest is preset for three to four minutes, but it is adjustable.

High-pressure safety control. This is a manual reset control located below the control box. If the discharge pressure reaches 450 psig (air-cooled unit) or 350 psig (water-cooled unit), this safety control will shut down the ice maker.

Installation of equipment

An ice machine is actually a food-manufacturing unit in that it takes in a raw material (water) and turns it into a food product (ice). The purity of the water is key to obtaining the pure, long-lasting ice. Although there is no filter that will eliminate all water problems, filtering the water is recommended. For optimum performance, a good filter should be used along with a polyphosphate feeder.

Electrical connections. First check the nameplate on the ice maker to determine the wire size needed for electrical hookup. The unit requires a solid chassis-to-chassis earth ground wire. Be sure that the unit is connected to its own electrical circuit and is individually fused. The voltage variation should not exceed 10 percent of the nameplate rating, even under starting conditions. Low voltage can cause erratic operation, which might in turn cause serious damage to the ice maker.

Make the electrical connections at the rear of the ice maker, inside the junction box (Fig. 14-9). All external wiring should conform to the national, state, and local electrical codes. An electrical permit and services of a licensed electrician are recommended.

Plumbing. Be sure to follow all local codes when installing the ice maker. The recommended water supply line is a $3/8$-inch O.D. copper tube with a minimum operating pressure of 20 psig and a maximum pressure of 80 psig. Connect the unit to a cold water supply line with standard fittings (Fig. 14-10). The shut-off valve should be installed in

Ice Makers, Water Coolers, and Food-Service Equipment   411

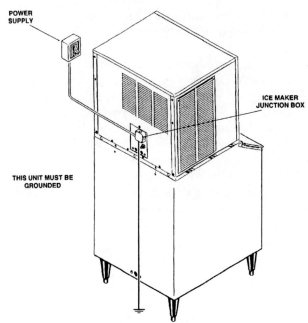

Figure 14-9 Typical electrical connections for ice-making machines. *(Courtesy of Scotsman®, Vernon Hills, IL)*

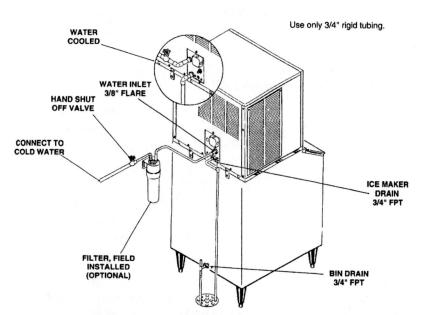

Figure 14-10 Typical water inlet and drain connections for ice-making machines. *(Courtesy of Scotsman®, Vernon Hills, IL)*

an accessible area somewhere between the water supply and the ice maker. Depending on the location of the unit, a plumber's services might be required.

All drains are the gravity type and should have a minimum of ¼-inch fall per foot on horizontal runs. Drains should be installed in accordance with the local plumbing codes. To ensure good drainage, install a vertical open vent on the drain line high point. A trapped and vented floor drain is the ideal type of receptacle. The recommended bin drain is ⅝-inch O.D. copper tubing. The tubing should be vented and run separately. Use insulation in high-humidity areas.

Drains should terminate above the drain receptacle with an air gap that follows local codes. The ice maker's sump drain is ¾-inch female pipe thread (FPT). There should be a vent at this connection for proper sump drainage.

Water-cooled condensers. On water-cooled units, a separate cold water inlet is needed. The inlet should be connected to a ⅜-inch FPT fitting at the rear of the cabinet. An additional drain line must be used to drain the water-cooled condenser; it must not connect with any other drains.

Final checks. When installing ice makers, the following are typical final checks:

- The water supply pressure should be checked to ensure a minimum of 20 psig and a maximum of 80 psig operating pressure.
- The reservoir should be properly secured to the bottom of the evaporator plates.
- All electrical, water, and drain connections should be inspected to see if they have been properly made.
- The water supply line shut-off valve should be installed with the wiring properly connected.
- All refrigerant and conduit lines should be inspected to guard against rubbing and potential failure.
- The cabinet should be in a room where the ambient temperatures are within the minimum and maximum temperatures specified.
- There should be at least six inches of clearance at the left and back sides of the cabinet for proper air circulation (air-cooled models).
- The cabinet should be level.
- Show the owner the service manual. Give the owner instructions on how to operate and maintain the ice maker, along with the name and phone number of the nearest authorized service agency.

Initial start-up

Once the ice maker has been delivered and installation is complete, the unit is ready for initial start-up. Begin by removing the front panels; remove all screws at the base and pull the panels out. Remove two screws and the control box cover. Check that the two switches on the control box—the ICE-OFF-WASH rocker switch and the COMPRESSOR-ON-OFF switch—are both in the off position. Open the water supply line shut-off valve.

The shaft of the timer and switch assembly is located inside the control box. Rotate the shaft of the timer and switch assembly clockwise to start the timer. The timer will start when the actuator arm on the microswitch drops off the outer cam into the cam slot. Move the ICE-OFF-WASH switch to the ICE position.

The water-fill cycle operates as follows: When the water pump starts, the water inlet solenoid valve opens. As incoming water flows from the valve through the tubing, the reservoir fills and excess water overflows into the stand pipe. This process continues through the harvest cycle. When the freeze cycle begins, the timer and relay close the water inlet solenoid valve, completing the water-fill cycle.

At this point, if the sump is not full, rotate the timer once again to repeat the harvest cycle. The water pump should continue pumping water through the tygon tubes up to the water distributor found at the top of each evaporator plate. It is here that the water is uniformly dispensed, cascading down both sides of the evaporator plates. The water drains back into the sump assembly for recirculation. When the sump fills, move the COMPRESSOR-ON-OFF toggle switch to the on position.

The freezing cycle operates as follows: The compressor is running, and the metal parts of the evaporator plate are cold to the touch. The tubing becomes frosted at the top of the evaporator plate. Ice is now beginning to form. The first cycle is the longest. Freezing time normally ranges from 12 to 15 minutes, or slightly longer if the outside temperature is above 70°F. The average complete cycle time is about 16 minutes.

Controlling cube size. The reservoir holds just enough water to make one full-sized batch of cubes. The water pump might pick up some air at the end of the freezing cycle, which is normal. If the water pump runs out of water before the freeze cycle ends, however, the cube-size control might be set too cold, or the water system could be leaking.

For a smaller sized cube, look for the cube-size control knob on front of the control box. Rotate the knob one eighth of a turn counterclockwise. Check the cube size in the next ice harvest. If it is still too large, adjust the knob in a one-eighth turn and check the cubes again. Repeat this procedure until the cubes are the desired size.

For a larger sized cube, look for the cube-size control knob on front of the control box. Rotate the knob one-eighth of a turn clockwise. Check the cube size in the next ice harvest. If it is still too small, adjust the knob in a one-eighth turn and check the cubes again. Repeat this procedure until the cubes are the desired size.

Controlling ice bin level

With the ice maker in the harvest cycle, place something directly below the transducer socket. The machine should switch off, indicating the end of the harvest cycle. Remove the obstruction, and within seconds the unit will start again. This control has two positions: full and partial. It will not fill up closer than eight inches from the bottom of the ice maker.

When all adjustments have been made to the unit, wash out the bin with clean, potable water. Replace the control box cover and all cabinet panels and screws. Answer any questions from the owner about start-up and operation.

Cleaning and sanitizing

An ice system represents a sizable investment of time and money for any company. To receive the best return for that investment, it must receive periodic maintenance. Stress to your customers that it is their responsibility to see that the unit is properly maintained. It is always preferable, and less costly in the long run, to avoid possible down time by keeping it clean; adjusting it as needed; and by replacing worn parts before they can cause failure. The following is a list of recommended maintenance tips that will help keep the machine running with a minimum of problems.

Maintenance and cleaning should be scheduled at a minimum of twice per year. Electrical power will be on when doing in-place cleaning.

Ice-making system: In-place cleaning

1. Check and clean any water treatment devices, if any are installed.
2. Pull out and remove the front panel.
3. Move the ON/OFF switch to off.
4. Remove all the ice from the storage bin.
5. Remove the cover to the water reservoir and block the float up.
6. Drain the water reservoir and freezer assembly using the drain tube attached to the freezer water inlet. Return the drain tube to its normal upright position and replace the end cap.

7. Prepare the cleaning solution: Mix eight ounces of Scotsman ice-machine cleaner with three quarts of hot water. The water should be between 90 to 100°F.

    Warning: Manufacturer's commercial ice-machine cleaners might contain harmful acid compounds, which might cause burns. In case of external contact, flush with water. Keep out of the reach of children!

8. Slowly pour the cleaning solution into the water reservoir until it is full. Wait 15 minutes, then switch the master switch to on.
9. As the ice maker begins to use water from the reservoir, continue to add more cleaning solution to maintain a full reservoir.
10. After all of the cleaning solution has been added to the reservoir, and the reservoir is nearly empty, switch the master switch to off.
11. After draining the reservoir, as in step 6, wash and rinse the reservoir.

    To sanitize the unit, repeat steps 8 to 11, except use an approved sanitizing solution in place of the cleaning solution. To mix a sanitizing solution, add one ounce of household bleach to two gallons of warm (95 to 115°F) water.

12. Remove the block from the float in the water reservoir.
13. Switch the master switch to on.
14. Continue ice making for at least 15 minutes, to flush out any cleaning solution. Check ice for acidic taste, and continue ice making until the ice tastes sweet.

    Caution: Do not use any ice produced from the cleaning solution. Be sure no ice remains in the bin.

15. Remove all ice from the storage bin.
16. Add warm water to the ice storage bin, and thoroughly wash and rinse all surfaces within the bin.
17. Sanitize the bin interior with an approved sanitizer using the directions for that sanitizer.
18. Replace the front panel.

Preventive maintenance

In many ice machines, the bin control uses devices that sense light; therefore these devices must be kept clean enough so that they can "see." At least twice a year, remove the bin control sensors from the base of the ice chute, and wipe the inside clean, as illustrated in Fig. 14-11.

The ice machine senses water level with a probe located in the water reservoir. At least twice a year, the probe should be removed from the reservoir and the tip wiped clean of mineral buildup (Fig. 14-12).

416 Chapter Fourteen

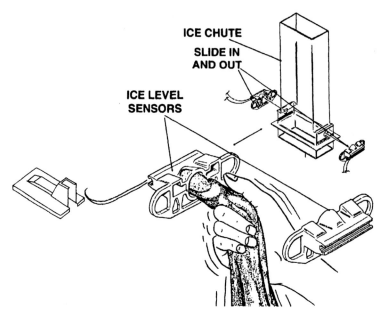

Figure 14-11 Cleaning an ice-level sensor. *(Courtesy of Scotsman®, Vernon Hills, IL)*

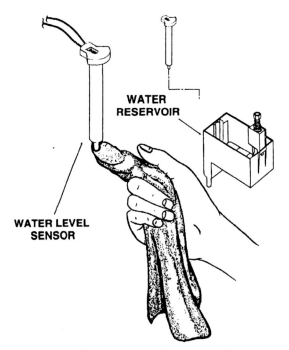

Figure 14-12 Cleaning a water-level sensor. *(Courtesy of Scotsman®, Vernon Hills, IL)*

The bearing in the breaker should be also checked at least twice a year. To check the bearing: remove the ice chute cover; unscrew the ice sweep; remove the water shed; unscrew the breaker cover; and unscrew the auger shield. Inspect the assembly, looking for wear. See the section, *Removal and replacement* to replace bearing or seals. Reverse to reassemble.

Clean or replace the air-cooled condenser filter. Clean the air-cooled condenser. Use a vacuum cleaner or coil cleaner if needed. Do not use a wire brush. Check and tighten all bolts and screws.

Water Coolers

Water coolers are another common item that rely on mechanical refrigeration systems. Common water cooler designs include self-contained, free-standing, or wall-mounted units with fountain tops, bottled water coolers, and remote chillers. Figure 14-13 illustrates the internal components of a typical free-standing unit.

Refrigeration cycle

The refrigeration cycle of a typical water cooler operates in the following manner. First, the compressor pumps hot, high-pressure refrigerant through its discharge tube. The refrigerant vapor passes through the condenser, releasing some of its heat into the atmosphere and condensing it into a liquid. The liquid refrigerant travels out the bottom of the condenser, goes through the drier and capillary tube, then flows into a larger coiled tube that is connected to the water storage tank.

A temperature control bulb sits in a control well that is positioned outside the water tank. This position allows the control to respond to the temperature of both the water and the refrigerant. As the liquid refrigerant removes heat from the water, it begins to boil off into a low-pressure vapor. The cycle ends as the vapor flows through the suction tube back to the compressor.

Water cooling cycle. The water cooling cycle begins as the water supply enters the cooler through its inlet water connection. Once inside the cooling tank, the water is kept cool by continuous contact with the refrigerated walls of the tank.

Installation and start-up procedure

1. Ensure proper ventilation by following manufacturer's recommended clearance from cabinet louvers to the wall on each side of the cooler.

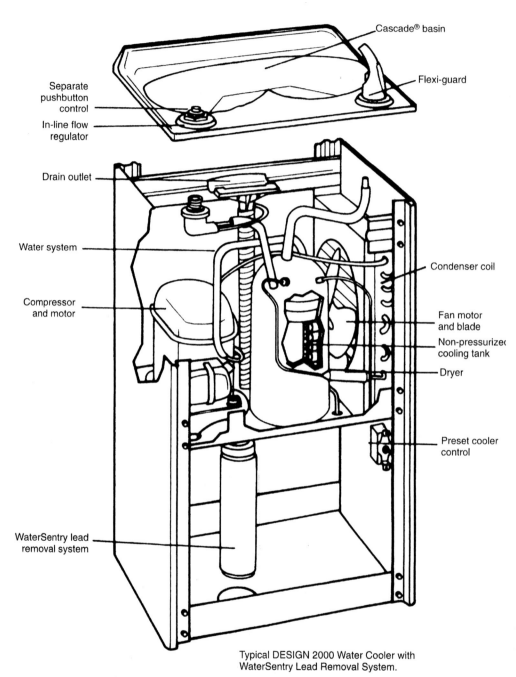

Figure 14-13 Components of a typical free-standing water cooler or fountain. *(Courtesy of Elkay Manufacturing Company)*

2. The water inlet should be ⅜-inch copper tube, with a waste 1¼-inch O.D. tailpiece for the slip trap. The contractor should supply the waste trap in accordance with local codes.
3. Connecting lines should be made of copper, thoroughly flushed to remove all foreign matter before being connected to the cooler. If flushing cannot remove all particles, a water strainer should be installed in the supply line.
4. Connect the cooler to the building supply line with a shut-off valve, installing a union connection between the valve and the cooler.
5. Check to be sure that the electrical power supply meets the specifications found on the cooler serial plate.

Note: As a preventive measure, a water line strainer should be installed ahead of the inlet water connection on both new and old plumbing. The strainer prevents clogging of water lines and bubbler valves.

Grounding considerations. Grounding of electrical equipment such as telephones, computers, etc., to water lines is a common procedure that can occur either inside or outside a building. This grounding creates electrical feedback into the cooler resulting in an electrolysis that increases the metal content of the water which can produce a metallic taste. To avoid this problem, install the cooler using the proper materials.

The water cooler should be connected to the water supply using a dielectric coupling. Your cooler should come equipped with a nonmetallic strainer that meets this requirement. The drain trap provided by the installer should be plastic to further isolate the cooler from the building's plumbing system. To start the unit:

1. Release air from the tank by holding the button down. Watch for a steady stream flow indicating that all air is removed.
2. Stream height is factory set at 35 psi. If the supply pressure varies greatly, the stream height will need adjustment. (See the section on cleaning and maintenance.) The water coolers are designed to operate between 20 psig and 105 psig supply line pressure. If inlet pressure is above 105 psig, a pressure regulator must be installed in the supply line. Any damage resulting from connecting the product to a supply line with pressure either less than 20 psig or more than 105 psig is not covered under warranty.
3. On hot tank models, depress the lever to ensure full stream flow. Do not connect electrical power until there is a full water flow.
4. Rotate the condenser fan blade (when used) to ensure proper clearance and free fan action.

Cleaning and maintenance

Clean stainless-steel and bronzetone surfaces with either a mild detergent or a vinegar-and-water solution. Rinse all surfaces thoroughly with clean water and wipe dry with a soft cloth. Never use steel wool or other abrasives as they can damage the finish.

Clogged water lines. If the unit in question has a low bubbler stream or no stream at all, the cause might be an obstruction in the water line. It's also possible that the bubbler assembly itself is malfunctioning or, at the very least, needs to be thoroughly cleaned. Inspect and repair the unit as follows:

1. Remove the hex nut and button.
2. With a screwdriver try to adjust the stream height by turning the bubbler cartridge adjusting screw toward the right to raise it. Push down on the cartridge stream housing to see if the stream height has increased any.
3. If there is no noticeable change, turn off the water inlet valve. Unscrew the bubbler cartridge retainer nut, then lift out the cartridge.
4. Open the water inlet valve, keeping in mind that you might need to catch excess water at the cartridge body housing.
5. If the water flow from the valve body is full, turn off the water inlet valve. Disassemble and clean the bubbler cartridge assembly, replacing it if needed.
6. If there is little or no water flow, disconnect the supply line to the cooler. Remove the strainer (located inside the $\frac{3}{8}$-inch female fitting) then clean or replace the strainer as needed. Reinstall the supply line, checking and adjusting the stream height to specifications.

Cleaning the condenser. If lint or dust is allowed to gather on the condenser, surface cooling capacity is reduced, making it less economical to operate the unit. To rid the condenser surface of lint or dust, you can use a vacuum cleaner, stiff bristle brush, or compressed air. To clean, proceed as follows:

1. Remove the front panel and unplug the service cord from the electrical outlet.
2. On models with a fan, insert a vacuum cleaner nozzle or a brush between the blades to clean. Turn the blades slowly so that they do not become bent or damaged.

3. Thoroughly clean the cabinet-side louvers located at the outlet side of the condenser.

Preparing for shipment or storage. All water in the cooler itself, the connecting tubing, and the fittings should be drained before the unit is shipped or stored. This is done to eliminate the possibility of serious damage that could result from freezing. To drain water from the cooling tank, do the following:

1. Disconnect all power supply and plumbing connections. If necessary move the cooler to an area where water-spills will not cause damage.
2. Remove the plug from the tank drain.
3. Depress the push-button to open the bubbler valve. This will allow air into the tank, speeding up the drainage process.

To drain water from the hot water tank, do the following:

1. Remove the plug from the tank drain.
2. Depress the push-button on the bubbler valve to help the water drain quicker.

Troubleshooting

Before making any repairs to a refrigeration system, first check the cooling load of the unit. If the load is greater than the capacity of the unit, it should be reduced. Next, examine the unit to ensure that it has sufficient ventilation. Any obstruction of air flow or dirt on the condenser will reduce cooling capacity. Finally, check the electrical service to be sure that the supply voltage is identical to the voltage listed on the data plate. If, after making your inspection, you find that the unit is still not working properly, consider the procedures listed below.

The refrigeration system is regulated by the cold control. If the cold control contacts remain closed, yet the refrigeration system continues to run, any of the following conditions could be the cause:

- The cold control is defective.
- The refrigerant charge has either partially or completely leaked out.
- There is a restriction in the refrigerant circuit.
- The compressor is defective.
- There is dirt in the condenser.

If the refrigeration system does not start, look for one of the following problems:

- Fault in the electrical power supply
- A defective cold control
- Defective compressor electrical component(s)
- A defective compressor

Once the type of refrigeration system failure has been identified, individual components should be checked in the order they are listed above. For specific information on checking individual components, refer to the following paragraphs.

Thermostat. The thermostat that controls the coldness of the water can fail with the contacts open or closed. Failure in the open position means the compressor and condenser fan motor will not run, and warm water is dispensed. Failure in the closed position (where contacts fuse together) means the compressor and condenser run continuously, producing frozen water. In either case, the control should be replaced.

Fan motor. To service, check the voltage at the motor terminals. If the motor is inoperable at the rated voltage, replace it. If the motor runs but does not reach normal operating speed, disconnect the electrical supply. Oil the fan, then run the motor again. Replace the motor if the fan does not turn freely after oiling.

Capacitor. Thoroughly inspect the electrical system before turning on the power. An incorrectly wired capacitor can burn out compressor windings in one minute or less. The capacitor terminal that is marked with either a red dot or a plus sign should be connected to the line side of the circuit. The replacement capacitor should have an equal or higher voltage rating. The replacement should have a capacitance rating (MFD) equal to that of the original unit. A factory-ordered replacement is recommended.

The bleed resistor-type capacitor is an excellent replacement because it prevents arcing across potential relay contacts. Where a bleed resistor capacitor is not available, a 2-watt, 15,000-ohm resistor can be soldered across the terminals of a correctly rated capacitor for relay protection.

Compressor relay. Each relay is designed specifically for one type of compressor, and should only be used as specified.

Compressor overload protection. This component will trip whenever it detects excessive temperature or current. Use only the manufacturer's recommended replacement.

Refrigerant system troubleshooting

When a refrigeration unit is operating with little or no cooling effect, inspect it as follows before opening up the system. Opening up the refrigeration system will void the warranty unless this intrusion has been specifically authorized by the factory.

Feel the compressor discharge line to see if it is warm, an indication that the compressor is pumping. Examine the capillary tube at both inlet and outlet connections. A frosted line could mean there is a partial loss of refrigerant or a restriction in the line. Apply a small amount of heat to the line, then observe system operation. If the unit functions normally, this would show that the restriction was caused by excess moisture in the system. Moisture in the unit can result in acid buildup and eventual system failure. The unit should be opened, evacuated, and recharged.

Check the pressure of the system using the proper gauges. A low reading indicates there is a loss of the refrigerant charge or a restriction somewhere in the circuit. Add a small amount of the type of refrigerant specified on the data plate in vapor form to the circuit. Compare gauge readings with the refrigerant pressure-temperature chart. An increase in pressure indicates either an initial low charge or a leak in the system that will need to be repaired. A continuous low pressure indicates a restriction, which can be located as previously described.

Leak testing. Once the unit has been pressurized with the correct amount of refrigerant, check all joints and components with a leak detector (one that detects a leak of ½ ounce per year is recommended). Once the leak has been located, bleed the system and repair it using silver solder.

Component replacement. All replacement components must be identical to the original parts in order to ensure that the refrigeration system maintains peak performance. Replacement parts lists are typically supplied by the manufacturer along with the unit to ensure that the correct components will be used.

Food-Service Equipment

In addition to the reach-in and walk-in refrigerators and freezers discussed in Chapter 12, the food-service industry uses many other types of refrigerated equipment for specialized purposes. These include under-counter refrigerator and freezers, back-bar storage coolers, pizza preparation tables, sandwich/salad tables, glass and plate chillers, dough retarders, reach-in horizontal bottle coolers, ice merchandisers, beer dispensers, down-sized dessert and food merchandisers, beverage dispensers, bulk milk dispensers, and slush ice machines (Fig. 14-14).

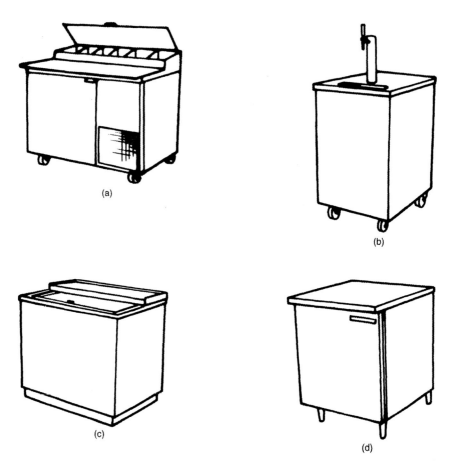

Figure 14-14 Food and beverage service equipment: (a) sandwich prep table, (b) glass and plate chiller, (c) draft beer dispenser, (d) horizontal bottle cooler, and (e) under-counter refrigerator. *(Courtesy of True Food Service Equipment, Inc.)*

With a few exceptions, most under-counter refrigerators, preparation tables, and chillers use a basic refrigeration system to cool an enclosed space. For example, a sandwich preparation table is basically an under-counter refrigerator with metal bins built into the top of the unit. These bins hold the sandwich or salad ingredients and protrude down into the refrigerated area. Airflow is directed under the pans to keep them cold. The refrigerated storage space is equipped with wire-rack shelving to keep meats, cheese, and other ingredients cold and fresh. Compressor size is small, usually in the ⅕ to ½ horsepower range. No external electrical or plumbing connections are required. The unit simply plugs into a 115-V wall outlet and is ready to run.

Horizontal beverage coolers are bench-high units accessed through sliding doors on the top. Operation is essentially the same as reach-in display coolers. The major design difference is the evaporator coils that run along the length of the horizontal cooler walls to provide even cooling throughout the unit.

In fact, just as with supermarket and ice-making equipment, the evaporator design and configuration is often the only unique feature of specialty food service equipment. The evaporator can be the standard coil, a flat plate, or a drum wrapped in coils. Forced air circulation across the evaporator might or might not be used.

Chapter

15

System Troubleshooting

This chapter covers a number of basic troubleshooting, service, and maintenance procedures common to many commercial and residential refrigeration systems. Table 15-1 lists some general troubleshooting problems and causes.

Compressor Failures

Replacement new or remanufactured compressors fail at four times the rate of original compressors. This increased failure rate is due to problems in the refrigeration system, not with problems in the compressor itself. After a compressor failure, field examination of the failed compressor will often reveal symptoms of system problems. Correct these system problems to avoid future compressor failures. The following are common compressor problems. The symptoms and remedies to these problems are summarized in Table 15-2.

Liquid slugging. Liquid slugging occurs when the compressor is trying to compress liquid in the cylinders. This liquid can be either refrigerant or lube oil, but is likely a combination of both. Slugging is most often caused by refrigerant migration into the oil on the compressors off cycle.

Liquid washout. Washout occurs when refrigerant washes the lubricating oil off of compressor surfaces. It is caused by off-cycle migration of saturated refrigerant into the crankcase. When the compressor starts up, a large volume of foam is generated that is pumped through the system, cleaning lubricant from wear surfaces. Washout is a minor condition of slugging.

Liquid dilution. Liquid dilution occurs when refrigerant returns to the compressor during the running cycle. This dilutes the oil. While the oil

TABLE 15-1 General System Troubleshooting

Problem	Possible causes

Compressor will not start—no hum	1. Disconnect switch open
	2. Blown fuse
	3. Defective wiring
	4. Overload protector tripped
	5. Oil pressure control tripped
	6. Low-pressure control contacts open. Check for:
	   a. Unit location too cold
	   b. Low refrigerant charge
	   c. Defective control
	7. High-pressure control contacts open. Check for:
	   a. Excessive refrigerant charge
	   b. Inadequate condenser or excessive condensing temperature
	   c. Defective control
	8. Open thermostat
	9. Burned motor windings—open circuit

Compressor will not start—hums but cycles on overload protector	1. Low voltage 2. Unit wired incorrectly 3. Starting capacitor defective 4. Starting relay contacts not closing 5. Compressor motor windings shorted or grounded 6. High head pressure 7. Bearings or pistons tight—inadequate lubrication 8. Run capacitor defective
Compresor starts, but starting winding remains in circuit	1. Low voltage 2. Unit wired incorrectly 3. Starting relay defective 4. Starting capacitor weak 5. Running capacitor defective 6. Compressor motor defective 7. High head pressure 8. Bearings or pistons tight—inadequate lubrication
Compressor starts and runs, but cycles on overload	1. Improperly wired 2. Low line voltage 3. Overload protector defective 4. Running capacitor defective 5. High head pressure 6. Start relay defective 7. Fan motor, pump, etc, wired to wrong side of overload protector 8. Compressor motor partially grounded or shorted 9. Unbalanced line voltage 10. Bearings or pistons tight—inadequate lubrication
Compressor tries to start when thermostat closes, but cuts out on overload. Finally starts after several attempts.	1. Low line voltage 2. Thermostat differential too close. Compressor attempts to short cycle. 3. Starting relay contacts badly pitted or burned 4. Starting capacitor weak 5. Slight restrictions in capillary tube, strainer, or drier 6. Air or noncondensible gases in system 7. Insufficient off time for system to balance—PSC compressor

TABLE 15-1 General System Troubleshooting (*Continued*)

Problem	Possible causes
Compressor starts but immediately cuts out on overload	1. Starting relay contacts welded together 2. Starting capacitor defective 3. Low line voltage 4. Excessive refrigerant charge 5. Insufficient off time for system to balance—PSC compressor
Starting relays burn out	1. Low line voltage 2. High line voltage 3. Compressor short cycling 4. Incorrect running capacitor being used
Starting capacitors burn out	1. Compressor short cycling 2. Incorrect starting capacitor being used 3. Relay contacts sticking 4. Compressor start winding held in circuit too long
Running capacitors burn out	1. Line voltage too low 2. High line voltage, light compressor load 3. Capacitor voltage rating too low
Head pressure too high	1. Unit overcharged 2. Air or other noncondensible gases in the system 3. Clogged condenser (air-cooled) 4. Defective condenser fan motor (air-cooled) 5. Air-cooled condenser location too hot 6. Restriction in water-cooled condenser 7. Water supply inadequate or shut off 8. Restriction in capillary tube, strainer or drier 9. Discharge valve partially closed 10. Restriction in discharge line 11. Excessive air temperature entering condenser
Head pressure too low	1. Insufficient refrigeration charge 2. Leaks in system 3. Cold location or cold water 4. Damaged valves in compressor 5. System operating without head pressure control when design requires
Compressor short cycles	1. Control differential set too close 2. Refrigerant undercharge 3. Refrigerant overcharge 4. Discharge valve leaking 5. Expansion valve leaking 6. Cutting out on high pressure control 7. Cutting out on overload protector because of tight bearings, tight piston, high head pressure, restrictive air or water cooled condenser, or interrupted water supply 8. Low head pressure
Running cycle too long, or unit operates continuously	1. Insufficient refrigerant charge, or leak in system 2. Dirty or restricted condenser 3. Unit location too warm 4. Control contacts stuck 5. Air or other noncondensible gases in system 6. Expansion valve blocked or defective

TABLE 15-1 General System Troubleshooting (*Continued*)

Problem	Possible causes
	7. Unit too small for application
	8. Fixture doors left open too long
	9. Insufficient, defective, or water-logged insulation
	10. Inefficient compressor
	11. Defrost heaters on
	12. Evaporator coil blocked with ice or dirt
	13. Thermostat bulb not in tight contact with evaporator
Refrigerated space temperature too high	1. Shortage of refrigerant, or leak in system
	2. Restricted expansion device
	3. Refrigerant lines too small
	4. Control setting too high
	5. Restricted strainer or drier
	6. Expansion valve too small
	7. Evaporator coil too small
	8. Unit too small for application
	9. Evaporator coil plugged with ice or dirt
	10. Evaporator oil logged
	11. Inefficient compressor
	12. Defrost heaters on
	13. Partially closed service valves
Noisy units	1. Compressor oil charge low
	2. Fan blade bent, causing vibration
	3. Fan motor bearings loose or worn
	4. Tube rattle
	5. Loose parts on condenser unit
	6. Discharge lines restricted or too small
Evaporator freezes but defrosts while unit is running	1. Moisture in the system. Clean the system and install a new drier.
Liquid line hot	1. Unit overcharged, or air (noncondensables) in system
	2. Expansion valve opened too far
	3. High head pressure
	4. Inadequate condenser
Liquid line frosted	1. Restriction in drier
	2. Shut-off valve on receiver partially closed or restricted
	3. Solenoid valve leaking
Suction line sweating or frosted	1. Expansion valve open too wide
	2. Unit overcharged (capillary tube expansion system) Note: Sweating suction lines are normal when the suction temperature design is below the ambient dewpoint. Frosted suction lines are normal any time the suction temperature is below 32°F.
Contacts stick on start relay	1. Short running cycle
	2. No bleed resistor on start capacitor
Loss of oil pressure	1. Loss of oil from compressor. Check for: a. insufficient oil in system b. oil trapped in system c. compressor short cycling

TABLE 15-1 General System Troubleshooting (*Continued*)

Problem	Possible causes
	2. Excessive liquid refrigerant returning to compressor
	3. Defective oil pump
	4. Restriction in oil pump inlet screen
	5. Extremely low suction pressure
	6. Excessive pressure drop in suction lines
	7. Suction lines too large or improperly trapped
	8. Expansion valves not feeding properly
	9. Compressor short cycling
Differential oil pressure control not functioning properly	1. Operating with cover removed
	2. Low or high voltage to control
	3. Compressor shutting off due to tripping of internal line-break motor
	4. Defective time delay switch
	5. Low side capillary connected to service valve instead of crankcase
	6. Time delay not in accordance with compressor manufacturer's recommendations
	7. Control wired improperly
	8. Control unable to respond to oil pressure changes due to low ambient conditions and high oil viscosity

pump and end bearing can receive adequate lubrication, the crankshaft rods and main bearings might not be properly lubricated.

High discharge temperature. When the temperature in the compressor head becomes too hot, the oil loses its ability to lubricate properly.

Lack of lubrication. This results when there is not enough oil in the crankcase to lubricate the running gear.

Electrical problems. While many motors fail as a result of mechanical or lubrication problems, electrical troubles also cause their share of failures. Windings can burn out in one or all phases or in the start winding of a single phase motor.

Compressor burnout

Compressor burnouts occur when the windings of a hermetic or semi-hermetic compressor overheat. The windings can burn and the heat can force a reaction with refrigerant and refrigerant oil to form acids and sludge in the system. Heat buildup leading to burnout can be caused by system overload, poor internal cooling, or low refrigerant charge. High compression temperatures and motor bearing failure are two other causes.

Ideally the compressor's internal protective device will shut down the compressor before major burnout damage occurs, but unfortunately this does not always occur. Compressor burnouts are classified

TABLE 15-2 Compressor Problems

Problem	Symptoms	Remedy
Liquid slugging	Broken reeds, rods, or crankshaft. Loose or broken discharge bolts. Blown gaskets.	1. Check pumpdown cycle operation. 2. Is TXV sized and operating properly? 3. Is suction line sized properly? 4. Check unloading.
Liquid washout	Worn pistons and/or rings. Cylinders worn. Scored pins. Scored or broken rods. Scored crankshaft. Worn bearings.	1. Check TXV bulb and super-heat setting. 2. Is TXV oversized? 3. Check crankcase heater (on during off cycle).
Liquid dilution	Rotor drag. Worn bearings. Scored and/or broken rods. Scored crankshaft.	1. Check thermal expansion valve bulb. 2. Check superheat setting. 3. Check defrost cycle.
High discharge temperature	Discoloration of valve plate that will not rub off. Overheated or burned valve reeds. Worn rings and pistons. Worn cylinders. Scored rods, bearings, and crankshaft. Spot burns in stator.	1. High compression ratio; check for low suction and high discharge pressures. Low load and evaporator problems. 2. Check low pressure control setting. 3. Check for dirty condenser, inoperative condenser fan, and ambient temperature. 4. Check for air flow across compressor.
Lack of lubrication	Scored bearings. Broken rods. Scored crankshaft. Low oil in crankcase.	1. Check oil failure switch. 2. Check pipe sizing and also for oil traps. 3. Poor defrost. 4. Low load. 5. Eliminate short cycling.
General or uniform electrical burn	Entire winding is uniformly overheated or burned.	1. Check for low voltage. 2. Rapid cycling of compressor. 3. Loose terminal connection. 4. Unbalanced voltage.
Single phase burn	Two phases of a three-phase motor are overheated or burned.	1. Check contacts in starter and contact slide mechanism for binding. 2. Terminal connections on compressor. 3. Unbalance voltage. 4. Blown fuse.
Half winding single phase burn	One-half of the motor has a single phasing condition on a PART WIND MOTOR with a two-contactor system.	1. Check both contactors as one will likely be defective. 2. Check timer for proper time delay.
Start winding burn	Only start winding burned in a single-phase motor due to excessive current flow through winding.	1. Check C, S, and R wiring. 2. Starting capacitor and/or start relay. 3. Compressor overloaded.
Primary single phase burn	One phase burned only. Result of losing one phase in the primary of a delta to wye or wye to delta transformer.	1. Check transformer for proper voltage incoming and outgoing.

as mild, moderate, and severe. Symptoms and corrective action for all types of compressor burnouts are summarized in Table 15-3.

Compressor replacement

If a compressor must be replaced, install an exact replacement unit if possible. This ensures the inlet, discharge, and electrical connections will be located in the same locations as the original. Unfortunately, exact replacement units are not always available for older compressors.

If your supply house cannot cross-reference a suitable unit, select the replacement compressor by matching the voltage (single or three phase), frequency, current rating, and BTU capacity of the original parts. Back pressure ratings and starting torque requirements should also match original specifications.

Refer to the compressor nameplate for electrical information the refrigeration unit nameplate for the required BTU rating. When

TABLE 15-3 Compressor Burnout

Classification	Symptoms	Corrective action
Mild	1. Motor windings open, shorted, or grounded. 2. Little or no acid or sludge formed. 3. Slight or no burned odor.	1. Replace compressor. 2. Install new liquid line filter-drier. 3. Evacuate and recharge.
Moderate	1. Motor windings open. 2. Some acid and/or sludge formed. 3. Slight discoloration of refrigerant. 4. Slight to moderate burned odor.	1. Replace compressor. 2. Install new liquid line filter-drier. 3. Install new suction line filter. Evacuate and recharge. 4. Run for 24 hours. 5. Test for acid. 6. Replace filters if acid is found. 7. Run for 24 hours and retest.
Severe	1. Large amounts of sludge formed. 2. Acid deposits in discharge and suction lines. 3. Strong acidic odor.	1. Replace compressor. 2. Install new liquid line filter-drier. 3. Install new suction line filter. 4. Evacuate and recharge. 5. Run for 24 hours. 6. Test for acid. 7. Replace filters if acid is found. 8. Run for 24 hours and retest. 9. Removal of all acid might take several days to a week.

replacing a compressor, the overload relay and capacitor must be matched to the new compressor. Never use the original parts if the new compressor is a different model number than the original.

Removal. Perform electrical lock-out/tag-out before beginning. Wear eye protection. Completely discharge the system of all refrigerant using a refrigerant recovery system before breaking any line connections to the compressor. Large commercial refrigeration systems normally employ shutoff valves which are used to isolate the compressor from the rest of the system. In this way the majority of refrigerant can stay in the system and be reused provided it has not been contaminated by a severe burnout.

Install a manifold gauge set to monitor system pressure as the refrigerant is discharged to zero pressure. Allow sufficient time for all liquid refrigerant in the system to vaporize and be recovered. This will be indicated by a zero pressure gauge reading when the manifold gauge valves are temporarily set to the closed position.

On smaller systems, an access valve for removing refrigerant might have to be installed. Use a piercing valve in this case. Install it in the line between the evaporator and suction inlet of the compressor. Locate the valve where the heat of the brazing operation during compressor replacement will not destroy its heat-sensitive gaskets or O-rings.

To avoid pressure buildup, keep the manifold gauge valves open during disconnection of the tubing from the compressor. Use an acetylene or Mapp gas torch to heat the compressor connections. This makes the tubes easier to pull out. Once the tubing is removed, allow them to cool and then cap the openings to prevent dirt and moisture from entering the system.

The open connections of the old compressor can be pinched shut and brazed to avoid spilling any oil that might remain in the compressor. If the failed compressor has burned out, the oil will produce a harsh, acidic odor. In this case, purge the system of contaminants after removal and prior to compressor replacement. See Chapter 8 for details on purging and evacuation.

Before installing the replacement compressor, double-check the location of all connections and be certain that all necessary fittings and tubing are available. New replacement compressors are usually supplied already filled with the proper lubricating oil. If in doubt, check to be sure. Any compressor operated without the proper oil charge will quickly fail.

The refrigeration system should be equipped with a new filter-drier when the compressor is replaced. It is recommended to install two filters: one in the compressor suction line and another in the liquid line between the condenser and evaporator. Two filters ensure that any residual moisture present in the system will be absorbed by desiccant and any solid contaminants will be trapped.

Before brazing the tubing connections at the replacement compressor, thoroughly clean the ends of the tubing to prevent leakage at the finished joints. Connect the manifold gauge set to the system and open the valves during brazing to allow pressure buildup from the heat to dissipate. Allowing an inert gas such as nitrogen to flow through the system during brazing will prevent the formation of oxides on the interior of the refrigerant lines. Avoid heating the tubing close to the compressor case to prevent melting the factory-brazed connection.

When the brazing operation is completed, allow the tubing to cooldown completely. Use water to clean any residual flux from the joints not only to prevent corrosion, but also to remove any flux that might be masking a leak.

After the compressor replacement is completed, all new joints must be leak tested before evacuating and charging the system. It is best if the system is first pressurized to only about 10 psi so that any large leaks will be easily detected. If none is found, the system pressure can be increased to 50 psi, or more. Carefully check at all locations where the parts were reassembled with a leak detector.

If no leaks are found, the standard evacuation and recharging procedure is followed. When evacuating the system, be sure that it will hold vacuum with all valves closed.

Evaporator Troubleshooting

Evaporators must be kept clean, both inside and out. The air stream being passed over the evaporator must not be blocked. Check air flow using an air velocity meter. Air-in and air-out temperatures can also be checked. Typically there is about a 15°F drop in air temperature as it moves across the evaporator.

Check the evaporator inlet and outlet pressures and temperatures. Frost accumulating near the TEV often indicates the superheat setting is too high. It can also be an indicator of low suction pressure. Adjust the superheat setting of the TEV as needed.

Frost accumulation on the evaporator coils reduces air flow and acts as an insulator against effective heat transfer. Spotting frost accumulation on the evaporator coils might indicate some of the defrost heating elements are not working. Check circuit continuity. Spotty frost can also be the result of uneven air movement across the coils.

A leaking evaporator can be removed and repaired by brazing or replacing the defective run of tubing. Perform a system pumpdown to move all refrigerant out of the evaporator before breaking system connections. Follow the brazing and annealing procedures outlined in Chapter 6. Leak-test prior to installing the repaired unit.

Air-Cooled Condensers

The most common problem with air-cooled condensers is dirt accumulation on the coils. This is particularly true when the condenser is located in the base of a reach-in cooler or in a poorly maintained machine room. Dirt acts as an insulator, making it more difficult for the condenser to dissipate heat. As a result the head pressure and high-side temperature both increase and the system compressor must work harder to provide the same amount of cooling capacity. The air moving across a clogged condenser will feel hot, rather than warm. To estimate the vapor temperature inside of a condenser, add approximately 35°F to the air temperature around the condenser. Use the pressure-temperature table for the refrigerant being used to determine the approximate pressure inside the condenser and compare it to the desired head pressure for the system.

A clogged and dirty air condenser is often the first step to a burned-out compressor. Always check and clean the condenser on any service call. Stress the importance of keeping the condenser clean to your customers. Small condensers can be cleaned using soft brushes or similar tools. Large condensers can be cleaned using high-pressure air- or water-cleaning equipment. Always wear goggles and the recommended protective clothing when using high-pressure equipment.

High condenser temperatures can also be caused by air in the system that accumulates in the condenser. Systems that are overcharged with refrigerant will also run hot on the high side. Both the air and/or liquid refrigerant displace refrigerant vapor in the condenser, effectively reducing its cooling ability.

Leaks in air-cooled condenser coils can be repaired by welding. The repaired unit should be pressure tested before being reinstalled.

Water-Cooled Condensers

Common problems with water-cooled condensers include insufficient cooling water flow, poor heat transfer, and leaks. Insufficient cooling water flow can be caused by a failed recirculating pump, a bad regulating valve, blocked strainers or filters, or problems with the cooling tower.

Poor heat transfer is almost always the fault of mineral or scale buildup in the water transfer tubes of the condenser. If this is a problem, the condenser can be disassembled and cleaned using wire bushes or special power equipment. An alternate cleaning method involves flushing with a special cleaning solution that dissolves mineral buildup.

Because refrigerant vapor pressure surrounding the cooling tubes is normally greater than the water pressure inside the cooling tubes, leaks in the tubes result in water entering the refrigerant flow. Leaks in condenser cooling tubes are not easily detected unless you are specifically looking for them. You normally have to check for traces of refrigerant in the cooling water. The easiest way to do this is to inspect a sample of water taken from the condenser drain line.

Although leaky water-cooled condensers can be repaired by careful welding, it is usually good practice to replace leaking units.

TEV and Other Valves

Many of the common problems associated with TEVs are outlined in Table 15-4. Do not be in a rush to condemn the thermostatic expansion valve or adjust it. Check system pressures and make certain the unit is correctly charged. Work through the problems outlined in the table one by one to eliminate them from consideration.

If the valve is leaking, it must be replaced. If the screen stainer is dirty or clogged, it can be removed and replaced or cleaned. Cleaning is best done by burning the dirt away with the tip of the torch. However, care must be taken not to burn the screen. Air pressure and/or solvents can also be used to clean the screen. Never reinstall a TEV with the strainer screen removed.

CDA valves

CDA valves are used in electronically controlled refrigeration systems. Their operation is discussed in Chapter 5. Troubleshooting CDA problems is summarized in Table 15-5.

EPR valves

Problems with EPR valves are covered in Table 15-6.

Solenoid valves

Troubleshooting solenoid valves is summarized in Table 15-7.

Fans and Motors

Tips on troubleshooting electric motors and capacitor start systems are given in Chapter 10. Also refer to Table 15-1.

Periodic Maintenance

Periodic maintenance of reach-in, walk-in, and display-type refrigeration equipment is normally the responsibility of the equipment owner.

TABLE 15-4 Troubleshooting the TEV

Malfunction	Cause	Action
Evaporator starved	Superheat adjusted too high	Adjust superheat to correct level.
	Excess moisture	Dehydrate; replace liquid line filter-drier.
	Dirt plugging strainer or valve mechanism	Remove and clean or replace.
	Wax buildup	Clean valve; install wax trapping drier.
	Internal equalizer misapplied	Replace with externally equalized TEV.
	External equalizer is: plugged capped restricted incorrectly positioned	Correct as follows: clear any obstructions install properly correct or re-pipe re-pipe
	Flash gas occurs upstream of the TEV	Raise head pressure if too low. If liquid supply line is not exiting branch from the bottom, re-pipe. If liquid supply line is too small, re-pipe.
	Inadequate pressure drop across valve	Raise head pressure. Install new, higher capacity valve at the reduced pressure.
	Dead bulb	Replace.
	Undersized TEV	Replace with correct size.
Evaporator flooded	Superheat adjusted too low	Adjust to correct level.
	Bulb not receiving adequate thermal pick-up	Check bulb and correct for good contact, location, and ambient influence.
	Excess moisture or dirt is holding the valve open	Clean and dehydrate; install new liquid line filter-drier.
	Oversized TEV	Replace with correct size.
	Valve damage or valve seat leak	Repair or replace as needed.
Poor performance	Uneven circuit load on multiple evaporators	Balance the load. (The lightly loaded evaporator is controlling the TEV, starving the loaded evaporator.)
	Excess moisture or dirt	Clean and dehydrate; install a new liquid line filter-drier.
	TEV improperly sized	Replace with correct size.

Stress the importance of periodic maintenance to your customer and instruct them to set up a routine cleaning program. Make sure the customer is familiar with the recommended guidelines for cleaning display cases and other types of units.

Caution: Always make sure the electricity to the unit or display case is shut off before cleaning or inspecting electrical components or connections.

TABLE 15-5 Troubleshooting the CDA Valve

Malfunction	Diagnostic board light status	Cause	Action
Valve stays closed	Defrost—ON	In defrost mode	Adjust or repair defrost clock.
	Sensor—OFF	Sensor or circuit open	Replace sensor or repair open in circuit.
	24 V—ON	Plug-in thermostat defective or wrong range	Test component and replace.
Valve stays open	24 V—OFF	No 24 V to panel	Check transformer circuit, correct open.
		Panelboard damaged or defective	Replace panelboard.
	Sensor—OFF	Sensor or circuit shorted	Replace sensor or repair short in circuit.
		Frost on sensor	Move sensor away from evaporator.
	Coil—OFF	Coil or circuit open	Replace coil or repair open in circuit.
	Defrost—OFF	Voltage to A thru D terminals below 80 Vac	Check input supply circuit and correct voltage.
	24 V—ON	Plug-in thermostat defective or wrong range	Test and replace as needed.
		Discharge to suction differential below 45 psi	Raise discharge pressure.
		Blocked orifice or restricted strainer	Clean or replace (exact replacement parts are required).
		Panelboard damaged or defective	Replace panelboard.
No temperature control	Sensor—OFF	Sensor or circuit open or shorted	Replace sensor or repair circuit.
	24 V—ON	Plug-in thermostat defective	Test and replace as needed.
		Valve damaged	Repair or replace.

1. When cleaning, do not overlook such areas as fan blades and coils. If dirt accumulates on the fan blades, it could lead to early motor failure. Coils should be free of dust as well.

2. Keep the drains free of debris, as any clogs will rob the case of needed refrigeration.

3. *Do not* use ammonia or ammonia-based cleaners on or around electronic components.

4. Check electrical connections periodically for loose connections or loose or frayed wiring.

5. It is also very important to inspect insulation around the suction lines for loose or missing insulation.

6. Check the environment around the case; misdirected air from fans, open windows, or doors can affect performance. Extra lighting can also raise the case temperature.

TABLE 15-6 Troubleshooting EPR Valves

Malfunction	Cause	Action
Fails to open	Dirt holding pilot port open Solenoid not operating because: a) defective b) circuit is open c) stuck in defrost mode	Disassemble pilot port and clean. Make adjustments: a) replace defective solenoid b) find and repair the open c) repair the defrost clock
Fails to close	High-pressure inlet strainer plugged	Clean or replace.
	High-pressure line pinched, shut off, or plugged	Clean or replace. If oil filled, re-pipe so oil is not trapped in the line.
	Sleeve or piston scored, causing high-pressure leak from the piston chamber	Replace.
	Dirt in piston chamber causing drag	Clean or replace.
	High-pressure supply lower than defrost vapor pressure	Re-pipe.
	Faulty T-seal or gasket	Replace.
Does not regulate temperature	High-pressure supply is low. A minimum of 50 psig difference is needed between high-pressure and down-stream suction	Increase pressure supply.
	Condensation in long high-pressure line	Insulate and/or relocate to a higher pressure source.

TABLE 15-7 Troubleshooting Solenoid Valves

Malfunction	Cause	Action
Fails to open	Dirt plugging valve port or equalizing port	Disassemble and clean.
	Solenoid is not operating because —defective —circuit open	Correct by —replacing the solenoid —locating and repairing the open
Fails to close	Dirt in valve port or equalizing port	Clean or replace.
	Barrel or piston is scored	Replace.
	Dirt in piston chamber causing drag	Clean or replace.

7. Make sure the case is loaded correctly. Do not use large plaque cards, or let product cover up air ducts. Keep product stacked neatly inside of the "load lines."

Supermarket display case maintenance. Modern display and merchandising cases are designed for easy and efficient cleaning. The cases have removable front ducts and lower trays. All models have fully accessible one- or two-inch waste outlets. Hinged fan panels release with quarter turn fasteners, allowing access to the lower coil and drain pan. The fiberglass reinforced plastic drain pan is waterproof and contoured for easy cleaning and rapid drainage.

The exterior surfaces of cases are typically finished with high-quality baked-on polyester base refrigerator enamel. Apply a coat of appliance wax shortly after installation to protect the finish. Daily cleaning with a damp cloth will keep the surface shining. Never use abrasives or scouring pads on painted surfaces.

The interior surfaces of meat and produce cases should be thoroughly cleaned at least once a week. Dairy cases should be cleaned every two to four weeks, and frozen food cases in two- to six-month intervals. The low temperatures in the frozen food cases will minimize growth of bacteria and mold. However, dirt and debris must still be removed to ensure efficient operation.

On cases with white, rigid PVC card molding, use a water-soluble solvent or general-purpose kitchen spray cleaner. Simply spray the clean on and wipe clean.

Caution: Never use abrasives or scouring pads on any surface, exterior or interior.

Standard cleaning. The following cleaning procedure is for a store with minimal cleaning equipment. The cases should be properly installed. There should be adequate caulking between the cases and sufficient drainage (one inch minimum diameter drip pipe size and ¼ inch per one-foot of drip pipe slope).

1. Shut off the refrigeration unit and all power to case fans, anti-sweat wires, and internal lights.
2. Remove all product, storing it in another case or a suitable walk-in area.
3. Remove any screens, trays, bottom pans, and ducts and clean separately. Be sure that lighted shelves are removed and cleaned separately. Do not soak lighted shelves.
4. Remove any loose debris that is clogging the drain.
5. Clean all surfaces with a germicidal detergent at recommended concentration using warm water. A brush or cleaner pad will aid in penetrating dirt. Never soak electrical wiring and fans. Keep them out of the water as much as possible.

   If detergent alone was used in step 5, it should be rinsed with warm water, then followed by a sanitizer. The germicidal agent/sanitizer should be rinsed off with clean water, using a hose for the rinse. Allow the surfaces to air dry, because wiping them would defeat the purpose of the sanitizer. (Because several gallons of water have already been used to clean the case, check now for any drain line problems.)

6. Replace all internal parts, making sure that they seat properly. This is important for proper case operation.
7. Start the refrigeration machine, turn on the case electrical circuits, and replace the product.

Caution: Never turn the case electrical power on until all components are dry. Do not use ammonia-based cleaning products on cases with electronic and solid-state components, as the ammonia could cause serious damage to these components. When cleaning, be sure to check that water is not pouring into the case faster than it can be carried out through the drain.

High-pressure/low-volume spray. The standard cleaning process just described is a good way to clean the easy-to-reach areas of a display case. But there is a way to clean those coil and duct areas that are often inaccessible by conventional methods. A thorough cleaning can be done on these hard-to-reach areas with high-pressure/low-volume spray equipment. This spray equipment is similar to the equipment found at a typical car wash. Spray nozzles are calibrated to spray about two to three gallons per minute at about 500 pounds per square inch of pressure. Use the following procedure for case cleaning:

1. Follow steps 1 to 4 from the standard cleaning procedure.
2. Use plain water first to loosen and soften any dirt or grime, which will also save detergent. A range of lower water temperatures can be used with the spray equipment; however, if the water temperature exceeds 160°F, it could bake the soil onto the case surface. In addition, the excessive heat creates steam that can fog the work area, making it hard to see the surfaces while working.
3. Use germicidal detergent and water at recommended concentrations. Be sure to wash down any air ducts and coils that are within reach. The case coil area can be reached through coil-and-drain inspection panels, depending on the construction of the case. Coils in the back might be harder to reach than those in the well area.
4. Aim the spray so that dirt is directed to the drain if possible. (A two-foot-long spray wand is best for cleaning open cases.)
5. Follow with a plain water rinse. Allow the surfaces to air dry.
6. Finish with steps 6 and 7 of the case cleaning procedure.

Warning: Do not misuse high-pressure spray equipment by directing the spray on the following electrical equipment: fan motors, light sockets, glass joints, or case joints. Cases are constructed with moisture-resistant, sealed components and connections; however, the seals can be penetrated if directly sprayed. Light sockets are particularly vulnerable and should be avoided if at all possible.

Chapter

16

Calculating Refrigeration Loads

Calculating refrigeration loads and matching and rating condensing units and evaporators for applications are essential skills for a well-trained service technician.

Refrigeration Loads

Refrigeration load, simply put, is the heat that must be removed from a refrigerated space to maintain the desired temperature. Knowing how to calculate the refrigeration load for a particular size walk-in cooler or supermarket display case is needed when considering a new or replacement unit. It is also often useful when troubleshooting an old one.

Heat load charts

Both equipment manufacturers and trade organizations, such as the American Society of Heating, Refrigerating, and Air-Conditioning Engineers (ASHRAE), publish heat load data designed to help service technicians and system designers accurately calculate what size refrigeration system is needed for a particular application.

The data in the heat load chart is calculated based on a certain set factors, such as operating hours per day (16 or 18), normal or heavy usage, and the K factor of the unit's insulation. If your application does not match these specifications, your calculation might have to be adjusted.

Note: All of the charts and data listed in this chapter are for example purposes only. This data is typical for the calculations being discussed. Always refer to current reference sources.

K factor. The *K factor* is the thermal conductance of an insulating material. The K factor is the reciprocal of the R factor (resistance to heat flow) you might be more familiar with from working with home insulation products. The K factor is calculated using the following formula:

$$K = \text{BTU} \times \text{inches of material/hour/square feet/degree F}$$

Example. Load tables allow you to quickly spec out a refrigeration system. For example, if a customer asks for the specifications for a 20' × 10' walk-in refrigerator, the manufacturer's heat load chart shown in Table 16-1 indicates a cooling capacity of 11,300 BTU/h is needed during normal operation.

TABLE 16-1 Condensing Unit Capacities for Walk-In Coolers*

Outside dimensions (in ft.)	Demand in BTU/h (18 hours operation)	
	Normal service	Heavy service
10 × 8	5,500	6,900
10 × 9	6,080	7,520
10 × 10	6,630	8,150
11 × 6	4,820	5,910
11 × 7	5,380	6,600
11 × 8	6,000	7,350
11 × 9	6,500	8,050
11 × 10	7,100	8,800
12 × 6	5,150	6,350
12 × 8	6,400	7,700
12 × 10	7,590	9,380
12 × 12	8,800	10,800
14 × 8	7,300	9,050
14 × 10	8,650	10,900
14 × 12	9,720	12,300
14 × 14	10,800	13,700
16 × 8	8,140	10,000
16 × 10	9,350	12,000
16 × 12	10,700	13,450
16 × 14	12,000	15,000
16 × 16	13,100	16,600
18 × 10	10,300	13,000
18 × 12	11,700	14,800
18 × 14	13,100	16,400
18 × 16	14,400	17,400
18 × 18	15,800	19,600
20 × 10	11,300	13,700
20 × 12	12,800	15,700
20 × 14	14,300	17,600
20 × 16	15,600	19,400
20 × 18	17,000	21,100
20 × 20	18,700	22,700

* 35°F temperature, 9 feet in height, 95°F ambient temperature, 4-inch insulation.

Calculating Refrigeration Loads

You can now compare this cooling capacity to compressor manufacturers specifications. Typical cooling capacities for various size compressors are listed in Table 16-2. For example, a 1½-horsepower medium-temperature compressor or condensing unit delivers approximately 12,000 BTU/h of cooling capacity. This may meet the needs during normal service loads, but as Table 16-2 states, heavy load conditions can require up to 14,000 BTU/h. It is easy to see that a 2-horsepower unit is the correct selection.

Basic components of refrigeration load

The heat load entering the refrigeration unit or space can be classified in one of four ways.

Transmission. Heat enters a unit through the walls, floor, and ceiling by *transmission*. Warm air outside the unit raises the temperature of the walls, floor, and ceiling. This heats moves through to the inside of the unit, where it warms the refrigerated air.

The coefficient of heat transfer or U factor of a given material states the number of BTU per hour that will pass through a square foot of wall, ceiling, or floor space for each degree of difference between the inside and outside unit temperature. If you know the U factor, the inside and outside unit temperatures, and the unit surface area (walls, floor, and ceiling), the following equation can be used to calculate transmission:

$$\text{Heat transmission } (HT) = \\ \text{U factor} \times \text{Area } (A) \times \text{Temperature difference } (TD)$$

Example: A medium temperature (35° F) walk-in refrigerator measuring 20′ long × 15′ wide × 10′ high is located in an area that has a maximum summertime temperature of 95°F. The walls and ceiling are constructed of 5″ foamed-in-place urethane insulation panels with a U factor of 0.025. The unit will be placed directly on a concrete slab foundation, but 5″ insulated floor panels with steel sheathing will provide added insulation in the floor.

TABLE 16-2 Typical Medium-Temperature Compressor Capacities

Horsepower	Capacity (BTU/h)
½	4,400
¾	6,000
1	8,500
1½	12,000
2	14,000
2½	18,000
3	24,000

The basic calculation is as follows:

$$\text{Heat transmission} = U \times A \times TD$$

For the two long walls:

$$HT = 0.025 \times (2 \times 20 \times 10) \times (95 - 35) = 600 \text{ BTU/h}$$

For the two short walls:

$$HT = 0.025 \times (2 \times 15 \times 10) \times (95 - 35) = 450 \text{ BTU/h}$$

For the ceiling:

$$HT = 0.025 \times (20 \times 15) \times (95 - 35) = 450 \text{ BTU/h}$$

For the floor:
The concrete slab floor has an average temperature of 55°F.

$$HT = 0.025 \times (20 \times 15) \times (55 - 35) = 150 \text{ BTU/h}$$

The total amount of heat entering the unit from transmission is calculated as follows:

$$\text{Total HT} = \text{Long walls} + \text{Short walls} + \text{Ceiling} + \text{Floor}$$

$$= 600 + 450 + 450 + 150$$

$$= 1650 \text{ BTU/h}$$

Air infiltration and change out. This type of heat load occurs whenever the door to a unit is opened. Warm air from outside the unit displaces the cooler air inside the unit. Poorly constructed or leaking walk-in units also experience air infiltration.

Table 16-3 lists average air change outs for both medium and low temperature units, based on the volume of the unit. Let's start by calculating the volume of our sample unit:

$$\text{Volume} = \text{Length} \times \text{Width} \times \text{Height}$$

$$= 20 \times 15 \times 10$$

$$= 3000 \text{ cubic feet (cu.ft.)}$$

The air infiltration rate found in Table 16-3 for 3000 cubic feet at temperature above 32°F is 9.5. The 9.5 refers to the number of times that the air within the unit is changed in a 24-hour period. For our 3000-cubic-feet example, the amount of air entering the unit is:

TABLE 16-3 Average Air Change Outs

Volume (cubic feet)	Air changes/24 hours	
	Above 32°F	Below 32°F
200	44.0	33.5
300	34.5	26.2
400	29.5	22.5
500	26.0	20.0
600	23.0	18.0
800	20.0	15.3
1,000	17.5	13.5
1,500	14.0	11.0
2,000	12.0	9.3
3,000	9.5	7.4
4,000	8.2	6.3
5,000	7.2	5.6

$$\text{Air infiltration} = \text{Air changes per 24 hours} \times \text{Volume} / 24$$

$$= 9.5 \times 3000 / 24$$

$$= 1187.5 \text{ cubic feet per hour}$$

Table 16-4 lists the number of BTUs that each cubic foot of air infiltration adds to the total heat load for various air temperatures, unit temperatures, and relative humidities. If we have a 95°F outside air temperature, a 35°F unit temperature, and a relative humidity of 50 percent, then the corresponding value from Table 16-4 is 1.99 BTU/cubic foot. Knowing this data, we can now calculate the amount of BTU/h added on to the total heat load through air infiltration:

$$\text{Heat infiltration} = \text{Air infiltration} \times \text{Factor}$$

$$= 1187.5 \times 1.99$$

$$= 2363 \text{ BTU/h}$$

Product load. Product load involves heat that is carried in to the unit by the products to be cooled, or the heat that comes from the product itself once in the unit. When a beverage distributor stocks his walk-in cooler with cases of soft drinks previously stored at room temperature or just unloaded from a hot delivery truck, he is introducing product load into the cooler.

Heat that enters a refrigeration system through product load can be either sensible heat or latent heat. As covered in Chapter 3, when sensible heat is removed from a product, the temperature of that product is lowered. When a product is cooled from room temperature to 35°F,

TABLE 16-4 BTU/Cubic Foot of Heat Removed as Air Is Cooled to Storage Room Conditions

Storage room temperature °F	Temperature of outside air, °F							
	85		90		95		100	
	Relative humidity, percent							
	50	60	50	60	50	60	50	60
65	0.45	0.64	0.68	0.91	0.93	1.20	1.21	1.51
60	0.66	0.85	0.89	1.12	1.14	1.41	1.42	1.71
55	0.85	1.04	1.08	1.31	1.33	1.60	1.61	1.91
50	1.03	1.22	1.26	1.49	1.51	1.78	1.79	2.09
45	1.19	1.39	1.43	1.66	1.68	1.94	1.95	2.25
40	1.35	1.55	1.59	1.61	1.83	2.10	2.11	2.41
35	1.50	1.70	1.74	1.96	1.99	2.25	2.26	2.56
30	1.64	1.84	1.88	2.10	2.13	2.39	2.40	2.70

the heat removed is sensible heat. As the product reaches the freezing point, any additional heat removed causes the product to solidify or freeze. This heat is latent heat. Once the product is frozen, if additional heat is removed, it is sensible heat, and the product's temperature once again is lowered.

Table 16-5 lists just a small sampling of the vast array of information on food product storage available from the American Society of Heating, Refrigerating, and Air-Conditioning Engineers (ASHRAE). Note that the table gives both sensible and latent heat values, which can be used to determine product load. An explanation of significant terms found in Table 16-5 follows:

Specific heat. This is sensible heat, or the number of BTUs needed to change the temperature of a product 1°F. The table lists this value for both frozen and unfrozen products. As you can see, there is quite a difference between the amount of heat needed to lower an unfrozen product 1°F compared to lowering a frozen one 1°F. For example, 0.91 BTU must be removed from 1 pound of beans to lower its temperature 1°F. One pound of frozen beans, however, requires only 0.47 BTU to lower the temperature 1°F.

Highest freezing. The temperature that a product, through removal of sensible heat, must reach before it begins to change state or freeze. Note that this temperature varies; beans have a freezing point of 30.7°F while tuna will freeze at 28.0°F.

Latent heat. When a product temperature is lowered to the freezing point, latent heat is the number of BTUs that must be removed from the product in order for it to freeze. Once beans reach their 30.7°F freezing point, latent heat is removed as they cool even further or freeze. It takes 127 BTUs to completely freeze a pound of beans once they reach the freezing point.

TABLE 16-5 Storage Requirements for Perishable Foods

Product	Storage temperature °F	Highest freezing °F	Specific heat above freezing BTU/lb °F	Specific heat below freezing BTU/lb °F	Latent heat BTU/lb
Vegetables					
Beans	40 to 45	30.7	0.91	0.47	127
Corn, sweet	32	30.9	0.79	0.42	106
Peas	32	30.9	0.79	0.42	106
Pumpkins	50 to 55	30.6	0.92	0.47	130
Tomatoes	55 to 60	31.0	0.94	0.48	133
Yams	61		0.79	0.42	105
Fruits					
Apples	32 to 38	29.3	0.87	0.45	121
Blueberries	31 to 32	29.7	0.86	0.45	118
Kiwifruit	31 to 32	29	0.86	0.45	118
Raspberries	31 to 32	30.9	0.87	0.45	120
Strawberries	31 to 32	30.6	0.92	0.47	129
Seafood					
Haddock, cod, and perch	31 to 34	28	0.85	0.44	117
Salmon	31 to 34	28	0.72	0.39	92
Scallop meat	32 to 34	28	0.84	0.44	114
Shrimp	31 to 34	28	0.81	0.43	109
Tuna	32 to 36	28	0.72	0.39	92
Meat					
Beef, fresh	28 to 34	28 to 29	0.48 to 0.57	0.30 to 0.33	46 to 63
Chicken	28 to 32	27	0.80	0.42	106
Frankfurters	32	29	0.66	0.37	80
Ham (74% lean)	32 to 34	29	0.66	0.37	80
Liver	32	29	0.77	0.41	100
Pork, fresh	32 to 34	28 to 29	0.48 to 0.57	0.30 to 0.33	46 to 63
Sausage, links	32 to 34		0.53	0.31	54
Turkey	28 to 32	27	0.72	0.39	92
Veal, lean	28 to 34		0.74	0.40	94
Dairy					
Butter	32	−4 to 31	0.36	0.25	23
Eggs, whole	29 to 32	28	0.73	0.40	96
Ice cream	−20 to −15	21	0.70	0.39	86
Milk, grade A	32 to 34	31	0.92	0.46	125
Other food products					
Bread	0	32 to 37	0.70	0.34	46 to 53
Coffee, green	35 to 37		0.32 to 0.35	0.23 to 0.24	14 to 21
Honey	50		0.35	0.26	26
Orange juice	30 to 35		0.91	0.47	127

SOURCE: Abstracted with permission of the American Society of Heating, Refrigerating, and Air-Conditioning Engineers from the *1994 ASHRAE Handbook—Refrigeration.*

For example, consider 2000 pounds of beans at 72.0°F that have been cooled to the freezing point (a decrease of approximately 41°F). Since the beans lose 0.91 BTU for every temperature decrease of 1°F, then 2000 pounds will lose 2000 × 0.91 or 1820 BTUs. For a decrease of 41°F, 41 × 1820 or 74,620 BTUs are lost. Once the beans reach the freezing point, every pound will then require 127 BTUs removed for freezing. Freezing 2000 pounds would mean a loss of 254,000 BTUs (2000 × 127). As you can see, much more refrigeration energy is used for freezing the product than is needed to reduce its temperature. So to cool and freeze the 2000 pounds of beans, a total of 328,620 BTUs must be removed. If we want the freezer to do this over an 8-hour period, the rate of product load is:

$$328{,}620 \text{ BTUs}/8 \text{ hours} = 41{,}070 \text{ BTU/h}$$

Heat of respiration. All fruits and vegetables generate heat as they ripen. This heat must be included when calculating heat loads for large amounts of stored produce. Heat of respiration is stated as BTU per ton of product over a 24-hour period. For example, a ton of green beans will generate approximately 12,000 BTUs over a 24-hour period or about 500 BTUs per hour.

Miscellaneous load. This type of load refers to any additional source of heat that must be removed so the unit can reach its desired operating temperature. For example, lights inside a large walk-in cooler radiate heat into the refrigerated space. This is a miscellaneous heat load that must be removed by the system.

The BTUs per hour generated by a light that is on continuously equals 3.4 times its wattage. Motors that run inside a refrigerated space are another miscellaneous heat source. A ⅓-hp evaporator fan motor can add about 1500 BTUs per hour.

People entering or working in the refrigerated space on a continuous basis, such a workers in a supermarket prep room, must also be considered a heat source. In a 40°F preparation room, each worker can add up about 850 BTUs per hour.

Total refrigeration load

Total refrigeration load is the sum of all the different types of heat loads placed on the system:

- Transmission = 1650 BTU/h
- Air infiltration = 2360 BTU/h
- Product load = 41,070 BTU/h
- Respiration = 500 BTU/h
- Miscellaneous load = 850 BTU/h for one worker

- 682 BTU/h for 200 watts in lighting
- 1500 BTU/h for a evaporator fan motor

Total Load equals the sum of the above or 48,612 BTU/h.

A final safety percentage of 20 percent must be factored in. This allows for the possibility of operation at greater loads, also for the fact that some of the figures used were estimates. Adding in the 20 percent safety factor gives a final figure of 58,334 BTU/h. Any unit with a condensing/evaporating system that provides refrigeration equal to or slightly greater than this amount would be acceptable.

When a refrigeration system is first started up, the load on the evaporator is heavy due to the warm air passing over it. Additionally, head pressure, suction pressure, and compressor current will be quite high. As the air temperature drops, the load decreases as pressure and current levels taper off. When the temperature setpoint is reached, the compressor shuts off.

The compressor then cycles on and off at intervals, based on the amount of heat entering the unit due to infiltration and transmission. In our example, infiltration and transmission combined equal just over 4000 BTU/h, so the 58,000 BTU/h cooling unit should quickly, almost effortlessly, remove this heat.

However, if a fresh 2000-lb. shipment of beans at 72°F are loaded into the unit, the load would once again be quite large-about 41,700 BTU/h. The 58,000 BTU/h walk-in refrigerator could cool the load down in eight hours, as specified. While the unit would not run constantly, it would have to work much harder when the product is first loaded into it. Once the beans have been cooled to the desired temperature, the system need only work to remove heat due to transmission, filtration, and respiration.

Walk-in freezer load calculations

The same type of load calculations can be performed for walk-in freezers. To keep our sample calculation simple, let's change the walk-in refrigerator in the previous example into a walk-in freezer with a setpoint temperature of –5°F. The product in this example is 2000 pounds of chicken, which must be lowered in temperature from 65°F to –5°F in four hours, then stored until it is sold.

Transmission. Transmission heat loss increases because the holding temperature is lower, now –5°F instead of 35°F. The temperature difference is now 95°F – (–5°F) = 100°F.

The basic calculation is as follows:

$$\text{Heat transmission} = U \times A \times TD$$

For the two long walls:

$$HT = 0.025 \times (2 \times 20 \times 10) \times 100 = 1000 \text{ BTU/h}$$

For the two short walls:

$$HT = 0.025 \times (2 \times 15 \times 10) \times 100 = 750 \text{ BTU/h}$$

For the ceiling:

$$HT = 0.025 \times (20 \times 15) \times 100 = 750 \text{ BTU/h}$$

For the floor:
The concrete slab floor has an average temperature of 55°F.

$$HT = 0.025 \times (20 \times 15) \times [55 - (-5)] = 450 \text{ BTU/h}$$

The total amount of heat entering the unit from transmission is calculated as follows:

$$\text{Total HT} = \text{Long walls} + \text{Short walls} + \text{Ceiling} + \text{Floor}$$

$$= 1000 + 750 + 750 + 450$$

$$= 2950 \text{ BTU/h}$$

Infiltration. The heat contributed by air infiltration for the walk-in freezer changes in the following ways. As shown in Table 16-3, the air changes in 24 hours will decrease from 9.5 to 7.4 when the unit temperature falls below 32°F.

$$\text{Air infiltration} = \text{Air changes per 24 hours} \times \text{Volume} / 24$$

$$= 7.4 \times 3000 / 24$$

$$= 925 \text{ cubic foot per hour}$$

Table 16-4 lists the number of BTUs that each cubic foot of air infiltration adds to the total heat load for various air temperatures, unit temperatures, and relative humidity. If we have a 95°F air temperature, a −5°F unit temperature, and a relative humidity of 50 percent, then the corresponding value from the table would be 2.76 BTU/cubic foot. We can now calculate the amount of BTU/h added on to the total heat load through air infiltration:

$$\text{Heat infiltration} = \text{Air infiltration} \times \text{Factor}$$

$$= 925 \times 2.76$$

$$= 2553 \text{ BTU/h}$$

Product load. The product load must be calculated in several steps because the 2000 pounds of chicken will be cooled, then frozen. The chicken must be reduced to its freezing point, which is 27°F. The chicken will be cooled from 65°F to 27°F.

To calculate the energy needed to cool the load to the freezing point:

Load = Product weight × Specific heat × Temperature difference

= 2000 × 0.80 × (65 − 27)

= 2000 × 0.80 × 38

= 60,800 BTU/h

To determine the energy needed to freeze the load:

Load = Product weight × Latent heat of fusion

= 2000 lbs × 106 BTU/lb

= 212,000 BTU/h

To calculate the energy needed to reduce the temperature of the load, once frozen, to −5°F:

Load = Product weight × Specific heat × Temperature difference

= 2000 × 0.42 × [27 − (−5)]

= 2000 × 0.42 × 32

= 26,880 BTU/h

There is no heat of respiration, so the total product load is 60,800 + 212,000 + 26,880 = 299,680 BTU. The hourly load over a four-hour time period would be 74,920 BTU/h. We will use the same amount of miscellaneous heat gain as in the previous example. So total heat load for this freezer application with 2000 pounds of chicken is:

Transmission = 29,500 BTU/h

Air infiltration = 2553 BTU/h

Product load = 74,920 BTU/h

Miscellaneous = 3032 BTU/h

Total = 110,005 BTU/h

Once the 20 percent safety factor is added in, the final figure would be 132,000 BTU/h. Using the value of 4200 BTU/h per 1 horsepower for low-temperature equipment, this would indicate that a 32 horsepower low-temperature walk-in freezer would be a good choice.

Equipment Selection

Once the total heat load on the system is determined, you can determine the size of the evaporator and condensing unit needed. It's a good idea to consult the manufacturer's literature. The manufacturer lists this information under capacity allowance. Although the condensing unit might be rated in horsepower, it is becoming more common to see BTU/h ratings for various evaporator and outdoor temperatures.

An evaporator is typically rated in BTU/h per degree difference between air passing over the evaporator and the refrigerant temperature. An evaporator that is too cold could dry out or dehydrate the product, which is undesirable. A 10°F difference is not too large, which is why a 25°F evaporator is a good choice. (35°F unit temperature − 10°F = 25°F evaporator.) For product that would not dry out, such as prepackaged frozen foods, a 20°F temperature difference would be allowed.

As mentioned, the manufacturer's literature might rate the system in horsepower instead of BTU/h. A medium-temperature system typically removes about 8000 BTU/h per 1 horsepower. To determine the horsepower for a walk-in unit with a capacity of 58,000 BTU/h, divide this figure by 8000. The result is 7.25 horsepower, which is not a common size. In this case simply go to the next larger available size.

Choosing an evaporator

Many types of evaporators are used with refrigeration units, but the two most popular designs are air-cooled and liquid-cooled.

Air-cooled. This evaporator cools the air within the unit directly. An air-cooled evaporator transfers heat from the air circulating over it to the liquid refrigerant as follows. Warm air contacts the evaporator, its molecules striking the fins, slowly releasing their energy (or heat) to the fins. This heat energy travels quickly past the fins into the tubing where it contacts liquid refrigerant, which slowly yet effectively absorbs the heat.

Air-cooled evaporators are also called dry evaporators. They might be of the frosting, defrosting, or nonfrosting type. Some air-cooled evaporators have forced circulation.

Liquid-cooled. The evaporator cools a liquid, which might then be used to cool other substances. Liquid-cooled evaporators can be either submerged or a tube-within-a-tube design.

Air circulation. Most refrigerator cabinets are specifically designed to run efficiently with evaporators. As a service technician, you should have a basic understanding of how air circulates in a cabinet and what the air-flow patterns should be when the unit is operating.

Baffles are surfaces, or air ducts, designed to increase the efficiency of air flow from the evaporator to all areas of the refrigeration unit. A well designed baffling system will direct air throughout the unit with no dead or warm air spots.

The air must move freely with no objects blocking its flow. In many cases, the colder air flows from the evaporator down into the center of the unit, while the warmer air moves up the walls to the evaporator.

Horizontal baffles or drain pans are often exposed to cold air on the topside at the same time that warm air is blowing on the underside. This temperature difference could cause undesirable condensation to collect on the baffling. Insulation is recommended to prevent this condensation buildup. Eddy currents (small, circular air flow) could also disturb the overall air flow in the cabinet.

With natural convection evaporators air flow in the unit is based on the temperature of the air within. The warmer air is lighter or less dense, so it rises while cooler, denser air falls. Baffling is used to speed up the natural circulation, which will remain effective as long as the airflow is not blocked in any way. Baffling can also be used as drain pans.

Balancing evaporator and condensing units

When trying to decide which size evaporator a customer or installation will need, keep in mind that the evaporator will remove heat from the cabinet only when the condenser is running. Also, the unit will only run 16 to 20 hours in a 24-hour period. This means that the unit must provide enough refrigeration in 16 hours to cover the total heat load in 24 hours.

The size of the evaporator you choose will depend on the following three conditions: cabinet temperature, refrigerant temperature, and the space allowed for the evaporator.

The size of the condensing unit is affected by low-side pressure, the condensing cooling agent—air or water—and the total size of the compressor and the condenser.

Evaporator capacity should be balanced against the condenser capacity—not against the heat load of the cabinet. When balancing these two components, be sure that the calculations are based on the same low-side pressure for each. This is important because the capacity of the evaporator increases as its temperature drops, and the capacity of the condensing unit lowers as its low-side pressure drops.

Manufacturers of evaporators and condensers typically rate their capacities in BTU for 1, 16, or 18 hour time periods. Table 16-6 lists condensing unit capacities at two different condensing temperatures—110°F and 120°F. Four evaporator temperatures are also listed—–30°F, –15°F, 20°F, and 40°F.

TABLE 16-6 Compressor Capacities (BTU/h) at Various Evaporating and Condensing Temperatures

Compressor horsepower	Evaporator temperature °F			
	−10		+20	
	Condensing temperature °F			
	110	130	110	130
½	2,600	2,200	5,100	4,400
¾	3,900	3,400	7,300	6,500
1	5,100	4,400	9,100	8,000
1½	7,400	6,600	13,700	12,000
2	13,100	11,500	16,100	14,000
3	20,100	17,200	26,500	24,000

TABLE 16-7 Condensing Unit and Evaporator Capacity Variations Due to Temperature and Pressure

Condensing unit		Evaporator	
Low-side temperature °F	BTU/h	Temperature difference °F	BTU/h
40	6,500	1	350
30	5,500	5	1,750
20	4,550	10	3,500
10	3,750	12	4,200
0	2,800	15	5,250
−10	2,200	—	—
−20	1,400	—	—
−30	875	—	—

Notes: All values are approximate. Evaporator temperature difference is difference between evaporator coil temperature and refrigerated space temperature.

Choosing a condensing unit

The type of condenser you choose will depend on your specific installation. Let's consider as an example a typical external drive air-cooled condenser with a 1-horsepower compressor motor. The evaporator chosen has a 15°F temperature difference and a 16-hour running time. For the refrigerator to run at 35°F, the refrigerant temperature should be 35 − 15 or 20°F. For an HFC-134a refrigerant, the condensing unit capacity would be matched with the evaporator capacity at 18.4 psig. Table 16-7 lists the average hourly capacity of different horsepower condensing units. The table shows an increase in condensing unit capacity as low-side pressure (and temperature) increases, assuming that head pressure is constant. It also shows the effect of condensing pressure and temperature on the overall capacity of the condenser.

Glossary

absolute pressure (psia) The pressure that exists in a perfect vacuum. It is 14.696 psi less than gauge pressure.

absolute temperature Temperature measurement equal to absolute zero, $-460°F$ or $-273°C$.

absorbent A substance with the ability to take up or assimilate another substance. (*See also* drying agent.)

accumulator Component in a refrigeration system that stores liquid refrigerant that did not vaporize in the evaporator.

air infiltration Occurs when air passes into or out of an enclosed area through cracks or other openings in the structure.

alternating current (ac) A type of electrical energy that reverses direction at a set rate.

ambient switch Control that opens or closes a circuit based on existing air temperature.

ambient temperature The prevailing or existing air temperature.

ammeter An electrical instrument that measures current flow in a circuit. It is usually calibrated in amperes.

ampere A unit of current flow equal to one coulomb per second.

ANSI American National Standards Institute.

ARI Air-Conditioning and Refrigeration Institute.

ASHRAE American Society of Heating, Refrigerating, and Air-Conditioning Engineers, Inc.

atmospheric pressure Absolute air pressure at a given altitude. Under standard conditions at sea level, the pressure is 14.696 psi absolute, or 29.92126 inches of mercury.

azeotropic mixture A blend of two or more refrigerants that produces desired pressure/temperature characteristics without combining chemically.

back pressure Pressure existing at the suction inlet of a compressor.

back-seating valve An access valve designed to close off refrigerant flow at each end of its adjustment travel.

barometer An instrument used to measure absolute atmospheric pressure. At standard conditions, absolute atmospheric pressure at sea level is 29.92126 inches of mercury (760 mm), 14.696 psia, 33.94 feet of water, or 101.3 kilo-Pascals (kPa).

bimetal Two metals with different rates of expansion fastened together. When heated or cooled, they will warp; they can be made to open a switch or valve.

bimetallic strip Temperature regulating or indicating device that works on the principle that two dissimilar metals with unequal expansion rates, welded together, will bend as temperatures change.

boiling point The temperature that a liquid changes to a vapor at a given pressure.

brazing A method of joining two metals using a silver-based filler metal; often called silver soldering.

BTU (British Thermal Unit) One BTU is the amount of heat energy needed to raise 1 pound of water 1 degree Fahrenheit.

bulb A chamber placed at the end of a capillary tube that stores liquid and gas refrigerant. It provides temperature sensing through the expansion and contraction of the refrigerant.

burnout Electrical failure of a motor or compressor that results in deteriorated wire insulation and/or refrigeration oil.

capacitance The measure of the charge-storing ability of a capacitor, usually specified in mircrofarads.

capacitor A component used in alternating circuits that draws a current that leads the applied voltage by 90 electrical degrees.

capacitor run motor An induction motor that uses a capacitor and start winding, which is energized whenever the motor is running. Sometimes called a permanent split capacitor (PSC) motor.

capacitor start motor An induction motor that uses a capacitor and start winding to provide the initial starting torque. The capacitor is switched out to the circuit when the motor reaches running speed.

capillary tube A small-diameter length of tubing that restricts the flow of liquid refrigerant. They are carefully sized by inside diameter and length for each individual application.

carbon dioxide An inert gas that can be used as a purging agent or to pressurize a system for leak testing.

Celsius (C) A temperature scale where water freezes at 0 degrees and boils at 100 degrees.

centigrade Another name for the Celsius temperature scale.

CFC (Chlorofluorocarbon) A family of chemicals that includes halogen elements like chlorine and fluorine. Freon 12 and 22 are examples of CFCs.

change of state Condition in which a substance changes from a solid to a liquid or a liquid to a gas due to the addition of heat. Or the reverse condition, where a substance changes from a gas to a liquid or a liquid to a solid due to the removal of heat.

charge Amount of refrigerant placed into a refrigerating unit.

charging The process of placing refrigerant into a system.

charging station A portable unit containing a manifold gauge set, vacuum pump, and refrigerant container. It is used to evacuate and charge a refrigeration system.

check valve A valve that allows the flow of refrigerant in one direction only.

chemical instability The tendency of a refrigerant or refrigerant oil to decompose when exposed to contaminants and heat in a sealed system.

chlorofluorocarbons A family of chemicals (such as Freon) that contains chlorine and fluorine halogen elements.

cold The absence of heat.

comb A plastic tool used to straighten out bent fins on evaporator and condenser coils.

compound gauge A pressure gauge designed to measure both pressure and vacuum.

compression cycle A refrigeration system that uses the mechanical compression of a circulated refrigerant to cool a space.

compression ratio The ratio of the absolute high-side pressure to the absolute low-side pressure in a compressor; sometimes called a pumping ratio.

compressor Pump of a refrigerating unit that draws a low pressure on the cooling side of the refrigerant cycle and squeezes or compresses the gas into the high pressure or condensing side of the cycle.

compressor seal A rotary seal used in open-type compressors that allows the externally driven crankshaft to pass through the housing wall while preventing loss of refrigerant.

condenser The part of the refrigeration unit that receives hot, high-pressure refrigerant gas from the compressor and cools the gaseous refrigerant until it returns to a liquid state.

condenser fan Forced-air device used to move air through the air-cooled condenser.

condensing unit The part of the refrigeration unit that pumps vaporized refrigerant from the evaporator, compresses it, liquifies it in the condenser, and returns it to the refrigerant control.

conduction The transfer of heat through a medium.

contactor A large electromagnetic-actuated relay. Typically, this is the relay that closes the circuit to the compressor.

contaminant Any solid, liquid, or gas found in a sealed system other than the refrigerant or refrigerant oil—such as air, water, or dirt.

convection The transfer of heat by means of circulation of a gas or liquid, due to gravitational force.

cooling coil The evaporator in a refrigeration system.

cooling tower A structure designed to cool water by evaporation as it is sprayed into the atmosphere.

coupling A connecting device that joins two refrigerant lines.

crankcase heater An electrically operated heating element placed around or inside a hermetic compressor. It prevents the migration of refrigerant and dilution of lubricating oil.

current relay A mechanical device that closes a pair of electrical contacts in response to current flowing through a coil.

cut-in The temperature at which a thermostatic switch activates a circuit.

cut-out The temperature at which a thermostatic switch deactivates a circuit.

cycle The complete course of operation of a refrigerant back to its starting point in the system. Also used to describe alternating current through 360 space degrees.

cylinder A container used to store a gas or liquid, usually under pressure.

defrost cycle A process that melts accumulated ice on the evaporator coil of a refrigeration system.

defrost timer A device connected to an electrical circuit that shuts the unit off long enough to allow any ice and frost accumulation on the evaporator to melt.

desiccant A moisture-absorbing substance that traps and holds residual water in a refrigeration system.

dewpoint The temperature at which the relative humidity in the air becomes 100 percent, causing the air to begin to condense.

diagnosis The process of analyzing the operation of a refrigeration system.

differential The difference between two levels of temperature or pressure.

direct current (dc) An electrical current that always flows in the same direction.

discharge The output port of a compressor, or the process of removing refrigerant from a system.

discharge line The high-pressure, high-temperature output line of a compressor.

discharge pressure The pressure read at the compressor outlet. Also called head pressure or high pressure.

drier Any substance or device used to remove moisture from a refrigeration system.

dry bulb thermometer An ordinary thermometer used to indicate ambient air temperature.

dry evaporator An evaporator in a refrigeration system designed to vaporize all liquid refrigerant before returning it to the compressor.

drying agent The desiccant placed in a receiver/drier or accumulator to absorb moisture.

duty cycle The percentage of operating time of a device as compared with the total time of one complete on-off cycle.

electronic leak detector An instrument that measures the change in the electrical characteristic of air when contaminated with refrigerant.

equalizer line A tube that connects one part of a component to another, to allow the pressures to become equal.

evacuate To remove all air, gas, moisture, and other contaminants from a sealed system, typically by means of a vacuum pump.

evaporation The process of a liquid changing to a vapor.

evaporative cooling The cooling effect of vaporization of a liquid in a moving air system.

evaporator Part of the refrigeration unit in which the liquid refrigerant absorbs heat as it vaporizes and becomes a gas.

evaporator superheat The actual temperature of the refrigerant vapor at the evaporator exit as compared to the saturated vapor temperature indicated by the suction pressure.

expansion valve An automatic metering device that controls the flow of liquid refrigerant and separates the high-pressure and low-pressure sides of a refrigeration system.

Fahrenheit A temperature scale (abbreviated F) where water freezes at 32 degrees and boils at 212 degrees.

filter-drier A component in a refrigeration system that contains a fine mesh screen and desiccant to trap small particles and moisture. It can be found in either a liquid or a gas line.

flare An enlargement at the end of a refrigeration tube held leak-tight by a fitting.

flare fitting Tapered end of flared tubing designed to grip the flare for a vapor-tight seal.

flash gas The phenomenon of instantaneous vaporization of liquid refrigerant in an evaporator or capillary tube.

flooded evaporator An evaporator in a refrigeration system where not all liquid refrigerant vaporizes to gas before reaching the outlet.

flush To purge a system of contaminants, usually with a liquid.

flux A heat-activated chemical that absorbs and prevents the formation of oxides during brazing or soldering.

frequency The number of cycles per second (Hertz) in an alternating current power source.

frost back A condition where the suction line of a refrigeration system develops frost or sweating due to liquid refrigerant in the line.

full-load amperes (FLA) The current drawn by a motor when operating at full load.

gas The vaporous state of a substance.

gauge pressure A differential pressure, referred to as psig, that is the difference between the absolute value of the pressure being measured and atmospheric pressure.

gauge set A pair of pressure gauges assembled into a manifold. This allows the simultaneous monitoring of the high-pressure and low-pressure sides of a refrigeration system, to evacuate or charge it.

halide leak detector This detector uses the property of refrigerant combustion to cause a change in flame color whenever a leak is present.

hard start Occurs when the compressor does not start under load. This condition happens if suction and discharge pressures have not equalized.

hard-start circuit Additional components added to a compressor's electrical circuit to increase starting torque.

HCFC (Hydrochlorofluorocarbon) A family of chemical refrigerants that pose little or no threat to the ozone layer and are being used as replacements for CFC refrigerants.

head pressure Pressure that exists at the condensing side of the refrigeration system at the compressor outlet.

head pressure control Pressure-operated control designed to open the electrical circuit if high-side pressure becomes too high.

heat exchanger A device that transfers heat energy between two mediums, such as refrigerant and air or refrigerant and water.

heat gain The amount of heat energy that enters or is produced within an enclosed area. Common sources of this energy are solar power, electrical power, infiltration, and habitation.

heat load The amount of heat energy, in BTU/h, that enters the evaporator heat exchanger in an air-conditioning system.

hermetic compressor In this type of compressor, the induction motor and compressor are both sealed in the same housing.

Hertz The unit of cycles per second in alternating current systems.

HFC (Hydrofluorocarbon) A family of chemical refrigerants that pose little or no threat to the ozone layer and are being used as replacements for CFC refrigerants.

high-pressure cutout A pressure control that interrupts compressor operation when discharge pressure exceeds a safe level.

high-pressure line The refrigerant tubing that carries either the high-pressure compressor discharge gas or high-pressure liquid refrigerant.

high side Refrigeration system components that are under condensing or high-side pressure.

humidistat A control switch that reacts to changes in the level of relative humidity.

humidity Relative humidity, or the amount of moisture in the air, expressed as a percentage of the maximum amount that can be contained at that temperature.

hydrocarbons Chemical compounds composed of carbon and hydrogen.

impedance protected A type of motor designed to withstand continuous excitation under locked rotor conditions without sustaining damage.

inches of mercury Atmospheric pressure is equal to 29.92 inches of mercury.

inches of water column One inch of water column equals 0.036 psi, commonly used to measure pressure levels of gas or static pressures.

induction motor An ac-operated motor that contains stator windings that are excited by an ac power source, and a rotor that carries induced current.

Infiltration The normal exchange of air that occurs in an enclosed area through cracks and crevices due to the pressure and temperature differences between the outside and inside air.

Kelvin An absolute temperature scale (abbreviated K) using the same divisions as the Celsius (C) scale and referenced to absolute zero. Degrees K equals degrees C plus 273.

kilowatt A unit of electrical power equal to 1000 watts.

latent heat The heat energy that changes the state (gas to liquid, liquid to gas) of a substance without changing its temperature. Latent heat is given up when a gas condenses to a liquid, and is absorbed when liquid vaporizes to a gas.

leak detector An instrument that reacts to a refrigerant as it is escaping from an opening in a sealed system to pinpoint the source of leakage.

limit control A pressure or temperature control that shuts down system operation when a preset limit has been reached.

liquid line Tube that carries liquid refrigerant from the condenser or liquid receiver to the refrigerant control mechanism.

liquid receiver Cylinder or container connected to the condenser outlet for storage of liquid refrigerant in a system.

locked rotor amperes (LRA) The current drawn by a motor when the rotor's shaft is not rotating.

low-pressure line The refrigerant line that connects the evaporator outlet to the suction inlet of the compressor.

low side That part of a refrigerating system that is below evaporating pressure.

manifold gauge set Consists of low-pressure and high-pressure gauges that are mounted on an assembly with hand-operated valves and hoses. It is used for diagnosing, evacuating, and charging a refrigeration system.

manometer Instrument used to measure pressure of gases and vapors.

megger An instrument that measures high values of insulation resistance of a compressor by impressing 500 volts or more between the electrical terminals of the unit under test and ground or frame.

megohm A unit of resistance equal to 1,000,000 ohms.

microfarad A unit of capacitance equal to 1/1,000,000 farads.

micron A unit of length equal to one millionth of a meter. Often used to specify the vacuum obtainable from high-grade vacuum pumps that can create a vacuum of less than 10 microns (a near perfect vacuum).

microprocessor A solid-state integrated circuit component made up of transistors, diodes, and resistors. It performs calculations or computer operations according to preprogrammed instructions and data input.

moisture indicator A chemical that changes color in the presence of moisture.

muffler A device attached to the discharge line of the compressor to minimize pumping noises.

NEMA National Electrical Manufacturers Association.

noncondensible gas A gas that will not change to liquid under the normal operating temperatures and pressures of a refrigeration system.

OEM Original equipment manufacturer.

off-cycle The part of the operating cycle of a system when the system is not running.

ohm A unit of resistance. A 1-ohm load will draw one ampere from a 1-volt power source.

ohmmeter An instrument that measures resistance.

oil bleed line Capillary tube that allows refrigerant oil that has collected in a component to return to the compressor.

oil charge A quantity of oil, under pressure or not, that is added to a refrigeration system. Also refers to the amount of oil that is contained in a sealed system.

oil separator A device that removes oil from gaseous refrigerant.

orifice A fixed opening designed to meter the flow of liquid refrigerant according to the pressure differential across it.

overload An electrical device that monitors current and temperature, and opens the circuit when a safe level has been exceeded.

phase change The process where latent heat is absorbed or released as a substance changes from liquid to vapor, or vapor to liquid.

POA valve Pressure-operated absolute valve. It automatically maintains a specified minimum evaporator pressure to prevent the buildup of frost on the coils.

pole Part of the stator of an induction motor. The number of poles is always an even number, and determines the rpm rating of the motor.

polyphase A multiphase electrical system, usually either two- or three-phase.

potential relay Electrical switch that opens on high voltage and closes on low voltage.

power element The part of a thermostatic control that reacts to temperature changes and causes a mechanical motion.

power factor The ratio of watts to volt-amperes in an alternating current system. Defined as the cosine of the angle of lead or lag of the current with respect to the voltage.

pressure Force per unit area. Can be measured in pounds per square inch, inches or millimeter of mercury, inches of water, or kilograms per square centimeter.

pressure drop Loss in pressure from one part of a refrigeration circuit to another.

pressure-enthalpy diagram Graphic illustration of the thermodynamic properties of a given refrigerant. It can be used to predict refrigeration system performance and to troubleshoot problems. Also called Mollier diagrams.

pressure line Discharge line of a compressor.

pressure regulator A device that maintains a predetermined pressure level at its output, when input level exceeds that level.

pressure switch A switch operated by a rise or drop in pressure.

pressure-temperature relationship The change affected in temperature when pressure is changed or vice versa. Only used at saturated conditions. An increase in pressure results in a temperature increase. A decrease in temperature results in a pressure decrease.

process tube A short stub of refrigeration tubing, usually located on the compressor, that evacuates and charges the system.

psi Pounds per square inch.

psia Pounds per square inch absolute. The pressure of a gas or liquid specified with reference to a perfect vacuum.

psig Pounds per square inch gauge. The pressure of a gas or liquid specified with reference to the atmospheric pressure at sea level (14.96 PSIA).

pumpdown The process of evacuating a refrigeration system.

purge Use of a pressurized gas or liquid to force contaminants out of a component or system.

Rankine An absolute temperature scale (abbreviated R) that uses the same divisions as the Fahrenheit scale and is referenced to absolute zero. Degrees R equals degrees F plus 460.

receiver/drier A component in an expansion-valve controlled refrigeration system that stores liquid refrigerant. It contains a desiccant to trap moisture.

reclaim To remove refrigerant in any condition from a refrigeration system and then transport the recovered refrigerant to a dedicated refrigerant reclamation center. At the center, the refrigerant is reprocessed to new specifications, which might include distillation. Reclamation requires that any reprocessed refrigerant be certified through a chemical analysis.

recover To remove refrigerant in any condition from a refrigeration system and store it in a department of transportation (DOT) rated or otherwise approved external container. Recovery does not provide for any cleaning or filtration of the refrigerant.

recycle To remove refrigerant in any condition from a refrigeration system, then reduce its contaminants through oil separation, filtration, and/or other methods. Recycling is typically done at the job site or a local service shop. Recycled refrigerant is not always suitable for reuse.

refrigerant Substance used in a refrigerating unit. It absorbs heat in the evaporator by changing state from a liquid to a gas, then releases heat in the condenser as it changes state from a gas to a liquid.

refrigerant control Device that meters the flow of refrigerant between two specific areas of the refrigeration system. It also maintains a pressure difference between high-pressure and low-pressure sides of the unit while it is running.

refrigeration Transfer of heat from a place where it is not wanted to a place where its presence is not undesirable.

refrigeration cycle The process of transferring heat by means of compression and expansion of a refrigerant.

relative humidity The amount of moisture in the air, expressed as a percentage of the maximum amount of moisture that can be contained in the air at the same temperature.

relay An electromechanical device that controls the opening or closing of its contacts via current flow in its coil.

reverse cycle defrost Method of heating an evaporator for defrosting. Valves move hot gas from the compressor to the evaporator.

reversing valve Device that reverses the direction of refrigerant flow depending on whether heating or cooling is desired.

rotor The rotating portion of an induction motor that carries the induced current.

RPM Revolutions per minute. In an induction motor this is the equivalent of (120)(frequency)/(# of poles).

saturated vapor A vapor at any given temperature or pressure that would begin to condense to liquid if its temperature fell or its pressure increased.

Schrader valve A spring-loaded valve that permits access to the high and low sides of a refrigeration system. The valve automatically opens when the manifold gauge set hose is connected.

sealed unit A hermetic compressor or refrigeration system.

semi-hermetic compressor A serviceable hermetic compressor with service valves.

sensible heat   Heat energy that causes a temperature change in a substance without changing its state.

service factor   A factor less than or greater than one by which the HP rating of a motor is multiplied to determine the maximum safe continuous load rating.

service valve   An access valve used to diagnose, evacuate, and charge a refrigeration system.

shaded pole   An induction motor design that uses an induced current path around the stator to generate starting torque.

short cycling   Occurs when a compressor fails to start due to high-side and low-side pressure differences.

sight glass   A glass tube or window that allows the visual examination of liquid refrigerant, indicating the presence of refrigerant oil or gas bubbles in the system.

silver brazing   Process where silver bearing filler material is used to join two metals at temperatures above 800°F (427°C) and below the melting point of the materials being joined; also called silver soldering.

single phase   An alternating power source that is supplied by two wires.

slugging   This condition occurs in a compressor when liquid enters the suction inlet, causing hammering of the moving parts.

soldering   Process where tin alloys are used to join two metals at temperatures of about 450°F (232°C).

solenoid   A wire-wound coil that creates a magnetic field when energized. The magnetic field pulls a movable iron core within the coil.

solenoid valve   A valve mechanically actuated by means of an electric current passing through a coil in the solenoid.

solid state   Electric components made of silicon, germanium, or other semiconductor materials, that are used in refrigeration control circuitry.

specifications   Data supplied by the manufacturer of a refrigeration system that specifies proper operating temperature, pressure, power consumption, etc.

split-phase motor   An induction motor with a start winding that is magnetically displaced from the main winding.

split-system   A refrigeration system designed so that the compressor/condenser assembly is located away from the evaporator assembly. Usually large capacity units.

standard conditions   A temperature of 68°F (20°C), atmospheric pressure of 29.92 inches of mercury, and 36 percent relative humidity.

starting relay   A relay that connects a motor-starting capacitor or winding, into a circuit, only during the starting sequence.

subcooling   Cooling of a liquid refrigerant below its condensing temperature, accounting for system pressure.

suction line   Tube or pipe used to carry refrigerant gas from the evaporator to the compressor.

suction pressure The pressure read at the inlet side of the compressor. Also called back pressure or low-side pressure.

suction side The part of the refrigeration system located between the refrigerant metering device and the inlet part of the compressor.

suction throttling valve An evaporator pressure control valve that automatically maintains sufficient evaporator pressure to prevent frost buildup.

superheat 1) Temperature of a vapor above its boiling temperature as a liquid at that pressure. 2) Difference between the temperature at the evaporator outlet and the lower temperature of the refrigerant evaporating in the evaporator.

superheated vapor Any gas that reaches a temperature above the boiling point of the substance at a given pressure.

swaging Enlarging one tube end so the end of another tube of the same size will fit into it.

synchronous speed The theoretical maximum rpm of an induction motor equal to (120)(frequency)/(# of poles).

system All of the components that make up the refrigeration unit.

temperature Heat energy as measured by the Celsius, Fahrenheit, Kelvin, or Rankine scales.

thermal cut-out A protective device that monitors temperature and opens its set of contacts when a predetermined temperature limit has been reached.

thermal relay A relay that is actuated by means of an element heated by electrical energy.

thermistor A resistor that has a positive or negative coefficient of resistance so that its value changes with temperature.

thermometer An instrument that measures temperature.

thermostat A device that reacts to ambient temperature changes, causing a pair of switch contacts to close or open.

thermostatic expansion valve Control valve operated by temperature and pressure within the evaporator. It controls the flow of refrigerant utilizing a control bulb attached to the outlet of the evaporator.

thermostatic switch A temperature-activated device used to control the on and off times of a compressor.

three-phase An alternating current power source supplied in three-wire delta or four-wire configurations.

ton of refrigeration Refers to the cooling effect created by melting 1 ton of ice in a 24-hour period. The equivalent of 12,000 BTUs per hour.

transformer An ac-operated component that raises or lowers the supply voltage.

transistor A solid-state component, normally made of silicon, designed to amplify current.

TEV or TXV   An abbreviation for the thermostatic expansion valve.

vacuum   Pressure lower than atmospheric pressure or less than 14.7 psia at sea level.

vacuum gauge   An instrument that measures the level of vacuum or negative pressure.

vacuum pump   Mechanical device designed to transfer vapor and air and water in order to evacuate a refrigeration system.

viscosity   A specified property of a liquid that refers its resistance to flow. Normally refers to refrigeration oil used in refrigeration systems.

volt-amperes   The combined product of voltage and current in an electrical load.

voltage   The pressure or electromotive force of an electrical power source.

voltmeter   An instrument used to measure voltage level.

watt   A unit of electrical power. One watt is dissipated in a load that is driven by one volt and carries one ampere of current.

Index

Accumulators, 64
ARC tubing, 141–142
ARI/GAMMA, 9
Automatic expansion valves, 66–67

Brazing, 16–17, 150–157
  alloys, 151
  breaking joints, 155
  fluxing, 152
  techniques, 152–155
  troubleshooting tips, 156–157

Capacitors, 257–258, 282–284
  running, 258
  starting, 258
  troubleshooting, 258, 282–284
Capillary tubes, 17, 61–65, 301–302
  advantage of, 61–62
  bubble point, 63
  cleaning kit, 17
  domestic refrigerator, 301–302
  installation and service, 64–65
Careers opportunity, 1–10
CDA valves, 437, 439
Certification, 10
Charging (refrigerant), 28–29, 220–227
  equipment, 28–29
  liquid, 224–226
  pump, 29
  scales, 29
  vapor, 220–224
  weighing, 226–227
Check valves, 107
Circuit breakers, 252–254
Clean Air Act, 175
Compressors, 40, 80–89, 300, 319–322, 427–435

Compressors (*Cont.*):
  accessible hermetic, 82–83
  burnout, 433
  centrifugal, 87–88
  cooling methods, 88
  domestic refrigerator, 300, 319–320, 321–322
  failures, 427, 429, 432
  hard-start circuits, 258
  hermetic, 81–82
  open, 80–81
  reciprocating, 83–84
  replacement, 433–435
  rotary, 84, 86
  screw, 86
  scroll, 86–87
  service valve, 90
  starting torque, 88–89
  unloader systems, 130–134
Condensers, 40, 90–93, 107, 299–301, 322, 365–366, 436–437, 457–458
  air-cooled, 93, 436
  construction, 91–92
  domestic refrigerator, 299–301, 322
  pressure regulator, 107
  remote, 93
  selecting, 457–458
  supermarket units, 365–366
  units, 92–93
  water-cooled, 92, 365–366, 436–437
Constant pressure expansion valves, 66–67
Contactors (*see* Relays)
Control switches, 99–100
Cooling capacity, 55, 128–134
  reduction control systems, 128–134
Cooling towers, 366
Copper tubing, 141–142

Desiccants, 182
Differential control switches, 105–106
Discharge line, 90
Disconnect switches, 254
Domestic refrigerator/freezers, 291–323
  capillary tubes, 301–302
  compressors, 301, 319–320, 321–322
  condensers, 299–301, 322
  core valves, 318
  defrost systems, 302–304, 321
  design features, 291–294
  electrical circuitry, 304–309, 315–317
  evaporators, 299
  gaskets, 298
  hardware, 296
  ice service, 298
  installing, 309
  leak repair, 318–319
  shutting down, 323
  specifications, 291, 313–315
  structure, 294–296
  troubleshooting, 309–311

Electrical components, 249–261
Electrical test equipment, 29–33
  ammeter, 31
  circuit testers, 30
  compressor analyzer, 33
  continuity tester, 31
  megger, 32
  multimeter, 32
  power factor meter, 32
  test capacitors, 33
  wattmeter, 31
Electricity, 239–249
  basic concepts, 239–241
  circuit conditions, 247–248
  circuit types, 243–246
  single-phase, 241
  three-phase, 242–243
Electromagnetism, 248–249
Electronic controls, 7, 134–139, 261, 384–389
  solid-state, 261
  supermarket systems, 384–389
  temperature control systems, 134–139
Electronic expansion valves, 75
Enthalpy, 55
EPR valves, 437, 440
Evacuation (of refrigerant lines), 216–220
  basic setup, 217
  multi-step, 217–219

Evaporators, 40, 75–78, 106–107, 299, 322, 435–436, 456–457
  design of, 75–78
  domestic refrigerator, 299, 322
  multi-circuited, 78
  pressure regulator, 106–107
  selecting, 456–457
  temperature difference relationship, 78
  troubleshooting, 435–446

Files, 13
Filter-driers, 94–97, 320–321
  capillary tube, 96
  domestic refrigerator, 320–321
  installation of, 96–97
  suction line, 95
Fin combs, 17
Fittings, 147–148, 150
  compression, 150
  O-ring, 150
Flared connections, 14–15, 144–148
  double-thickness, 146–147
  fittings, 147–149
Float switches, 106
Float valves, 65–66
Flow controls (refrigerant), 40, 61–71
Flow switches, 105
Food service equipment, 423–425
Freezers (*see* Domestic refrigerator/freezers; Reach-in freezers; Walk-in refrigerators/freezers)
Fuses, 252

Graduated cylinders, 227

Hacksaws, 13
Heat, 42–44, 46–50
  latent, 46–47
  sensible, 46
  specific, 45
  transfer methods, 48–50
  units, 44
Heat-recovery systems, 377–380
High-pressure switch, 100
Hose, flexible, 142, 182
  permeation, 182
Hot gas defrost, 127
Humidstats, 104

Ice makers, 403–417
  cleaning, 414–415
  cuber system, 407–410
  flaker systems, 405–406

Ice makers (*Cont.*):
  installation, 410–412
  maintenance, 415–417
  operating cycle, 403–405
  start-up, 413–414
Inches of mercury, 51
Inches of water, 51–52
Inert gas, 28

Leak detectors, 23–28
  compound specific, 24
  electronic, 26–27
  halide torch, 25
  halogen selective, 24
  soap solution, 25
  UV light, 27–28
Leak testing, 219–220
Liquid line, 93
Low-pressure switch, 101–102
Lubricants (refrigerant), 188–191, 353–354
  alkylbenzenes, 189
  mineral oils, 188–189
  poly alkylene glycols, 189–190
  poyol esters, 189
  pressure protection controls, 353–354
  retrofit concerns, 190–191

Mallets, 12
Manifold gauge set, 19, 21
Manometer, 21
Mirrors, 17
Montreal Protocol, 174
Motors (electric), 254–255, 267–290
  capacitor-start, 274
  capacitor-start-capacitor-run, 275–276
  design, 267–270
  hermetic compressor, 277–279, 288–290
  permanent split-phase, 273
  replacement, 286–287
  shaded-pole, 270–271
  single-phase, 279
  split-phase, 271–273
  strength, 270
  three-phase, 276–277, 279
  troubleshooting, 279–286, 288–290
  wire terminal identification, 278

National Electrical Code, 252
Nutdrivers, 11

Oil pressure sensors, 127–128
Oil separators, 89–90

Oils (*see* Lubricants)
Overload devices, 252

Piercing valves, 17
Pliers, 12
Pressure, 50–54, 100–102, 106–107
  absolute, 50–51
  critical, 54
  gauge, 50–51
  regulators, 106–107
  switches, 100–102
  temperature relationship, 52–54
Pressure controls, 348–352
  installation checks, 350
  setting, 351
Pressure-enthalpy diagrams, 192–207
  key features, 192–195
  troubleshooting with, 199–207
  working with, 195–199
Purging (refrigerant lines), 215

Reach-in freezers, 329–335
  operation, 329–330
  hot-gas defrost, 330–331
  electric defrost, 331
  installing, 332
  maintenance, 332–333
  service, 333–335
Reach-in refrigerators, 325–329, 332–335
  electrical systems, 328–329
  maintenance, 332–333
  operation, 328
  service, 333–335
  types and construction, 325–328
Reamers, 13
Receiver, 93–94
Recovery and recycling (refrigerant), 28, 176–178, 209–214
  equipment, 28, 209–214
  working tips, 178–180
Refrigerant(s), 7, 169–191, 228–238
  alternative, 170–173, 181
  ASHRAE numbering system, 171
  azeotropes, 171–172
  CFC, 169, 174
  cylinders, 184–188
  disposal of, 178
  HCFC, 169–170, 174
  HFC, 170
  identification, 171
  lubricants, 188–190
  mixed, 180–181
  moisture concerns, 188

476 Index

Refrigerant(s) (*Cont.*):
  reclaim, 178
  recovery, 176–177, 209–214
  recycling, 177–178
  retrofits, 190–191, 228–238
  safety, 182–184
  weighing, 226–227
  zeotropes, 172, 181–182
Refrigeration, 1–33, 39–42, 56–97, 427–443
  apprentice programs, 8
  careers in, 1, 4–10
  history of, 1–3
  mechanical, 40
  principles of, 39–42
  skills needed, 5–6
  system components, 59–97
  tools, 11–33
  troubleshooting systems, 427–443
  types of systems, 4–5
  types of work, 6–7
  vapor compression cycle, 40–42, 56–58
  vocational programs, 8–9
Refrigeration loads, 445–458
  air infiltration, 448–449
  equipment selection, 456–458
  heat load charts, 445–446
  K factor, 446
  miscellaneous, 452
  product load, 449–452
  respiration, 452
  transmission, 447–448
  walk-in freezer calculation, 453–456
  walk-in refrigerator calculation, 447–453
Refrigerators (*see* Domestic refrigerator/freezers; Reach-in refrigerators; Walk-in refrigerators/freezers)
Relative humidity, 55–56
Relays, 256–257
  start, 258–260
Retrofits, 190–191, 228–238
  equipment needed, 228–229
  HFC-134a tips, 234–238
  overview, 228
  procedure, 229–233

Safety, 33–38, 152, 182–184, 186–188
  brazing, 152
  cold, 34
  cylinders, 186–188
  electrical, 34–35
  fire, 34

Safety (*Cont.*):
  lifting, 37
  material safety data sheets, 35–37
  personal, 37
  pressure vessel, 35
  refrigerant, 182–184
  right-to-know laws, 35
  tools and equipment, 37–38
Saturated vapor, 53–54
Screwdrivers, 11–12
Signal lights, 255
Soldering, 16–17, 151–157
Solenoid valves, 107–111, 437, 440
  high temperature, 111
  maxium operating pressure, 110–111
  operation of, 108–110
  suction line, 111
  troubleshooting, 437, 440
Specific gravity, 54
Specific heat, 45
Specific volume, 54
States of matter, 45–48
Steel tubing, 142
Suction line, 78–79, 107, 163–168
  filter-drier, 79
  pressure regulator, 107
  risers, 163–168
  service valve, 79
  sizing, 164, 167–168
Superheat, 71–73
  calculating, 72–73
Supermarket refrigeration, 355–402, 437–443
  air distribution, 381–382
  cooling requirements, 357–358
  condensers, 365–366
  control settings, 390–391
  control systems for, 382–389
  defrost controls, 369–376
  defrost systems, 367–369
  display case design, 355–357
  electrical connections, 393–394
  expansion valve adjustment, 391–393
  heat recovery in, 377–380
  installation of systems, 389–401
  multiplexed systems, 358–361
  night covers for, 394–395
  parallel compressor units, 361–364
  periodic maintenance, 437–443
  servicing, 401–402
  single-compressor units, 358–359
  start-up procedures, 396–401
  subcooling refrigerant in, 376–377

Swaging, 15, 149–150
  joints, 149–150
  tool for, 15
Switches, 255

Temperature, 44, 52–54
  controls, 65
  critical, 54
  pressure relationship, 52–54
Temperature contol schemes, 111–139, 345–348
  alarm control, 123
  condenser fan control, 125
  controller installation and troubleshooting, 344–347
  defrost termination, 126–127
  electronic, 134–139
  evaporator temperature, 115–117
  high limit control, 125
  ice thickness sensing, 118
  low pressure limit control, 124
  multiple compressor, 115
  multiple evaporator, 112–113
  product temperature, 118
  pumpdown control, 121–123
  return air temperature, 118
  single evaporator, 111–112
  suction pressure, 120
Thermometers, 18–19
Thermostatic expansion valves, 68–75, 437, 438
  adjustment, 73–74
  basic operation, 68–71
  effects of heat load on, 74–75
  refrigerant charge with, 75
  superheat calculations, 71–73
  troubleshooting, 437, 438
Thermostats, 65, 102–104
  basic design, 102–103
  bi-metal switch, 103
  thermistor, 103
Time clocks, 104–105, 369–376
  mechanical defrost timers, 369–371
  modular timers, 373–376
  setting mechanical timers, 371–373

Time-delay relays, 104
Torches, 16–17
Troubleshooting (general), 427–443
Tube benders, 13–14
Tube cutters, 13
Tubing, 141–168
  bending, 143–144
  brazing, copper, 141–142
  cutting, 143
  elbow design, 158
  expansion, 161–162
  insulating runs, 157–158
  joining, 144–158
  plastic, 142
  repairing leaks, 157
  steel, 142
  supporting runs, 159–160

Vacuum, 51–52, 216–219
  detecting leaks under, 219
  measuring, 216–217
Vacuum gauges, 22
Vacuum pumps, 21–22, 217
Vapor compression cycle, 40–42, 56–58

Walk-in refrigerators/freezers, 335–350
  compressors, 337–338
  condensers, 337–338
  construction, 335–336
  crankcase pressure regulators, 343–344
  defrost systems, 342
  liquid-suction heat exchanger, 342–343
  pressure controls, 347–350
  refrigerant flow controls for, 338–341
  temperature controls, 341, 345–347
Water coolers, 417–422
  operation, 417
  installation, 418–419
  maintenance, 420–421
  troubleshooting, 421–422
Wiring, 249–252, 261–266
  schematics, 261–266
Wrenches, 12

ABOUT THE AUTHOR

John A. Corinchock is a technical writer with extensive experience in the industrial trades. He researches and writes operator and maintenance manuals, training materials, and test procedures for a variety of technical subjects.

Printed in the United States
5258